W0044177

FRONTIERS IN CEREBRAL VASCULAR BIOLOGY
Transport and Its Regulation

ADVANCES IN EXPERIMENTAL MEDICINE AND BIOLOGY

Editorial Board:

NATHAN BACK, *State University of New York at Buffalo*

IRUN R. COHEN, *The Weizmann Institute of Science*

DAVID KRITCHEVSKY, *Wistar Institute*

ABEL LAJTHA, *N.S. Kline Institute for Psychiatric Research*

RODOLFO PAOLETTI, *University of Milan*

Recent Volumes in this Series

A Continuation Order Plan is available for this series. A continuation order will bring delivery of each new volume immediately upon publication. Volumes are billed only upon actual shipment. For further information please contact the publisher.

FRONTIERS IN CEREBRAL VASCULAR BIOLOGY

Transport and Its Regulation

Edited by

Lester R. Drewes

University of Minnesota
Duluth, Minnesota

and

A. Lorris Betz

The University of Michigan
Ann Arbor, Michigan

SPRINGER SCIENCE+BUSINESS MEDIA, LLC

Library of Congress Cataloging-in-Publication Data

Frontiers in cerebral vascular biology : transport and its regulation
  / edited by Lester R. Drewes and A. Lorris Betz.
      p.   cm. -- (Advances in experimental medicine and biology ; v.
331)
    Proceedings of a workshop held July 11-13, 1992 in Duluth, Minn.
    Includes bibliographical references and index.
    ISBN 978-1-4613-6267-8   ISBN 978-1-4615-2920-0 (eBook)
    DOI 10.1007/978-1-4615-2920-0
    1. Blood-brain barrier--Congresses.   I. Drewes, Lester R.
II. Betz, A. L. (A. Lorris)   III. Series.
    [DNLM: 1. Blood-Brain Barrier--physiology--congresses. 2. Brain-
-blood supply--congresses. 3. Biological Transport--physiology-
-congresses. 4. Cell Communication--physiology--congresses.   W1
AD559 v.331 1993 / WL 200 F935 1992]
QP375.5.F76 1993
612.8'24--dc20
DNLM/DLC
for Library of Congress   93-16860
                                                               CIP

Proceedings of CVB '92, a conference on Frontiers in Cerebral Vascular
Biology: Transport and Its Regulation, held July 11-13, 1992,
in Duluth, Minnesota

ISBN 978-1-4613-6267-8

© 1993 Springer Science+Business Media New York
Originally published by Plenum Press, New York in 1993

All rights reserved

No part of this book may be reproduced, stored in a retrieval system, or transmitted
in any form or by any means, electronic, mechanical, photocopying, microfilming,
recording, or otherwise, without written permission from the Publisher

PREFACE

During the past three decades, the cerebral vasculature and its role in blood-brain transport has been an increasingly active area of investigation and learning, particularly from an anatomical and physiological point of view. However, much less is known at the molecular and cellular level about the blood-brain barrier especially regarding the macromolecules responsible for transport, the roles played by vascular wall components (endothelial cell, pericyte, smooth muscle, basement membrane), and the mechanisms regulating brain vascular-specific protein expression and their molecular alterations during development and disease. Fundamental questions still unanswered include: What are the molecular constituents of brain endothelial cell tight junctions? What are the membrane proteins responsible for transport of specific substrates? What are the molecular signals that cause glucose transporter gene expression to be 20 to 100 times greater in brain endothelial cells in vivo than in vitro? What roles do pericytes, smooth muscle cells and basement membrane have in establishing or maintaining blood-brain transport characteristics? Are brain vascular transport systems responsible for edema following injury? Are transporter systems regulated via receptor-mediated events? Do hormones or neuromodulators regulate transporter expression? What is the molecular mechanism by which plasma proteins enter the extravascular space? Are transporters asymmetrically distributed between the luminal and abluminal endothelial cell membranes? Can prodrugs or pharmacologic agents be designed as substrate analogs and be delivered to the central nervous system via existing transporters or receptors? Can new and beneficial transporters be introduced into the brain vasculature? Can vascular cells be genetically engineered and introduced into the brain to create new functions or to modify existing barrier-specific properties? Can transport be monitored dynamically in vivo by non-invasive approaches? How do cells of the immune system interact with cells of the brain vascular system? Is junctional integrity regulated by receptor-mediated signals? What are the signaling mechanisms for capillary tube formation and angiogenesis? One of the most important challenges during the 1990s' Decade of the Brain will be to answer these and other provocative questions by innovative approaches from the newer fields of cell physiology, recombinant DNA technology and molecular biology.

The Cerebral Vascular Biology (CVB '92) conference was organized and held in Duluth, Minnesota, for three days during July 1992. The objectives of the conference were to examine results of research in this field using new approaches involving cellular and molecular technology; to provide a forum in a conducive environment for creation of new and fresh insights into the function and regulation of the brain vascular wall; to foster the participation of trainees, new investigators and experts in the field in an open and free exchange of their views and ideas; and to enhance the understanding of the advances in cerebrovascular research to physicians who treat patients with cerebrovascular disease. Nearly one hundred participants from the United States, Canada, Europe and Japan attended or contributed to the scientific program. With the aid and advice of an international scientific advisory panel, the conference program was developed with the aims of presenting the most recent insights into mechanisms of blood-brain transport and regulation during physiological or pathological perturbations and to characterize the roles of vascular cells and basement membrane in forming the vascular wall.

This volume is divided into five thematic sections, each with a title that represents a timely and central issue in cerebral vascular biology today. The authors represent an international collection of neuroscientists and authorities at the forefront of their field. It is anticipated that this volume will set the scene for exciting conceptual developments, collaborations, and future conferences. Furthermore, as progress is attained, the insights gained will be relevant to the clinical environment where new strategies are needed for treating neurological disease.

L.R.D.
A.L.B.

CONTENTS

GLUCOSE TRANSPORT

ELECTROLYTE TRANSPORT

RECEPTORS AND INTRACELLULAR MESSENGERS

INTERACTION OF BRAIN ENDOTHELIAL CELLS WITH OTHER CELLS

IMMUNOLOGIC FUNCTIONS

BARRIER FUNCTION AND CHARACTERISTICS

Glucose Transport

DEVELOPMENTAL EXPRESSION OF GLUCOSE TRANSPORTERS, GLUT1 AND GLUT3, IN POSTNATAL RAT BRAIN

Susan J. Vannucci, Lisa B. Willing and Robert C. Vannucci

Department of Pediatrics
Pennsylvania State University College of Medicine
Hershey, PA 17033

INTRODUCTION

Glucose is the predominant fuel for the developing, as well as the adult brain under physiologic conditions. However, cerebral energy requirements of most mammals are quite low at birth, reflective of the immaturity of the central nervous system at this stage of development. Several studies have reported values for cerebral glucose utilization in the fetal and newborn rat brain to be 10-30% of the adult value (3-5), although it is somewhat higher in the brain stem and other structures such as the thalamus (6,8).

The initial step in cerebral glucose utilization is the transport of glucose across the blood-brain barrier and then into the neurons and glia. The transport of glucose across most cell membranes is mediated by the facilitative glucose transporter proteins, designated GLUT1 - GLUT5 for the order in which they were cloned (for review, see Pessin and Bell, 7). To date, two glucose transporter proteins, GLUT1 and GLUT3, have been definitively identified in rat brain. GLUT1 is most concentrated in the microvessels of the blood-brain barrier where it is detected as a 55 kDa band on sodium dodecyl sulphate polyacrylamide gel electrophoresis (SDS-PAGE). A lower molecular weight form of GLUT1, 45 kDa, has been detected in vascular-free cerebral membranes and presumably reflects the form of this transporter present in the neuronal/glial membranes. GLUT3 appears to be expressed exclusively in neurons (9,10).

The purpose of this study was to determine the expression of both GLUT1 and GLUT3 glucose transporters in developing rat brain and to correlate these observations with the changes in cerebral energy requirements and glucose utilization that occur during the first few weeks of postnatal life.

Frontiers in Cerebral Vascular Biology: Transport and Its Regulation
Edited by L.R. Drewes and A.L. Betz, Plenum Press, New York, 1993

METHODS

Rats were studied at 1, 3, 7, 14, and 21 days of postnatal age and as adults. Pups were kept with their dams until immediately before sacrifice. Animals were killed by decapitation, and the brains were removed and placed in a pre-cooled brain slicer (Activational Systems, Inc.). Samples (5-10 mg) were taken from four forebrain regions (cerebral cortex, hippocampus, thalamus, and hypothalamus). Samples were homogenized in TES buffer (20 mM Tris-HCl, 1 mM EDTA and 255 mM sucrose, pH 7.4, in the presence of protease inhibitors, aprotinin, leupeptin, pepstatin and PMSF, 1 µg/ml). Total membranes were prepared by centrifugation at 150,000 x g for 20 min at 4°C. Protein was determined by the Pierce BCA method.

Membrane samples (15 µg) were subjected to one-dimensional SDS-PAGE and Western blot analysis using antipeptide antisera specific to the C-terminal 12 amino acid sequence of GLUT1 and to the C-terminal 20 amino acid sequence of mouse GLUT3. Both antibodies were generously provided by Dr. Ian A. Simpson, National Institutes of Health, Bethesda, MD. The secondary antibody was ^{125}I F'Ab fragments and the results were quantitated by laser densitometry. All blots contained equal amounts of membrane protein as well as an adult brain membrane (vascular-free) standard for normalization. Results are expressed as a percent of the standard.

RESULTS AND DISCUSSION

To determine the relative amounts of GLUT1 and GLUT3 glucose transporter proteins in the four forebrain regions of postnatal rats, samples of cortex (C), hippocampus (H), thalamus (T) and hypothalamus (HT) were taken from rats at 1, 3, 7, 14, and 21 days and prepared for SDS-PAGE and Western blot analysis as described in Methods. Comparable samples were taken from an adult rat. Representative Coomassie-blue stained gels and the corresponding autoradiograms obtained with 1-, 7-, and 21-day-old rats are shown in Figure 1 A-C. Figure 1D shows the results obtained with the adult rat brain. As can be seen from these autoradiograms, GLUT1 in whole brain microsomes is detected as two discrete bands, a 55 kDa vascular form, and a 45 kDa form, presumably from the neuronal/glial fraction. The brain standard, prepared without a vascular component, only presents the 45 kDa GLUT1 band. Not only does the total GLUT1 increase by 3-fold during the initial weeks of postnatal life, but as can be readily seen from these autoradiograms, the 45 kDa form steadily increases. This most likely relates to the ongoing glial proliferation during this period. The compilation of the data obtained from 1- and 3-day-old rats did not differ and these groups are combined in the tables. The low level of GLUT1 in the early newborn period correlates well with the reduced glucose transport and utilization at this age. There was no significant difference noted in GLUT1 expression among the regions studied.

The pattern for GLUT3 expression is quite different (bottom panels of Figure 1 A-D, Table 3). GLUT3 protein was barely detectable in the 1-day-old cortex and represented less than 10% of the adult brain standard, whereas more primitive areas such as the thalamus expressed 3 times this level of GLUT3. The GLUT3 levels doubled between 1 and 7 days and then again between 7 and 21 days, with 14 days being intermediate between these. This finding is in good agreement with a recent study utilizing in situ hybridization to detect GLUT1 and GLUT3 mRNA in development (2). This study demonstrated that neurons do not begin to express the GLUT3 protein until they have migrated to their final cortical destination and start to send out processes. This period of rapid increase in the levels of GLUT3 protein, as reported in this study, corresponds to the period of the most rapid rate of synaptogenesis in the newborn rat brain (1). The observation that near adult levels of GLUT3 protein, on a per mg of membrane protein basis, is achieved by 21 days is also interesting since this age marks the end of the suckling period in the rat. At this time, the

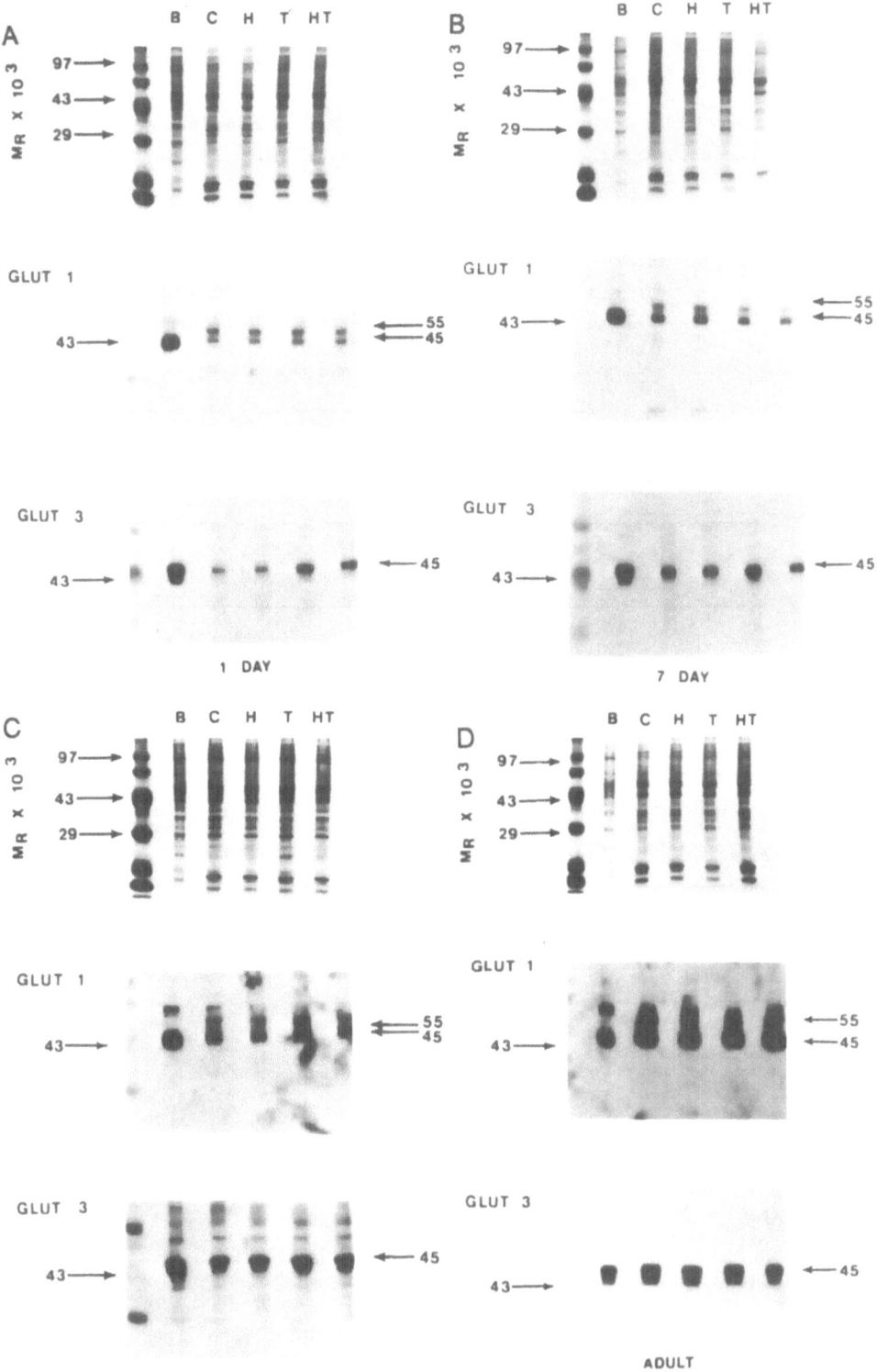

Figure 1 A-D: Coomassie-blue stained gels and autoradiograms of Western blot analysis for GLUT1 and GLUT3 proteins in brain regions from postnatal rats aged: (A) 2 day; (B) 7 days; (C) 21 days; (D) adult. The results of all of the analyses, 3-5 rats per age and inclusive of 14 days, are presented in Tables 1-3.

availability of alternate fuels, such as ketone bodies, declines and glucose becomes nearly the sole fuel for the brain under normal conditions.

Table 1. GLUT1 in Postnatal Rat Brain: % of Brain Standard

Age	Cortex	Hippocampus	Thalamus	Hypothalamus
1	20.3±2.4	24.3±4.1	20.9±4.4	22.0±2.5
7	78.1±6.6	68.8±4.7	66.2±6.9	53.9±8.8
14	63.5±3.1	50.9±0.4	65.0±5.9	59.7±4.1
21	72.3±3.9	58.3±3.4	69.7±3.6	57.1±2.1

Table 2. GLUT1 in Postnatal Rat Brain: 45kDa Form as a % of Total

Age	Cortex	Hippocampus	Thalamus	Hypothalamus
1	41.2±0.9	42.4±1.0	43.0±3.8	37.1±3.0
7	56.7±2.9	51.9±1.9	54.3±2.1	56.9±3.2
14	65.8±1.1	70.4±3.6	69.0±4.6	73.1±4.4
21	58.6±0.9	55.3±1.0	64.6±1.5	65.0±0.8

Table 3. GLUT3 in Postnatal Rat Brain: % of Brain Standard

Age	Cortex	Hippocampus	Thalamus	Hypothalamus
1	9.8±0.8	14.3±0.4	27.0±1.8	20.5±1.1
7	22.4±2.0	26.9±1.3	41.9±1.4	36.1±3.6
14	35.3±4.2	72.5±1.1	62.0±3.0	56.9±7.2
21	55.3±6.1	70.1±10.0	86.7±12.0	83.9±5.2

REFERENCES

1. Aghajanian, G.K., and Bloom, F.E., The formation of synaptic junctions in developing rat brain: A quantitative electron microscopic study, *Brain Res.* 6:716-727, 1967.
2. Bondy, C.A., Lee, W.H., and Zhou, J., Ontogeny and cellular distribution of brain glucose transporter gene expression, *Mol. and Cell. Neurosci.*, in press.
3. Cremer, J.E., Cunningham, V.J., Pardridge, W.M., Braun, L.D., and Oldendorf, W.H., Kinetics of blood-brain barrier transport of pyruvate, lactate and glucose in suckling, weanling and adult rats, *J. Neurochem.* 33:439-445, 1979.

4. Daniel, P.M., Love, E.R., and Pratt, O.E., The effect of age upon the influx of glucose into the brain, *J. Physiol.* 274:141-148, 1978.

5. Dyve, S., and Gjedde, A., Glucose metabolism of fetal rat brain in utero, measured with labeled deoxyglucose, *Acta. Neurol. Scand.* 83:14-19, 1991.

6. Nehlig, A., Pereira de Vasconcelso, A., and Boyet, S., Quantitative autoradiographic measurement of local cerebral glucose utilization of freely moving rats during postnatal development, *J. Neurosci.* 8:2321-2333, 1988.

7. Pessin, J.E., and Bell, G.I., Mammalian facilitative glucose transporter family: structure and molecular regulation: *in:* "Annu. Rev. Physiol." 54:911-930, J.K. Hoffman, ed., Annual Reviews Inc., Palo Alto, CA, 1991.

8. Vannucci, R.C., Christensen, M.A., and Stein, D.T., Regional cerebral glucose utilization in the immature rat: Effect of hypoxia-ischemia, *Ped. Res.* 26:208-214, 1989.

9. Nagamatsu, S., Kornhauser, J.M., Burant, C.F., Seino, S., Mayo, K.E., and Bell, G.I., Glucose transporter expression in brain: cDNA sequence of mouse glut3, the brain facilitative glucose transporter isoform, and identification of sites of expression by *in situ* hybridization, *J. Biol. Chem.* 267:467-472, 1992.

10. Maher, F., Vannucci, S., Takada, J., and Simpson, I.A., Expression of mouse-glut3 and human-glut3 glucose transporter proteins in brain, *Biochem. Biophys. Res. Comm.* 182:703-711, 1992.

ALTERATIONS IN BRAIN GLUCOSE TRANSPORTER PROTEINS, GLUT1 AND GLUT3, IN STREPTOZOTOCIN DIABETIC RATS

Fran Maher [1], Ian A. Simpson [1] and Susan J. Vannucci [2]

[1]EDMNS, Diabetes Branch, NIDDK, National Institutes of Health, Bethesda, MD 20892
[2]Department of Pediatrics, Pennsylvania State University College of Medicine, Hershey, PA 17033

INTRODUCTION

The brain is dependent on glucose for energy, and it requires glucose transporter proteins to facilitate the passage of glucose across the blood-brain barrier and across the plasma membranes of neurons and glia. Two glucose transporter isoforms, GLUT1 and GLUT3, have been identified in rat brain. GLUT1 is detected as a 55 kDa form associated with microvessels and a 45kDa form associated with microvessel-free cerebral membranes. GLUT3, the neuronal glucose transporter (2,6) is expressed in most brain regions and the neurohypophysis (3).

Several studies have examined blood-to-brain glucose transport during diabetic hyperglycemia and have reported it to be increased, decreased or unchanged (see 5 for review). Few studies have specifically examined the glucose transporter proteins, particularly in the neuronal or glial membranes. The concentration of glucose transporter in cerebral microvessels (GLUT1) has been variably reported to increase or decrease in diabetes. The expression and/or abundance of GLUT1 in the neuronal and glial cells in vivo is currently unclear, as it has been detected by immunohistochemistry only in glia, but is expressed in cultured neurons and glia. GLUT1 regulation in these cells during diabetes has not been specifically addressed. GLUT3 has only recently been cloned and recognized as a neuronal glucose transporter (2,3,6) and has not been studied in pathological states.

The purpose of this study was to examine the expression of GLUT1 and GLUT3 in cortical microvessels and several brain regions of streptozotocin-diabetic rats.

METHODS

Animals. Diabetes was induced in male Sprague-Dawley rats (200-300 gm) with streptozotocin (65 mg/kg, i.p.). Untreated controls and diabetic animals (blood glucose > 400 mg/dl) were studied 2 weeks after injection of streptozotocin. Animals were anesthetized

Frontiers in Cerebral Vascular Biology: Transport and Its Regulation
Edited by L.R. Drewes and A.L. Betz, Plenum Press, New York, 1993

with 70% CO_2 : 30% O_2 and decapitated. Brains were rapidly removed, the hippocampus, neurohypophysis and a slice of cerebellum dissected out and frozen on dry ice for subsequent homogenization and membrane preparation, and the cerebral cortex was transferred immediately into buffer for microvessel and cortical membrane preparation. Tissue from 3-5 animals was pooled for each group of control and diabetic animals in seven experiments.

Microvessels and Membrane Preparations. Cortical shells were cleaned of meninges and surface vessels and homogenized in 5 volumes of Krebs-Ringer-Hepes buffer (KRBH), pH 7.4, containing 5 mM glucose, 0.5% BSA and protease inhibitors (aprotinin, leupeptin, pepstatin and PMSF, 10 µg/ml each). The initial homogenate was centrifuged for 10 min at 1000 x g, 4°C, and the resulting supernatant was centrifuged at 150,000 x g for the recovery of vascular-free cortical membranes (neuronal/glial membranes). The initial pellet was resuspended in 17% dextran in KRBH and centrifuged at 10,000 x g for 15 min to separate the myelin and debris from the vessels. The pellet was resuspended in KRBH and passed through 125-µm mesh onto a column of 0.25 mm glass beads retained by 45-µm mesh. After extensive washing, the microvessels were recovered from the column and pelleted at 1000 x g. Microvessels and cortical membranes were washed and resuspended in 20 mM Tris / 1 mM EDTA / 233 mM sucrose, pH 7.4, with protease inhibitors (TES).

Hippocampus and cerebellum were homogenized in TES and centrifuged at 200,000 x g for 20 min at 4°C to recover total membranes. The neurohypophysis was sonicated in TES. Protein content of all preparations was determined with the BCA assay.

Immunoblotting and Antisera. Protein (25 µg of membrane/lane; 5-10 µg of microvessel protein /lane) was separated by 10% SDS-polyacrylamide gel electrophoresis and transferred to nitrocellulose. Glucose transporter proteins were detected on duplicate blots with rabbit polyclonal antisera specific to the GLUT1 and mouse-GLUT3 (6) C-terminal sequences and ^{125}I-protein A. Antiserum to human-GLUT3 C-terminus was also used. Quantitation of immunoblots was performed by excising the radioactive band and counting in a gamma-counter. Antiserum specificity was determined by competition with 10 µg/ml homologous peptide as previously reported (2,3).

RESULTS

Detection of Glucose Transporter Isoforms in Rat Brain and Isolated Microvessels

Previously we have reported the detection of GLUT1 in rat brain total membranes as two molecular weight forms (45 kDa and 55 kDa) and GLUT3 as a 45 kDa protein (3). In microvessels isolated from cerebral cortex, immunofluorescence analysis revealed strong GLUT1 immunoreactivity (Figure 1). Immunoblot analysis of isolated microvessels and cortical neuronal/glial membranes demonstrated the presence of only the 55 kDa form of GLUT1 in microvessels, and only the 45 kDa form in neuron/glial membranes (Figure 2). GLUT1 was detected in all regions of rat brain studied, but was most concentrated in microvessels. The 45 kDa form was the only form detected in microvessel-free cortical membranes and was the major form detected in all other regions. Because microvessels were not isolated from hippocampus and cerebellum, these regions contained both the 45 kDa and the 55 kDa forms of GLUT1. The neurohypophysis expressed only the 45 kDa form of GLUT1 (Figure 2).

Antiserum to mouse-GLUT3 C-terminal sequence detects a 45 kDa protein in rat brain, but does not cross react with the GLUT3 protein in human brain (3) or canine brain (unpublished data). GLUT3 was not detected in isolated rat cortical microvessels by immunofluorescence analysis with antisera to mouse-GLUT3 or human-GLUT3 (Figure 1) or by immunoblot analysis. GLUT3 was detected in the cortical neuronal/glial membrane

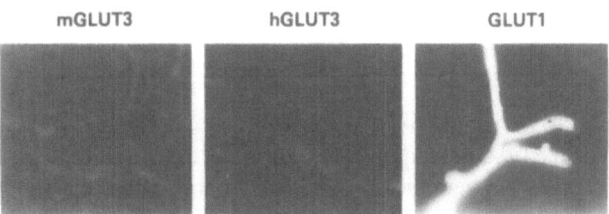

Figure 1. Immunofluorescence of isolated rat cortical microvessels with antisera to mouse-GLUT3, human-GLUT3 and GLUT1. Freshly isolated microvessels were to attached to poly-L-lysine/collagen coated slides and fixed with 4% paraformaldehyde (30 min) prior to incubation with antisera to GLUT1 and GLUT3. The secondary antibody was rhodamine-conjugated goat anti-rabbit IgG.

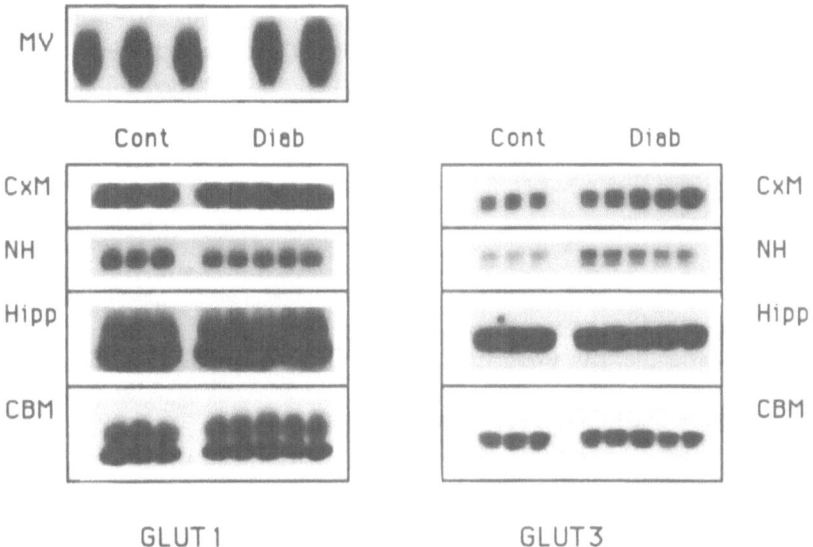

Figure 2. Representative immunoblots of GLUT1 and GLUT3 in control and diabetic animals in each of the brain regions studied. Cortical microvessels (MV); Cortical neuronal/glial membranes (CxM); Hippocampus (HIPP); Cerebellum (CBM); Neurohypophysis (NH).

fraction, and all other regions of the brain studied, with antiserum to mouse-GLUT3 (Figure 2) but not human-GLUT3.

Effects of Diabetes on GLUT1 and GLUT3 Expression

In diabetic animals, GLUT1 was increased in microvessels by 26±6% (% change ± S.E.M.) and in cortical membranes by 17±10%. Preliminary analysis indicated no significant change in hippocampus and cerebellum. In contrast GLUT1 was decreased in the neurohypophysis by 32±4%. GLUT3 in diabetics was increased in cortical membranes by 26±14% and in neurohypophysis by 45±12%, but appeared to be unchanged in hippocampus and cerebellum. Preliminary experiments indicated no clear differences in the alterations to GLUT1 and GLUT3 expression between 1 and 2 weeks of diabetes.

CONCLUSIONS

This study is the first direct demonstration of diabetes-induced changes in both GLUT3 and GLUT1 expression in brain as well as a comparison of transporters in cerebral microvessels and microvessel-free cortical membranes of the same animals. Comparable increases (20-30%) in transporter expression were seen for GLUT1 in microvessels as well as GLUT1 and GLUT3 in microvessel-free cortical membranes. These increases correlate with reported increases of 25-30% in cytochalasin B binding (1) and brain glucose uptake (4) in streptozotocin-induced diabetes. The absence of changes in transporter levels in hippocampus and cerebellum suggests regional specificity for regulation of glucose transporter number.

The greatest increase (45%) in GLUT3 levels was observed in the neurohypophysis, a structure lying outside the blood-brain barrier. In marked contrast to other regions, GLUT1 was decreased in this tissue. These results suggest alternate regulatory mechanisms for the expression of glucose transporters inside and outside the blood-brain barrier.

These increases in GLUT1 and GLUT3 glucose transporter expression suggest an enhanced capacity for brain glucose transport, both across the blood-brain barrier and into neurons, in response to the changes in blood glucose and/or insulin.

REFERENCES

1. Harik S.I., Gravina, S.A., and Kalaria, R.N., Glucose transporter of the blood-brain barrier and brain in chronic hyperglycemia, *J. Neurochem.* 51:1930-1934, 1988.
2. Maher, F., Davies-Hill, T.D., Lysko, P.G., Henneberry, R.C., and Simpson, I.A., Expression of two glucose transporters, GLUT1 and GLUT3, in cultured cerebellar neurons: evidence for neuron-specific expression of GLUT3, *Mol. Cell Neurosci.* 2:351-360, 1991.
3. Maher, F., Vannucci, S.J., Takeda, J., and Simpson, I.A., Expression of mouse-GLUT3 and human-GLUT3 glucose transporter proteins in brain, *Biochem. Biophys. Res. Commun.* 182:703-711, 1992.
4. Mans, A.M., DeJoseph, M.R., Davis, D.W., and Hawkins, R.A., Brain energy metabolism in streptozotocin-diabetes, *Biochem. J.* 249:57-62, 1988.
5. McCall, A.L., The impact of diabetes on the CNS, *Diabetes* 41:557-570, 1992.
6. Nagamatsu, S., Kornhauser, J.M., Burant, C.F., Seino, S., Mayo, K.E., and Bell, G.I., Glucose transporter expression in brain: cDNA sequence of mouse GLUT3, the brain facilitative glucose transporter isoform, and identification of sites of expression by in situ hybridization, *J. Biol. Chem.* 267:467-472, 1992.

GLUCOSE TRANSPORTER (GLUT1) EXPRESSION BY CANINE BRAIN MICROVESSEL ENDOTHELIAL CELLS IN CULTURE: AN IMMUNOCYTOCHEMICAL STUDY

Jay M. Hemmila and Lester R. Drewes

Department of Biochemistry and Molecular Biology
School of Medicine, University of Minnesota
Duluth, MN 55812

INTRODUCTION

Brain microvessel endothelial cells (BMEC) form a selectively permeable barrier known as the blood-brain barrier (BBB) (16). The barrier is maintained structurally by the formation of tight junctions between adjacent BMEC plasma membranes (1). As a result, metabolites exchanged between the blood and brain must pass through the BMEC and BMEC, are marked by transport systems that are highly expressed in comparison to endothelial cells of non-barrier origin.

BBB dysfunction has been linked to several pathological conditions, including anaplastic and metastatic brain tumors (7,8), multiple sclerosis (3), Alzheimer's disease (10) and acquired immunodeficiency syndrome (17). Also, a competent barrier is often an obstacle to efficient drug delivery to the brain. Therefore, studies that further our understanding of BBB maintenance and regulation are warranted. For this purpose, BMEC cultures as in vitro BBB models have received the most attention [reviewed by Joó (9)].

Glucose is supplied to the brain for energy metabolism via a carrier-mediated system (2). BMEC are rich in facilitative glucose transporters of the human erythrocyte/HepG2 (GLUT1) isoform as identified by biochemical (11), immunohistochemical (5,11,15) and molecular biological methods (15,19). In anaplastic astrocytomas (7) and in some experimental brain neoplasms (6), GLUT1 expression was found to be independent of barrier tightness, suggesting that these two BBB characteristics may be regulated by different mechanisms. GLUT1 mRNA is maintained at high levels in primary cultures of BMEC (19). We have begun an immunochemical study of this culture system using an antiserum (5) to the carboxyl terminus of GLUT1. A BMEC in vitro BBB model in which GLUT1 is highly expressed would facilitate the study of GLUT1 regulation at the BBB. Also, an immunocytochemical detection system would be an efficient method for screening for GLUT1 induction and/or maintenance factors.

Frontiers in Cerebral Vascular Biology: Transport and Its Regulation
Edited by L.R. Drewes and A.L. Betz, Plenum Press, New York, 1993

13

MATERIALS AND METHODS

Medium 199 (M199), antibiotic/antimycotic solution (penicillin, streptomycin, amphotericin), dextran, fetal bovine serum (FBS), gelatin and glycerol gelatin were purchased from Sigma (St. Louis, MO, USA); type IV collagenase from Worthington Biochemical (Freehold, NJ, USA); trypsin in a 2.5% solution from Boehringer Mannheim (Indianapolis, IN, USA); Mito+ serum extender from Collaborative Research (Bedford, MA, USA); 1,1'-dioctadecyl-3,3,3',3'-tetramethylindocarbo-cyanine perchlorate acetylated low-density lipoprotein (DiI-Ac-LDL) from Biomedical Technologies (Stoughton, MA, USA); and a peroxidase-antiperoxidase system in kit form (Dako PAP Kit System 40, universal rabbit) from Dako (Carpinteria, CA, USA).

Brain microvessels were isolated from canine cerebral cortex as described previously (4). Isolated microvessels were suspended in M199 and checked microscopically for quality. The preparations typically were found enriched in microvessels with only minor amounts of associated debris. The isolated microvessels were then resuspended in 1 mg/ml type IV collagenase in M199 and incubated at 37°C in a shaking water bath overnight. After collagenase digestion, the cell suspensions were washed once with M199, resuspended in 0.25% trypsin in M199 and incubated in a 37°C shaking water bath for 30 min. The cells were then washed twice in growth medium (M199, 10% fetal bovine serum, Mito+ serum extender) and seeded into 24-well plates containing gelatin-coated 12-mm glass coverslips and/or gelatin-coated T75 flasks. Usually one prep was plated into either 48 wells or 24 wells and one T75 flask for passaging cells. The media was changed twice weekly. Cultures reached confluence in 9 to 12 days. In some experiments T75 flasks were passaged 1:3 with 0.25% trypsin.

As a negative control, endothelial cells of non-barrier origin were cultured. Primary cultures of canine aortic endothelial cells (AEC) were prepared by collagenase digestion essentially as described by Madri et al. (12). AEC were seeded and maintained under the same conditions as BMEC. The endothelial character of the BMEC and AEC cultures was confirmed by DiI-Ac-LDL uptake (18). Briefly, cells were incubated in growth medium containing 10 µg/ml DiI-Ac-LDL for 3-4 hours at 37°C. Cultures were rinsed with phosphate buffered saline (PBS), fixed (10 min in 3.7% formaldehyde, PBS) and mounted with glycerol gelatin on glass slides. Phase contrast and epifluorescence photomicrography was performed using a Nikon Diaphot-TMD inverted microscope.

Coverslips with cells were removed daily for GLUT1 immunodetection. After rinsing well with PBS, cells were fixed and permeabilized (5 min in 3.7% formaldehyde, 70% ethanol) and immunostained using a peroxidase-antiperoxidase system. A primary antiserum to the carboxyl terminal of GLUT1 (5) was used at a dilution of 1:50 in PBS, or antiserum pre-incubated with 10 µM carboxyl terminal peptide was applied as a negative control. The chromogen was amino-ethylcarbazole. The cells were counterstained with Mayer's hematoxylin, and the coverslips were mounted onto glass slides with glycerol gelatin. Light photomicrography was performed with an Olympus photomicrographic system.

RESULTS

Freshly plated BMEC were seen mainly as aggregates of cells that spread into monolayers during the first 3-4 days of culture. These early colonies displayed spindle-shaped cells and well-defined borders. Aortic endothelial cells (AEC) were well dispersed when plated and formed confluent monolayers in six days. The endothelial nature of both cell types was confirmed by DiI-Ac-LDL uptake (Figure 1).

We found that immunocytochemistry was an efficient method to screen for GLUT1 expression in BMEC. One preparation of isolated canine BMEC was sufficient to seed 48 wells (two 24-well plates) with each well containing a 12-mm glass coverslip. Individual

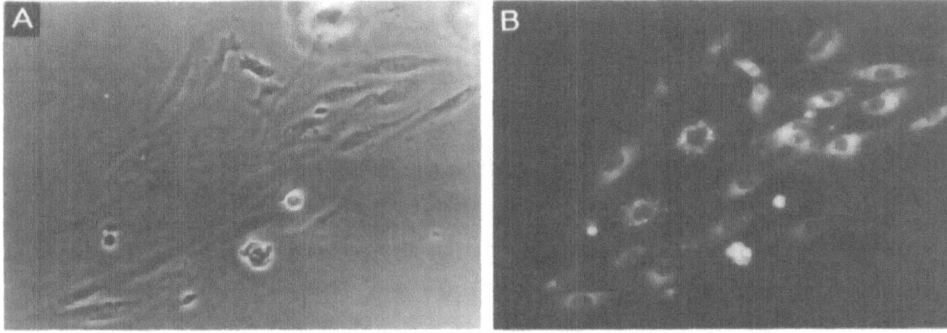

Figure 1. LDL-uptake by cultured BMEC. A primary BMEC colony after six days in culture is stained with DiI-Ac-LDL and shows the perinuclear staining pattern that is typical of endothelial cells. Phase-contrast (A) and fluorescence (B) photomicrographs.

coverslips with BMEC colonies attached were removed from the 24-well plates for GLUT1 staining at any desired time point, and the staining procedure required 3-4 hr. After three days in culture, BMEC displayed intense GLUT1 staining in their plasma membranes (Figure 2A). The antiserum pre-absorbed with GLUT1 carboxyl terminus peptide failed to stain BMEC, thus showing the specificity of the staining for GLUT1 (Figure 2B).

Cell division was rapid from 4 days after plating to confluence. At six days, most cells continued to stain for GLUT1, but staining appeared decreased in several of the cells (Figure 2C). Cells around the periphery of the colony were generally less stained than those in the center.

At confluence (9 to 12 days in culture), areas of positive GLUT1 staining remained evident, but most cells in the monolayer exhibited low or no staining (Figure 2D). When these cells were passaged by trypsinization, they tended to plate as aggregates of cells as did the primary cultures. However, first-time passaged cells spread more quickly than did the primaries. GLUT1 staining decreased further in first-time passaged BMEC, although a few GLUT1-positive cells could still be seen (not shown). By the third passage, BMEC exhibited typical polygonal morphology and dispersed mainly as single cells when plated. Third passage BMEC did not express GLUT1 in detectable amounts in any cells (Figure 2E). In contrast to BMEC, primary cultures of AEC did not express detectable levels of GLUT1 (Figure 2F).

DISCUSSION

Glucose transport is a critical function of BBB endothelial cells, and BMEC plasma membranes contain an abundance of facilitative glucose transporters (GLUT1 isoform) to perform this task. Changes in the BBB glucose transport system have been documented. For example, GLUT1 expression is lost in some brain tumors (6), and brain glucose transport decreases with age (14). Therefore, effectors and underlying mechanisms of regulation of GLUT1 expression and activity in the BBB need to be identified. An in vitro BBB model where GLUT1 is hyperexpressed would facilitate these studies. Cultured BMEC as in vitro BBB models have been studied extensively [reviewed by Joó (9)]. We used immunocytochemistry to examine GLUT1 expression in BMEC primary cultures. GLUT1 is hyperexpressed in BMEC in vivo, and our results indicate that this phenotype is retained in freshly plated primary cultures. However, as cells proliferate, many low or non-GLUT1-staining cells are seen, and by the second or third passage, no GLUT1 expression is detected immunocytochemically.

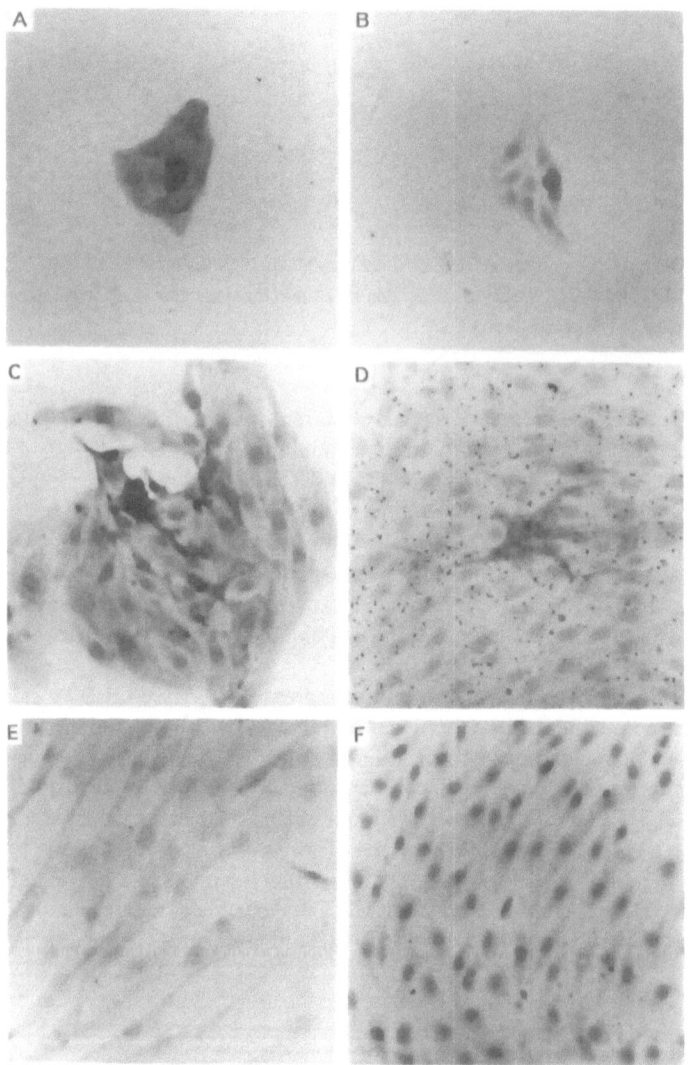

Figure 2. GLUT1 immunostaining in endothelial cell cultures. Cell cultures were routinely stained with GLUT1 carboxyl-terminal antiserum or antiserum pre-absorbed with GLUT1 carboxyl-terminal peptide and counterstained with hematoxylin. Three-day-old primary BMEC cultures showed intense plasma membrane staining for GLUT1 in all cells (A), and this staining was blocked by GLUT1 carboxyl-terminal peptide (B). After six days in culture, expanding BMEC colonies consisted of positive and negative GLUT1 staining cells (C). At confluence (9-12 days), most BMEC were negative for GLUT1 staining; however, islands of positive GLUT1 reactivity were present (D). By the third passage, no detectable GLUT1 staining could be found in BMEC (E). Endothelial cells from a non-barrier source (aorta) did not stain positively for GLUT1 at any time; GLUT1 staining of six-day-old primary culture of AEC is shown (F).

These results are in accordance with those of Meyer et al. (13) who showed similar decreases in the activities of two BBB markers, γ-glutamyl transpeptidase and alkaline phosphatase, as cells proliferated. In another study, Weiler-Guttler et al. (19) found abundant GLUT1 mRNA levels in primary cultures of porcine BMEC. These levels remained constant in rapidly dividing cultures and in confluent monolayers. In our study, we found a dramatic decrease in GLUT1 expression in confluent primary cultures. This apparent discrepancy may be a result of different initial seeding densities. Our primary cultures reached confluence in 9-12 days, whereas the primary cultures reported by Weiler-Guttler et al. (19) required only six days. GLUT1 expression was still significantly high at six days in our cultures.

In contrast to BMEC, primary cultures of AEC did not express detectable levels of GLUT1. This suggests that the GLUT1 expression seen in cultured BMEC was the result of the temporary maintenance in vitro of the in vivo GLUT1 marker and not an artifact of the culture method.

We found that immunocytochemistry was an efficient method to follow changes in the qualitative expression of GLUT1 in cultured BMEC. Since relatively few cells are required for immunostaining, several factors and growth conditions may be screened for qualitative effects on GLUT1 expression with each culture preparation. In contrast the quantitative techniques of Western and Northern blotting require many cells and are time consuming, making these methods inefficient for screening multiple factors. The effectiveness of this in vitro BBB model and GLUT1 immunostaining system for identifying effectors of GLUT1 expression in BMEC awaits further studies.

SUMMARY

We have found that immunocytochemical techniques can be used to monitor the qualitative expression of GLUT1 in cultured cells. Primary cultures of BMEC showed significant GLUT1 staining, and primary cultures of endothelial cells from a non-barrier source (aorta) were negative. BMEC express significant levels of GLUT1 in young primary cultures, thus affording an opportunity for short-term regulation studies of GLUT1 activity and expression.

Immunocytochemistry may be an efficient method for screening potential GLUT1 induction and maintenance factors at the BBB. The immunocytochemical technique used is rapid and requires relatively few cells and may provide a useful adjunct to the more quantitative techniques of Western and Northern blotting.

ACKNOWLEDGMENTS

This work was supported by grants from the National Institutes of Health (NS-27229) and the American Diabetes Association, Minnesota Affiliate.

REFERENCES

1. Brightman, M.W., and Reese, T.S., Junctions between intimately apposed cell membranes in the vertebrate brain, *J. Cell Biol.* 40:638-677, 1969.
2. Crone, C., Facilitated transfer of glucose from blood into brain tissue, *J. Physiol. London* 181:103-113, 1965.
3. Gay, D., and Esiri, M., Blood-brain barrier damage in acute multiple sclerosis plaques. An immunological study, *Brain* 114:557-572, 1991.

4. Gerhart, D.Z., Broderius, M.A., and Drewes, L.R., Cultured human and canine endothelial cells from brain microvessels, *Brain Res. Bull.* 21:785-793, 1988.

5. Gerhart, D.Z., LeVasseur, R.J., Broderius, M.A., and Drewes, L.R., Glucose transporter localization in brain using light and electron immunocytochemistry, *J. Neurosci. Res.* 22:464-472, 1989.

6. Guerin, C., Laterra, J., Drewes, L.R., Brem, H., and Goldstein, G.W., Vascular expression of glucose transporter in experimental brain neoplasms, *Am. J. Path.* 140:417-425, 1992.

7. Guerin, C., Laterra, J., Hruban, R., Brem, H., Drewes, L.R., and Goldstein, G.W., The glucose transporter and blood-brain barrier of human brain tumors, *Ann. Neurol.* 28:758-765, 1990.

8. Hasegawa, H., Ushio, Y., Hayakawa, T., Yamada, K., and Mogami, H., Changes of the blood-brain barrier in experimental metastatic brain tumors, *J. Neurosurg.* 59:304-310, 1983.

9. Joó, F., The cerebral microvessels in culture, an update, *J. Neurochem.* 58:1-17, 1992.

10. Kalaria, R.N., and Harik, S.I., Reduced glucose transporter at the blood-brain barrier and in cerebral cortex in Alzheimer disease, *J. Neurochem.* 53:1083-1088, 1989.

11. Kalaria, R.N., Gravina, S.A., Schmidley, J.W., Perry, G., and Harik, S.I., The glucose transporter of the human brain and blood-brain barrier, *Ann. Neurol.* 24:757-764, 1988.

12. Madri, J.A., Dreyer, B., Pitlick, F.A., and Furthmayr, H., The collagenous components of the subendothelium. Correlation of structure and function, *Lab. Invest.* 43:303-315, 1980.

13. Meyer, J., Mischeck, U., Vehyl, M., Henzel, K., and Galla, H.-J., Blood-brain barrier characteristic enzymatic properties in cultured brain capillary endothelial cells, *Brain Res.* 514:305-309, 1990.

14. Mooradian, A.D., Morin, A.M., Cipp, L.J., and Haspel, H.C., Glucose transport is reduced in the blood-brain barrier of aged rats, *Brain Res.* 551:145-149, 1991.

15. Pardridge, W.M., Boado, R.J., and Farrell, C.R., Brain-type glucose transporter (Glut-1) is selectively localized to the blood-brain barrier. Studies with quantitative western blotting and in situ hybridization, *J. Biol. Chem.* 265:18035-18040, 1990.

16. Reese, T.S., and Karnovsky, M.J., Fine structural localization of a blood-brain barrier to exogenous peroxidase, *J. Cell Biol.* 34:207-217, 1967.

17. Rhodes, R.H., Evidence of serum-protein leakage across the blood-brain barrier in the acquired immunodeficiency syndrome, *J. Neuropathol. Exp. Neurol.* 50:171-183, 1991.

18. Voyta, J.C., Via, D.P., Butterfield, C.E., and Zetter, B.R., Identification and isolation of endothelial cells based on their increased uptake of acetylated-low density lipoprotein, *J. Cell Biol.* 99:2034-2040, 1984.

19. Weiler-Guttler, H., Zinke, H., Mockel, B., Frey, A., and Gassen, H.G., cDNA cloning and sequence analysis of the glucose transporter from porcine blood-brain barrier, *Biol. Chem. Hoppe-Seyler* 370:467-473, 1989.

APPLICATION OF NOVEL PCR STRATEGIES TO AMPLIFY AND SEQUENCE GLUCOSE TRANSPORTERS IN CANINE BRAIN

Nancy D. Borson, Wilmar L. Salo, and Lester R. Drewes

Department of Biochemistry and Molecular Biology
School of Medicine
University of Minnesota
Duluth, MN 55812

INTRODUCTION

The transport of glucose in the brain of mammalian species is extremely important, not only for growth and development, but to provide energy for functional activity. Glucose is transported across plasma membranes in mammalian tissues in part by members of a family of glucose transporters. Five highly conserved isoforms of this family (Glut1-Glut5) have been identified in various species and tissues, and all five members are believed to be integral membrane proteins that facilitate glucose diffusion.[1,2] The Glut1 and Glut3 isoforms have been detected in brain in several species,[3-8] but a question arises as to whether or not other glucose transporter family members are also expressed. Also, the possibility exists that an as yet unidentified family member may be expressed in brain. Therefore, a rapid and reliable method of determining which glucose transporter isoforms are expressed in various species and tissues is desirable.

The usual method of identifying members of a multigene family is through cDNA cloning and screening at low stringency with a probe produced from a known family member. Another method is to employ the polymerase chain reaction (PCR) primed by pairs of consensus primers targeted at regions of a highly conserved nucleotide sequence. The template used in the PCR can be mRNA, total RNA, or cDNA produced by reverse-transcription of mRNA. However, because of the similarity in size of the glucose transporter isoforms (the range among the five known human isoforms is from 492 to 501 amino acids[2]), the production of differentiable PCR products for glucose transporters is difficult when using pairs of internal consensus primers (Fig. 1). These products could be cloned and screened for glucose transporters at low stringency, but the process again becomes time-consuming, and it would be desirable to acquire discrete PCR products that could be directly sequenced and identified.

An alternative approach using the PCR is to amplify the 3'-end products of each family member. The advantage to this approach is the increased probability of producing PCR

Frontiers in Cerebral Vascular Biology: Transport and Its Regulation
Edited by L.R. Drewes and A.L. Betz, Plenum Press, New York, 1993

19

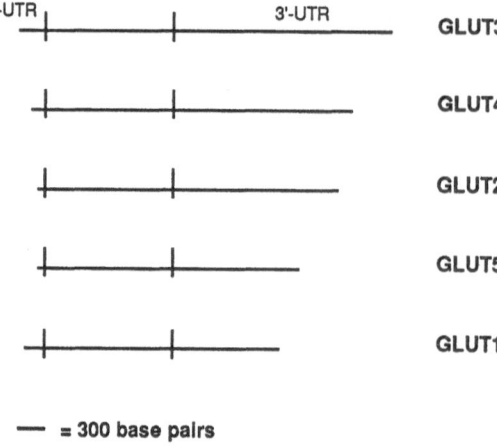

Figure 1. Variations in mRNA lengths of human glucose transporter isoforms. Each line depicts a representative isoform mRNA, and the 5'-untranslated, coding and 3'-untranslated regions are indicated. Information was obtained from Northern blots1 and from Gene Bank.

products that vary considerably in length due to the differences usually found in the lengths of 3'-untranslated regions (3'-UTRs) of glucose transporter mRNA transcripts (Fig.1).

An approach to amplify 3'-ends, known as 3'-RACE, has been described.[9] In this method, first-strand cDNA template is reverse-transcribed using an oligo(dT) primer that contains some multiple cloning sequence at its 5'-end. The PCR primer pair then consists of an upstream consensus primer based on a region of conserved sequence in the coding region of all, or many, family members, and a downstream primer composed of the multiple cloning sequence (MCS) only. However, because the oligo(dT) primer used for cDNA synthesis can bind at multiple sites along the length of the mRNA poly(A) tail, a heterogeneous population of products is usually obtained in the PCR. When these products are electrophoresed on an agarose gel and viewed under the UV light after ethidium bromide staining, a smear is often observed.[10] The resolution of these products then also requires Southern blotting, cloning, and screening before identification by sequencing can be done.

Thus, an objective of this study was to design a modified oligo (dT) primer for cDNA synthesis, that would subsequently enable the acquisition of discrete, first-round PCR products for 3'-cDNA ends of any reverse-transcribed poly(A) RNAs. These products could then be isolated from an agarose gel and sequenced directly, circumventing a need for Southern blotting and cloning. A primer that would lock in at the junction of the 3'-untranslated region and the poly(A) tail of mRNA was designed. This "lock-docking" primer was used for reverse transcription of canine cortex mRNA to produce single-stranded cDNA template (lock-docked cDNA) for use in the PCR.

The 3'-RACE procedure using lock-docked cDNA (lock-docking/3'-RACE system) was first tested in two separate PCRs to amplify the 3'-end products for canine Glut1 and Glut3[11]. Subsequently, two different canine-sequence based consensus primers for use as upstream primers in the PCR (Primers CO1 and CO2) were designed. When either of these primers is employed in the PCR, using the lock-docking/3'-RACE system with first-strand cDNA template derived from canine kidney, intestine, brain cortex, and brain microvessels, the 3'-ends of the more highly expressed glucose transporters are obtained as products. When these products are electrophoresed on an agarose gel, the multiple products are discrete and differentiable, and a display of the glucose transporters that are abundantly expressed in each tissue can be observed.

METHODS AND MATERIALS

Lock-docking Primer Design

CCGCATGCGGCCGCAGATCTAGATATCGA(T)$_{16}$(ACG)(ACGT)

The degeneracy at the 3'-end of this primer, which is used for cDNA synthesis, enables it to specifically lock in at the junction of the 3'-end sequence of a poly(A) RNA transcript and the beginning of polyadenylation.

MCS Primer Design

CCGCAGATCTAGATATCGA

Consensus Primer Designs

Consensus primer CO1 is based entirely on canine specific sequence from Glut1 and Glut3. The nucleotide sequence that correlates with the highly conserved (YF)KVPET amino acid sequence (amino acids 448-455 in human Glut1) was chosen for this primer.

Primer CO1:
(AT)CTTCAAAGT(TC)CCTGAGACC

Consensus primer CO2 is also based on canine specific sequence from the same region as primer CO1, but includes some nucleotide representation from human Glut5 near its 5'-end.

Primer CO2:
(AT)CTT(CG)A(AT)(AT)GT(TC)CCTGAGACC

Isolation of mRNA and cDNA Synthesis

The methods for isolation of mRNA and for cDNA synthesis have been previously described.[11]

PCR Protocols

The protocol used in amplification of Glut1 and Glut3 3'-ends is as described.[11] A single PCR protocol was used for all reactions using the consensus primers. The PCR thermal-cycling parameters were done using a "touchdown" protocol as described.[11] Reactions were performed in final volumes of 50 ul, using AmpliWax[TM] PCR Gems (Perkin-Elmer Cetus, Norwalk, CT). Following a typical PCR Gem protocol, a lower layer consisted of 1.25 ml 10X Stoffel Buffer (Perkin-Elmer Cetus), 6 ml of 25mM MgCl$_2$, 10 ml of a dNTP solution that was 1.25 mM in each dNTP (dATP, dCTP, dGTP, dTTP), and 20 pmoles of each primer. An AmpliWax[TM] PCR Gem was added, the tube was incubated at 80° C for 5 min, and then cooled to room temperature. An upper layer was composed of 25 ml of deionized H$_2$O, 5 ml Stoffel buffer (see above), 10 units of AmpliTaq Stoffel Fragment (Perkin-Elmer Cetus), and 1 ml of a 1:10 dilution of canine brain cortex cDNA (as described[11]). Final concentrations in the PCR reaction were 1.25X Stoffel buffer, 3mM MgCl$_2$, and 250mM in each dNTP. When using the CO2 primer (described above), the reactions were improved by adding dimethyl sulfoxide (Sigma, St. Louis, MO) at a final concentration of 2%.

Gel Electrophoresis

All PCR products were electrophoresed on 1% agarose gels (FMC Bioproducts, Rockland, ME) and viewed under UV light after ethidium bromide staining.

RESULTS

Initially, lock-docked cDNA template from canine brain cortex was used successfully in a 3'-RACE procedure to obtain discrete PCR products for the 3'-ends of canine Glut1 and Glut3.[11] A single, discrete PCR product of 1025 base pairs was obtained for Glut1, and a single, discrete 2400 base pair product was obtained for Glut3. Sequencing of these two products revealed the upstream sequence of canine Glut1 and Glut3, respectively, and both products yielded sequences that traversed the termination codons for these two transcripts.

When the CO1/MCS primer pair was used in a PCR with lock-docked cDNA template from canine upper jejunum, one prominent product of approximately 1425 bp was observed upon electrophoresis (Lane 2, Fig.2). Similarly, kidney lock-docked cDNA as a source of template yielded two major products of approximately 1200 base pairs and 1425 base pairs (Lane 3, Fig. 2), while brain microvessel lock-docked cDNA template thus used resulted in products of approximately 1030 and 560 base pairs (Lane 4, Fig. 2).

The consensus primer CO1 yielded several discrete products when used in a PCR in conjunction with the MCS primer and when using canine cortex lock-docked template (Lane 5, Fig. 2). The five most prominent products were approximately 2350, 1475, 1425, 1030, and 300 base pairs in length. The use of Primer CO2 in place of Primer CO1 with canine cortex lock-docked cDNA template yielded an identical pattern when the PCR products were electrophoresed on an agarose gel (data not shown). All products were observed after one round (31 cycles) of PCR.

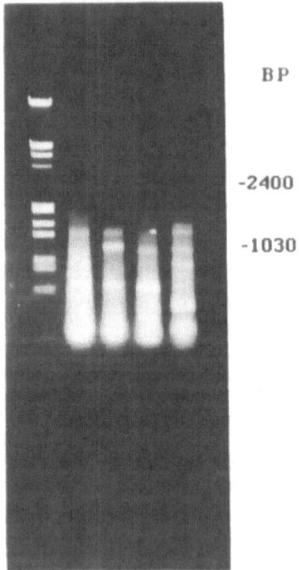

Figure 2. Electrophoresed PCR products that display the 3'-ends of the more abundantly expressed isoforms of the glucose transporter family for canine upper jejunum (Lane 2), kidney (Lane 3), brain microvessel (Lane 4), and brain cortex (Lane 5). Lane 1: Standard (lDNA/Hind III + EcoRI, Boehringer Mannheim Biochemicals, Indianapolis, IN).

DISCUSSION

The lock-docking primer/3'-RACE system proved to be an efficient means of amplifying the 3'-ends of canine brain cortex Glut1 and Glut3[11]. Each of the resultant products was verified by direct sequencing, giving credence to the integrity of the lock-docking system.

The consensus primer that was designed and employed in the lock-docking/3'-RACE system will amplify the 3'-ends of the mRNA transcripts for all of the more abundantly expressed members of the glucose transporter family. These products are obtained simultaneously, after a single round (31 cycles) of PCR. As expected, different band patterns are observed upon electrophoresis of the PCR products obtained from canine small intestine, kidney, brain cortex, and brain microvessel cDNA templates. For example, a single prominent band is observed at 1425 base pairs in upper jejunum. Because upper jejunum reportedly expresses Glut5[1], this PCR product is a likely candidate for the 3'-end of Glut5. A 1425 base pair band is also observed when kidney cDNA is used as template; kidney has also been shown by Northern blotting to express Glut5.[12] The major kidney glucose transporter is believed to be Glut2, so there is a strong possibility that a second, more prominent, kidney PCR product, at approximately 1200 base pairs, is a 3'-end product of Glut2. It is also possible, however, that the two kidney products are alternately terminated Glut2 transcripts. Northern blots of human Glut2 in kidney indicate more than one commonly expressed transcript.[1]

When the lock-docking/3'-RACE system is applied to canine brain cortex and microvessels, differing product patterns are observed upon electrophoresis. From canine brain cortex, several discrete bands are seen. The expected size band of approximately 1030 base pairs for the 3'-end of Glut1 is observed, as well as the expected size band of approximately 2350 base pairs for the 3'-end of Glut3. These two bands are not observed in kidney and small intestine where they are not abundantly expressed in human tissues.[1]

In contrast, from canine brain microvessels, two prominent bands are observed. The differences that are observed in this pattern from the brain cortex pattern would indicate that there was a successful isolation of microvessels from the surrounding cortex tissue. The band at 1030 base pairs is the known, expected size for Glut1. Glut1 has previously been shown by immunocytochemistry to be functionally expressed in the blood-brain barrier.[3,4] The second band at 560 base pairs is presently unidentified. Although Glut3 may be expressed in microvessels, it was below the level of detection by this method.

The remaining discrete PCR products that are observed in canine brain cortex (other than Glut1 and Glut3), and the prominent band at 560 base pairs observed from canine brain microvessels, potentially could be other glucose transporter isoforms (perhaps as yet unidentified), alternately terminated transcripts of expressed isoforms, or alternately spliced transcripts. All bands await sequencing.

The lock-docking/3'-RACE system should be helpful in delineating glucose transporter expression in any species or tissue. Also, this process should be applicable to other gene families when some species-specific sequence is known. The problems inherent in the detection of rarer transcripts using the PCR remain. However, many of the standard approaches that are now taken to overcome this problem should be applicable with this primer system as well. For example, it could be helpful to perform a linear first round PCR reaction with the species specific consensus primer in excess. When first-strand cDNA is used as template, this would result in increased copies of a transcript that was initially present in low copy number. The linear PCR products could then be used as templates in a regular round of PCR.

SUMMARY

We employ here a modified oligo(dT) primer called a "lock-docking" primer which enables the production of a cDNA template that can subsequently be amplified in the 3'-RACE PCR procedure to produce discrete, first-round products for 3'-cDNA ends of any reverse-transcribed poly(A) RNAs.

An upstream consensus primer targeted to a highly conserved region (amino acid region 448-455 in human Glut1) among all glucose transporter isoforms was used in the 3'-RACE PCR procedure with lock-docked cDNA template. Sources of lock-docked cDNA template were canine intestine, kidney, brain cortex, and brain microvessels. This procedure made it possible to delineate, in one round of PCR (31 cycles), all of the more abundantly expressed glucose transporter isoforms in each of these tissues. Other as yet unaccounted for products were also obtained. These products are potentially alternately terminated transcripts of glucose transporter isoforms, alternately spliced transcripts, or as yet uncharacterized isoforms. After electrophoresis on an agarose gel and purification of the DNA, each of these PCR products is available for direct sequencing.

ACKNOWLEDGMENTS

This work was supported by grants from the National Institutes of Health (NS-27229, L.R.D.) and the American Diabetes Association, Minnesota Affiliate (W.L.S. and L.R.D.).

REFERENCES

1. Burant, C.F., et al., Mammalian glucose transporters: structure and molecular regulation, *in* "Recent Progress in Hormone Research," Academic Press, Inc., 47:349-389, 1991.
2. Bell, G.I., et al., Molecular biology of mammalian glucose transporters, *Diabetes Care* 13:198-208, 1990.
3. Pardridge, W.M., Boado, R.J., and Farrell, C.R., Brain-type glucose transporter (Glut-1) is selectively localized to the blood-brain barrier, *J.Biol.Chem.* 265:18035-18040, 1990.
4. Gerhart, D.Z., LeVasseur, R.J., Broderius, M.A., Drewes, L.R., Glucose transporter localization in brain using light and electron immunocytochemistry, *J. Neurosci. Res.* 22:464-472, 1989.
5. Kayano, T., Fukumoto, H., Eddy, R.L., Fan, Y., Byers, M.G., Shows, T.B., Bell, G.I., Evidence for a family of human glucose transporter-like proteins: sequence and gene localization of a protein expressed in fetal skeletal muscle and other tissues, *J.Biol.Chem.* 263:15245-15248, 1988.
6. Gerhart, D.Z., Broderius, M.A., Borson, N.D., Drewes, L.R., Neurons and microvessels express the brain glucose transporter protein GLUT3, *Proc. Natl. Acad. Sci.* 89:733-737, 1992.
7. Nagamatsu, S., Kornhauser, J.M., Burant, C.F., Seino, S., Mayo, K.E., Bell, G.I., Glucose transporter expression in brain: cDNA sequence of mouse Glut3, the brain facilitative glucose transporter isoform, and identification of sites of expression by in situ hybridization, *J. Biol. Chem.* 267:467-472, 1992.
8. Maher, F., Vannucci, S., Takeda, J., Simpson, I.A., Expression of mouse-Glut3 and human-Glut3 glucose transporter proteins in brain, *Biochem. Biophys. Res. Comm.* 182:703-711, 1992.
9. Frohman, M.A., RACE: Rapid amplification of cDNA ends, *in:* "PCR Protocols: A Guide to Methods and Applications," M.A.Innis, D.H. Gelfand, J.J. Sninsky, T.J. White, eds., pp. 28-38, Academic Press, Inc., San Diego, 1990.
10. Jain, R., Gomer, R.H., Murtagh, Jr., J.J., Increasing specificity from the PCR-RACE technique, *Bio.Techniques* 12:58-59, 1992.
11. Borson, N.D., Salo, W.L., Drewes, L.R., *PCR Methods and Applications*, 1992, in press.
12. Kayano, T., et al., Human facilitative glucose transporters, *J. Biol. Chem.* 265:13276-13282, 1990.

ESTIMATION OF UNIDIRECTIONAL

CLEARANCES OF FDG AND GLUCOSE

ACROSS THE BLOOD-BRAIN BARRIER IN MAN

Steen Hasselbalch, Gitte Moos Knudsen, Johannes Jakobsen,
Søren Holm, and Olaf B. Paulson

Department of Neurology, Rigshospitalet
University State Hospital, Copenhagen, Denmark

INTRODUCTION

The deoxyglucose method for calculation of regional cerebral glucose metabolism using ^{18}F-2-fluoro-2-deoxy-D-glucose (FDG) as a tracer (1,2) requires knowledge of differences in the transport of FDG and glucose across the blood-brain barrier (BBB), as well as affinity for the enzyme hexokinase, in the first step of the glycolysis. These differences are incorporated in a correction term, the lumped constant. The unique feature of the PET-FDG method is the ability to measure the absolute metabolic rate of glucose, but the correct estimation of the lumped constant has been a major problem. Few methods for estimation of the lumped constant have so far been put forward (1,3,4). Simultaneous measurements of the transport of FDG and glucose across the BBB have been performed in animals (5,6), but never before in humans.

The double-indicator method adapted for studies in man (7) requires injection of the test and the reference substance into the carotid artery. Based on this method, a new method — the intravenous double-indicator method — has recently been developed. Here the test and reference substances are injected into a peripheral vein instead of the carotid artery and, therefore, the method can also be applied in normal volunteers.

The main purpose of the present study was to evaluate, for the first time in man, the relationship between the transport (unidirectional clearance) across the BBB of FDG and glucose using the intravenous double-indicator method, thereby determining the first correction term of the lumped constant.

METHODS

Twenty-four healthy subjects, 11 females and 13 males, were studied. Mean age was 23±4 years. All 24 subjects were studied under identical, normoglycemic baseline conditions (plasma glucose = 5.2±0.7 mM). Fifteen subjects were restudied during a hyper-insulinaemic, euglycaemic clamp (plasma glucose = 5.2±0.7 mM, plasma insulin = 500±91

Frontiers in Cerebral Vascular Biology: Transport and Its Regulation
Edited by L.R. Drewes and A.L. Betz, Plenum Press, New York, 1993

25

pM) and 9 during moderate hypoglycemia after 3.5 days of starvation (plasma glucose = 3.4±0.4). On each study day, a small polyethylene catheter was inserted into the radial artery and into two antecubical veins under local analgesia. Using the Seldinger technique, a catheter was placed in the internal jugular vein through percutaneous puncture 3 to 4 cm above the clavicle.

Blood-Brain Barrier Permeability Measurements

Theory. A single membrane (well-mixed) model was used because it has been found suitable for the analysis of glucose (8). With this model, the theoretical output values for the test substance may be calculated if the CBF and the reference output values are known. The estimates of the model parameters and their standard errors are obtained by minimizing the squares of the differences between the theoretical and the measured output test values.

The operational equation for the single membrane model is

$$C_{test}(t) = \int_0^{\theta_{max}} C_{ref}(\theta)\, C_c[L, t-\tau_1(\theta), \tau_c(\theta)]\, d\theta \qquad (1)$$

in which $C_{ref}(t)$ and $C_{test}(t)$ are the amounts of the reference and test substances as measured at the jugular vein at time t. The transit time through the large vessels is τ_1 and τ_c is the transit time through the capillary bed. The total capillary transit time is θ while θ_{max} is the value of θ for which $t-\tau_1(\theta) = 0$. $C_c(L, t, \tau_c)$ is the capillary concentration at time t, when L is the capillary length of a capillary with the transit time τ_c.

With intravenous injection, the integration of equation (1) is accomplished by replacing $C_{ref}(\theta)$ with the transfer function of the reference material between the artery and the jugular vein. The result is then convoluted with the arterial test concentration to yield theoretical venous values, which are then compared to the measured venous test values. From these measurements, the unidirectional extraction fraction E is calculated. Provided CBF is known, the unidirectional clearance from blood to brain (K_1) can then be calculated: $K_1 = E * CBF$.

Procedure. Cerebral blood flow (CBF) was determined using the ^{133}Xe intravenous injection method with stationary detectors (9). For the double-indicator study, a 5- to 10-ml bolus, containing two test substances (radioactive labeled glucose and FDG) and three BBB impermeable reference substances ([36]Cl, [24]Na, and ^{99m}Tc-DTPA) was injected intravenously through the antecubical catheter. Starting 2 to 3 seconds before injection, continuous samples of 1 ml of blood were collected at a fixed time interval of 1.3 seconds from the radial artery and jugular vein by means of a sample machine.

RESULTS

The mean absolute value of K_1 (glucose) was 0.098±0.021 during baseline studies and 0.104±0.038 during insulin clamp with no significant difference between the two values. It increased significantly to 0.135±0.022 during hypoglycemia. The same pattern was seen for FDG, where K_1^* remained unchanged from a baseline value of 0.141±0.028 to a hyperinsulinemic value of 0.142±0.038, but increased during hypoglycemia to 0.188±0.033. The ratio K_1^*/K_1 was 1.48±0.219 during baseline conditions. The mean value of this ratio did not change significantly during the hyperinsulinemic clamp (mean 1.524±0.376). No change in the ratio was observed during moderate hypoglycemia (mean 1.413±0.228), since both K_1 and K_1^* increased by the same magnitude during hypoglycemia.

DISCUSSION

In the present study, the ratio of the transport of glucose and FDG across the BBB was determined for the first time in man. The value of 1.48 is in the same order of magnitude as the values of 1.67 and 1.65 obtained in animal experiments (5,6). It thus seems that the ratio between the clearances of the two hexoses does not differ substantially between rats and humans. Most methodological errors in the present study would affect both hexoses identically, since both K_1 and K_1^* were determined in the same experiment, and the ratio is, therefore, well determined.

The ratio remained constant during hyperinsulinemia and during moderate hypoglycemia. This constancy is crucial for the FDG method (and for all tracer studies) because the BBB transport of the tracer (FDG) should reflect the BBB transport of the native molecule (glucose) during all conditions. In theory, it can be shown that the value of the lumped constant increases in a predictable manner as plasma glucose decreases (10). However, the theoretical relationship between plasma glucose and the lumped constant is only valid under the assumption that the ratio of BBB transport between FDG and glucose is constant. The present study has now confirmed those previous assumptions and confirmed the validity of the use of the FDG method during hyperinsulinemia and moderate hypoglycemia.

REFERENCES

1. Sokoloff, L., Reivich, M., Kennedy, C., et al. The C-14-deoxyglucose method for the measurement of local cerebral glucose utilization: theory, procedure, and normal values in the conscious and anesthetized albino rat, *J. Neurochem.* 28:897, 1977.

2. Reivich, M., Kuhl, D., Wolf, A., Greenberg, J., et al., The 18-F-fluorodeoxyglucose method for the measurement of local cerebral glucose metabolism in man, *Circ. Res.* 44:127, 1979.

3. Reivich, M., Alavi, A., Wolf, A., et al., Glucose metabolic rate kinetic model parameter determination in humans: the lumped constants and rate constants for F-18-fluorodeoxyglucose and C-11-deoxyglucose, *J. Cereb. Blood Flow Metab.* 4:179, 1985.

4. Kuwabara, H., Evans, A.C., and Gjedde, A., Michaelis-Menten constraints improved the regional lumped constant measurements with F-18-fluorodeoxyglucose, *J. Cereb. Blood Flow Metab.* 10:180, 1990.

5. Crane, P.D., Pardridge, W.M., Braun, L.D., and Oldendorf, W.M., Kinetics of transport and phosphorylation of 2-fluoro-2-deoxy-D-glucose in rat brain, *J. Neurochem.* 40:160, 1983.

6. Fuglsang, A., Lomholdt, M., and Gjedde, A., Blood-brain transfer of glucose and glucose analogs in newborn rats, *J. Neurochem.* 46:1417, 1986.

7. Lassen, N.A., Trap-Jensen, J., Alexander, S.C., et al., Blood-brain barrier studies in man using the double-indicator method, *Am. J. Physiol.* 220:1627, 1971.

8. Knudsen, G.M., Pettigrew, K.D., Paulson, O.B., et al., Kinetic analysis of blood-brain barrier transport of D-glucose in man: quantitative evaluation in the presence of tracer backflux and capillary heterogeneity, *Microvascular Res.* 39:28, 1990.

9. Obrist, W.D., Thompson, H.K., Wang, H.S., et al., Regional cerebral blood flow estimated by xenon-133 inhalation, *Stroke* 6:245, 1975.

10. Crane, P.D., Pardridge, W.M., Braun, L.D., et al., The interaction of transport and metabolism on brain glucose utilization: a reevaluation of the lumped constant, *J. Neurochem.* 36:1601, 1981.

RAT BRAIN GLUCOSE CONCENTRATION AND TRANSPORT KINETICS DETERMINED WITH ^{13}C NUCLEAR MAGNETIC RESONANCE SPECTROSCOPY

Graeme F. Mason[1], Kevin L. Behar[2], Margaret A. Martin[1], and Robert G. Shulman[1]

[1]Department of Molecular Biophysics and Biochemistry
[2]Department of Neurology
Yale University School of Medicine
New Haven, CT 06510

INTRODUCTION

Previously reported results indicate that under normal physiological conditions, brain glucose transport is faster than metabolism and, therefore, does not ordinarily limit glucose metabolism. However, glucose transport may be limiting when glucose consumption is increased as in seizures (2) or when glucose transport is decreased as reported in chronic hyperglycemia (7) and Alzheimer's disease (12).

Although glucose transport does not normally limit metabolism, it can influence kinetic tracer studies of glucose metabolic rates by delaying entry of the tracer into the brain. When not included in the data analysis, this delay may cause uncertainty in calculated metabolic rates when the tracer is labeled glucose (15).

Many methods have been used to study brain glucose transport. Chemical methods for determining glucose require fixation of the brain prior to the measurement and are, therefore, sensitive to post-mortem metabolism. Autoradiography and positron emission tomography (PET) use non-metabolizable glucose analogues such as [^{14}C]deoxyglucose (21), [^{14}C]glucose (11), [^{18}F]fluorodeoxyglucose (17) and [^{11}C]glucose (19). These methods yield regional information about glucose utilization. However, they require corrections for differences in transport kinetics between glucose and the labeled analogues or assumptions about the labeling of products beyond glucose. With the exception of PET, these methods also require removal of the brain. ^{13}C NMR is an alternate method that allows the direct, non-invasive detection of intracerebral ^{13}C-labeled glucose (1), which can be quantified using internal (14) or external standards (9).

Frontiers in Cerebral Vascular Biology: Transport and Its Regulation
Edited by L.R. Drewes and A.L. Betz, Plenum Press, New York, 1993

29

METHODS

Model of Glucose Transport

A symmetric Michaelis-Menten model of facilitated transport was used to determine the kinetic parameters K_m and V_{max} from the concentration of glucose in blood plasma and brain (6,14). Glucose was assumed to be uniformly distributed throughout the brain water space when the brain glucose concentration was at steady state. The model treats transport as two enzymes that obey Michaelis-Menten kinetics, with one for influx and the other for efflux. Thus, at steady state the concentration of brain glucose (G_i) can be described by the following equation (13):

$$\frac{dG_i}{dt} = \frac{V_{max}G_o}{K_m + G_o} - \frac{V_{max}G_i}{K_m V_d + G_i} - V_{gly'} \tag{1}$$

where G_o is the concentration of glucose in blood plasma (mM), K_m is the half-saturation concentration for transport (mM), V_{max} is the maximum transport rate (μmol g^{-1} min^{-1}), V_d is the distribution space of glucose in the brain (0.77 ml g^{-1}) (8), and V_{gly} is the rate of cerebral glucose consumption (μmol g^{-1} min^{-1}), which is assumed to be independent of G_o and G_i. When solved for G_i, Eq. 1 yields:

$$G_i = \frac{\dfrac{V_{max}/V_{gly}}{K_m + G_o} G_o - 1}{\dfrac{V_{max}/V_{gly}}{K_m + G_o} G_o + 1} K_m V_d \bullet \tag{2}$$

Animal Preparation

Male Sprague-Dawley rats weighing 170-240 gm and fasted 20-30 hr were prepared as described by Mason et al. (14). The scalp was retracted and a single-turn elliptical surface coil (8 x 12 mm) placed on the skull. The surface coil functions as a radio antenna for the purpose of exciting ^{13}C and ^{1}H spins in the brain and detecting the signals arising from these nuclei in glucose and other compounds.

Only 1.1% of naturally occurring carbon is the NMR-active isotope ^{13}C, so [1-^{13}C]glucose was infused through a tail vein. The infusion raised the plasma glucose ^{13}C fractional enrichment to a constant level according to the protocol described by Fitzpatrick et al. (5). Insulin was administered in some animals for measurements at low plasma glucose concentrations.

Spectroscopy

^{13}C NMR spectra were obtained using an AM360 wide-bore spectrometer with an 8.4 Tesla field (Bruker Instruments). The α and β anomers of glucose were detected at 92.8 and 96.6 ppm, respectively (Figure 1A). When both the plasma glucose level and the ^{13}C signal from brain glucose stabilized, data were accumulated for 20-60 min to achieve good signal-to-noise ratios. After detection of glucose in vivo, ischemia was induced by a cardioplegic dose of potassium chloride. The [1-^{13}C]glucose concentration was determined using the method detailed in Mason et al. (14). Briefly, the ratio of the NMR resonance intensities of [1-^{13}C]glucose (premortem) to [3-^{13}C]lactate (postmortem) was calculated after the lactate concentration stabilized (Figures 1A and 1B). [3-^{13}C]Lactate was then detected by ^{1}H-observed/^{13}C-edited NMR (20) and its ratio with the ^{1}H signal from total creatine (creatine + phosphocreatine) was determined. The concentration of total creatine was assumed to be 10.5 μmol g^{-1} wet tissue weight (Figures 1C and 1D) (18) and thus allowed the calculation

of the [3-13C]lactate and [1-13C]glucose concentrations from the NMR signal ratios in the
1H and 13C NMR spectra. The [1-13C]glucose concentration was divided by the plasma
fractional [1-13C]glucose enrichment and adjusted for the small brain blood volume to
calculate the total brain glucose concentration.

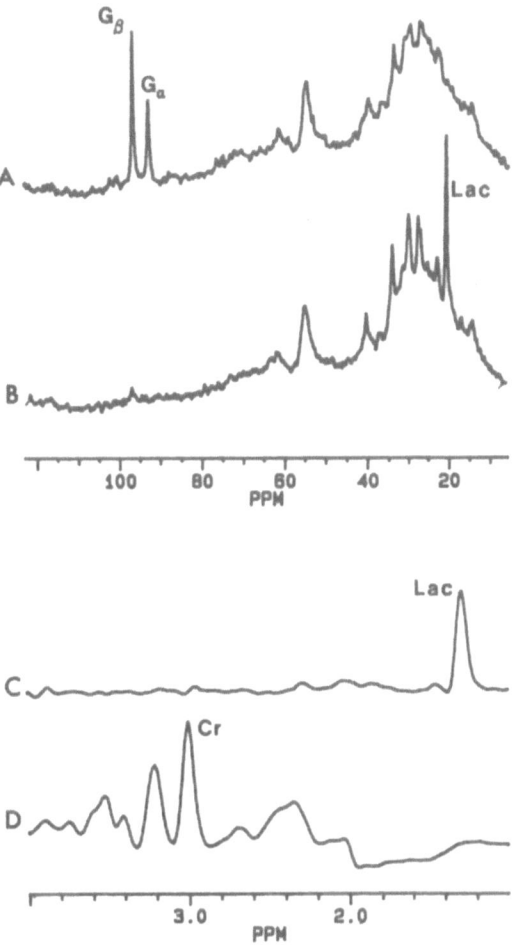

Figure 1. (A) 13C spectrum acquired during infusion of [1-13C]glucose. The
integrated 13C NMR signal intensities of the α and β anomers of [1-13C]glucose
(92.8 and 96.6 ppm) were averaged over the period of measurement. (B) 13C
spectrum of acquired post-mortem, showing the C3 resonance of lactate (20.7
ppm). (C) POCE spectrum acquired post-mortem, showing [3-13C]lactate (1.33
ppm). (D) post-mortem 1H spectrum showing creatine (3.0 ppm). [From
reference 14]

RESULTS

Following an intravenous infusion of [1-13C]glucose, the C1 of the α and β anomers of
glucose were detected in the 13C NMR spectrum of the rat brain (Figure 1A). The β:α ratio
of the areas of the NMR resonances was 3:2, as expected for a solution of glucose in
equilibrium. A plot of the brain glucose concentration as a function of the plasma glucose
concentration for each measurement is shown in Figure 2. For plasma glucose levels
between 3.1 and 62 mM, the concentration of brain glucose varied from 0.7 to 19 μmol g-1.

A symmetric Michaelis-Menten model of glucose transport (Eq. 2) was fit to the data as shown in Figure 2. The best fit was obtained with values of 13.9 ± 2.7 mM for K_m and 5.8 ± 0.8 for the ratio of the maximum glucose influx rate to the glucose utilization rate (V_{max}/V_{gly}).

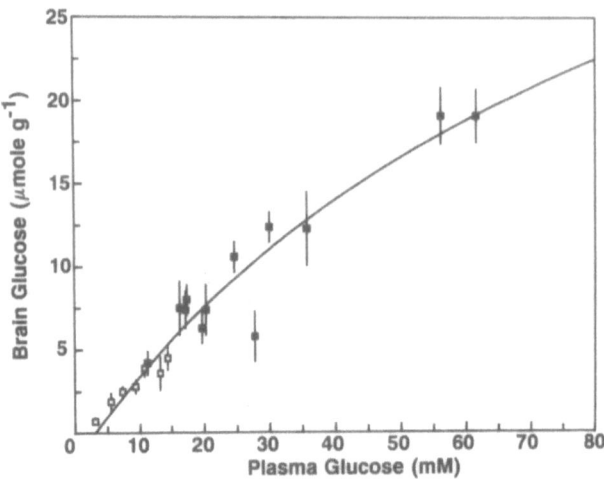

Figure 2. The concentration of intracerebral glucose as a function of the plasma glucose concentration. Each filled symbol represents an uninsulinized animal; open symbols represent insulinized animals. Error bars represent one standard deviation, calculated from the signal-to-noise ratios of all NMR measurements in each experiment. A least squares fit of the saturable symmetric Michaelis-Menten model is superimposed on the graph. [From reference 14]

DISCUSSION

The level of intracerebral glucose determined by NMR was less than the plasma glucose level, with a brain-to-plasma glucose concentration of ratio of ~0.2 to ~0.45. As described in Mason et al. (14), the ratios of brain to plasma glucose calculated from the present data are in close agreement with values obtained using focused microwave irradiation (3,4) and freezing in situ (18) to arrest glucose metabolism quickly.

Values of V_{max} and K_d were determined from the calculated ratios of V_{max}/V_{gly} and K_d/V_{gly}, using the glucose utilization rate of 0.8 μmol min^{-1} g^{-1} wet weight reported by Fitzpatrick et al. (5) for animals studied under conditions similar to the present study. Using the saturable Michaelis-Menten model, V_{max} was calculated to be 4.6 ± 0.6 μmol min^{-1} g^{-1}.

The results obtained from these studies indicate that glucose transport does not limit glucose utilization under conditions of light anesthesia. As shown previously (14), the values determined for and K_m and V_{max} are somewhat higher than in non-steady-state measurements. The transport parameters may be artifactually low in previous studies (13). In the present study the fractional uncertainty in K_m is considerably larger than that of V_{max}, showing that the ^{13}C NMR method is more sensitive to changes in V_{max} than in K_m.

We have shown that ^{13}C NMR can be used for the direct detection of intracerebral glucose in the rat brain. Large-bore magnets in conjunction with external concentration standards will allow the procedure to be performed entirely in vivo, as recently demonstrated in humans by Gruetter et al. (9). Extension of this method to localized spectroscopy (10,16) should permit regional determinations of glucose transport parameters in a host of human cerebral disorders involving suspected changes in glucose metabolism and/or transport, such as diabetes and Alzheimer's disease.

ACKNOWLEDGMENTS

This work was supported by National Institutes of Health Grant DK27121 (R.G.S.) and a First Independent Research and Transition Award NS-26419 (K.L.B.). G.F.M. was supported by National Institutes of Health Training Grant 5T32-GM07223.

REFERENCES

1. Behar, K.L., Petroff, O.A.C., Prichard, J.W., Alger, J.R., and Shulman R.G., Detection of metabolites in rabbit brain by ^{13}C NMR spectroscopy following administration of [1-^{13}C]glucose, *Mag. Res. Med.* 3:911, 1986.
2. Borgström, L., Chapman, A.G., and Siesjö, B.K., Glucose consumption in the cerebral cortex of rat during bicuculline-induced status epilepticus, *J. Neurochem.* 27:971, 1976.
3. Crane, P.D., Pardridge, W.M., Braun, L.D., and Oldendorf, W.H., Two-day starvation does not alter the kinetics of blood-brain barrier transport and phosphorylation in rat brain, *J. Cereb. Blood Flow Metab.* 5:40, 1985.
4. Cremer, J.E., Cunningham, V.J., and Vesille, M.P., Relationships between extraction and metabolism of glucose, blood flow, and tissue blood volume in regions of rat brain, *J. Cereb. Blood Flow Metab.* 3:291, 1983.
5. Fitzpatrick, S.M., Hetherington, H.P., Behar, K.L., and Shulman, R.G., The flux from glucose to glutamate in the rat brain in vivo as determined by ^{1}H-observed, ^{13}C-edited NMR spectroscopy, *J. Cereb. Blood Flow Metab.* 10:170, 1990.
6. Gjedde, A., and Christensen, O., Estimates of Michaelis-Menten constants for the two membranes of the brain endothelium, *J. Cereb. Blood Flow Metab.* 4:241, 1984.
7. Gjedde, A., and Crone, C., Blood-brain glucose transfer: repression in chronic hyperglycemia, *Science* 214:456, 1981.
8. Gjedde, A., and Diemer, N.H., Autoradiographic determination of regional brain glucose content, *J. Cereb. Blood Flow Metab.* 3:303, 1983.
9. Gruetter, R., Novotny, E.J., Boulware, S.D., Rothman, D.L., Mason, G.F., Shulman, G.I., Shulman, R.G., and Tamborlane, W.V., Direct measurement of brain glucose concentrations in humans by ^{13}C NMR spectroscopy, *Proc. Natl. Acad. Sci. USA* 89:1109, 1992.
10. Hanstock, C.C., Rothman, D.L., Jue, T., and Shulman, R.G., Volume-selected proton spectroscopy in the human brain, *J. Magn. Reson.* 77:583, 1988.
11. Hawkins, R.A., Hass, W.K., and Ransohoff, J., Measurement of regional brain glucose utilization in vivo using [2-^{14}C]glucose, *Stroke* 10:690, 1979.
12. Kalaria, R.N., and Harik, S.I., Reduced glucose transporter at the blood-brain barrier and in cerebral cortex in Alzheimer disease, *J. Neurochem.* 53:1083, 1989.
13. Lund-Andersen, H., Transport of glucose from blood to brain, *Phys. Rev.* 59:305, 1979.
14. Mason, G.F., Behar, K.L., Rothman, D.L., and Shulman, R.G. NMR determination of intracerebral glucose concentration and transport kinetics in rat brain, *J. Cereb. Blood Flow Metab.* 12:448, 1992.
15. Mason, G.F., Rothman, D.L., Behar, K.L., and Shulman, R.G., NMR determination of the TCA cycle rate and α-ketoglutarate/glutamate exchange rate in rat brain, *J. Cereb. Blood Flow Metab.* 12:434, 1992.
16. Ordridge, R.J., Connelly, A., and Lohman, J.A.B., Image-selected in vivo spectroscopy (ISIS). A new technique for spatially selective NMR spectroscopy, *J. Magn. Reson.* 66:283, 1986.

17. Phelps, M.E., Huang, S.C., Hoffman, E.J., Selin, C., Sokoloff, L., and Kuhl, D.E., Tomographic measurement of local cerebral glucose metabolic rate in humans with (F-18)-fluoro-2-deoxy-D-glucose: validation of method, *Ann. Neurol.* 6:371, 1979.

18. Pontén, U., Ratcheson, R.A., Salford, L.G., and Siesjö, B.K., Optimal freezing conditions for cerebral metabolites in rats, *J. Neurochem.* 21:1127, 1973.

19. Raichle, M.E., Larson, K.B., Phelps, M.E., Grubb, R.L. Jr., Welch, M.J., and Ter-Pogossian, M.M., In vivo measurement of brain glucose transport and metabolism employing glucose-^{11}C, *Am. J. Physiol.* 228:1936, 1975.

20. Rothman, D.L., Behar, K.L., Hetherington, H.P., den Hollander, J.A., Bendall, M.R., Petroff, O.A.C., and Shulman, R.G., [1]H-observed/^{13}C-decoupled spectroscopic measurement of lactate and glutamate in the rat brain in vivo, *Proc. Natl. Acad. Sci. USA* 82:1633, 1985.

21. Sokoloff, L., Reivich, M., Kennedy, C., Des Rosiers, M.H., Patlak, C.S., Pettigrew, K.D., Sakurada, O., and Shinohara, M., The [^{14}C]deoxyglucose method for the measurement of local cerebral glucose utilization, *J. Neurochem.* 28:897, 1977.

NON-INVASIVE MEASUREMENTS OF THE CEREBRAL STEADY-STATE GLUCOSE CONCENTRATION AND TRANSPORT IN HUMANS BY ^{13}C NUCLEAR MAGNETIC RESONANCE

Rolf Gruetter[1]*, Edward J. Novotny[2], Susan D. Boulware[2], Douglas L. Rothman[3], Graeme F. Mason[1], Gerald I. Shulman[3], William V. Tamborlane[2], and Robert G. Shulman[1]

[1]Department of Molecular Biology and Biochemistry, Yale University
[2]Department of Pediatrics and Neurology
[3]Department of Internal Medicine, Yale School of Medicine
Yale University, New Haven, CT 06510

INTRODUCTION

Glucose transport across the blood-brain barrier (BBB) has been shown to be a carrier-mediated, saturable process of the facilitated diffusion type (4). The cerebral steady-state glucose concentration depends on the kinetic properties of transport and is lower than in blood. Direct measurement of brain glucose is not possible in humans by non-invasive techniques using radioactive tracers because the signal of labeled glucose cannot be distinguished from that of its metabolic products.

Nuclear magnetic resonance (NMR) spectroscopy has been shown to be a non-invasive technique that can be used to study many metabolites in vivo. In particular, it has been shown that glucose can be observed in the mammalian brain by ^{13}C NMR after [1-^{13}C] glucose infusion (1).

Recently, we have shown the quantification of glucose in rat brain by ^{13}C NMR (14). Since the methods used in that study require postmortem measurements, we developed a new method for the non-invasive quantification of ^{13}C NMR signals in the human brain (9,10). This paper reviews the measurements of brain glucose that we have made in humans using this non-invasive technique and explores some of the alternatives that potentially can be provided by ^{1}H NMR measurements of glucose changes in the brain.

---

*Present address: Magnetic Resonance Center, Inselspital and University, Inselheimmatte, CH-3010 Bern, Switzerland.

Frontiers in Cerebral Vascular Biology: Transport and Its Regulation
Edited by L.R. Drewes and A.L. Betz, Plenum Press, New York, 1993

METHODS

Subjects

Seven healthy subjects aged 13-16 years were studied. One arm was cannulated for the administration of glucose and insulin and the other was used to sample blood for the determination of plasma glucose and isotopic enrichment. Measurements were performed in a 2.1 Tesla whole-body magnet with the subject placed supine on the NMR transceiver, a double-concentric $^{13}C(7cm)/^1H(14cm)$ surface coil. Sensory stimulation was minimized by having the subjects wear ear plugs and eye-patches.

Infusion Protocol

In order to increase the sensitivity of the ^{13}C NMR measurement, the isotopic fraction of [1-^{13}C]glucose in blood was increased to 50% by raising the fasting plasma glucose concentration acutely from approximately 5 mM to 10 mM using an infusate of 99%-enriched D-[1-^{13}C]glucose. Subsequent variation of plasma glucose as well as the isotopic enrichment of glucose in plasma were minimized by continuing the infusion using a 50% enriched infusate of exogenous glucose at a variable rate, based on plasma samples measured every 5 min. Plasma glucose was lowered to euglycemia by stopping the glucose infusion temporarily. In order to minimize hepatic output of unlabeled glucose during this period, a primed continuous insulin infusion (0.5 mU kg^{-1}min^{-1}) was begun during hyperglycemia.

Localized ^{13}C NMR Measurement of D-[1-^{13}C]glucose

The glucose signals were obtained from a 144-ml volume as described previously in Gruetter et al. (9). The pulse sequence is based on the ISIS technique as described by Gruetter et al. (10). Localization in the occipital lobe excluded major blood vessels and the ventricles as judged from magnetic resonance images obtained just prior to the infusion. Quantification of α- and β-glucose was achieved by comparing the in vivo measurement to that of an aqueous glucose solution placed in a two-liter bottle on the coil obtained immediately after the study.

The comparison with a solution measurement resulted in a volumetric measurement of ^{13}C labeled glucose, which was assumed to be equivalent to the per weight measurement. The brain glucose concentration was calculated from the brain D-[1-^{13}C]glucose concentration by dividing the latter by the steady-state isotopic enrichment in blood.

Kinetic Analysis

The steady-state relationship between brain and blood glucose concentrations can be analyzed with the symmetric Michaelis-Menten model of transport as in previous studies of animal brain (7,12,14). This model assumes glucose influx and efflux to be characterized by the same Michaelis-Menten kinetic parameters and that glucose is uniformly distributed in the brain water, in which the distribution volume is V_d=0.77 ml/g (6). At steady-state, dG_{brain}/dt equals zero, and the standard differential equation (6,7,12-14) can be rewritten to express the brain glucose concentration G_{brain} (μmol/g) as a function of the half saturation concentration K_t, the ratio of maximal transport rate to glucose consumption, T_{max}/V_{gly}, and the glucose concentration in plasma G_{plasma}, as

$$G_{brain} = V_d K_t \frac{(T_{max}/V_{gly}-1)G_{plasma}/K_t-1}{T_{max}/V_{gly}+G_{plasma}/K_t+1} \qquad [1]$$

Equation 1 was fitted to the steady-state glucose measurements using non-linear, iterative numeric methods (Levenberg-Marquardt algorithm). T_{max} was calculated from T_{max}/V_{gly} assuming $V_{gly}=0.3$ µmol g^{-1}min^{-1} reported for the human brain under sensory deprivation (15).

RESULTS

Localized ^{13}C NMR measurements were begun after the unlocalized NMR signal of glucose and the plasma concentration stabilized, which was typically the case ~30 min after the start of the infusion.

Figure 1 shows the glucose region from localized ^{13}C NMR spectra obtained in the same study at euglycemia (bottom) and at hyperglycemia (top). In both cases the measurement time and fractional enrichment in blood were the same, which makes the relative intensities directly comparable and Figure 1, thus, shows directly the profound effect of hyperglycemia on brain glucose concentrations.

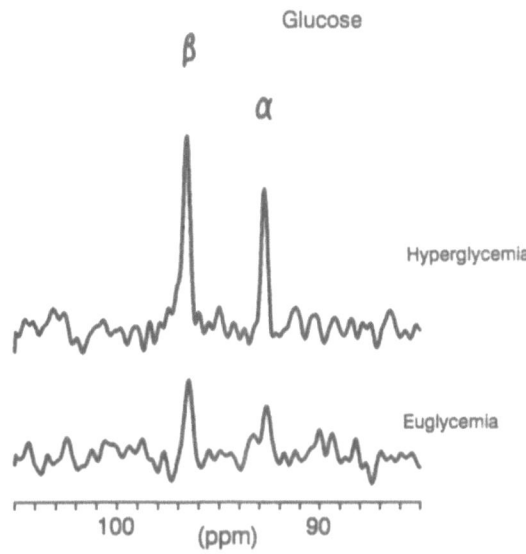

Figure 1. In vivo spectra of D-[1-^{13}C]glucose from a 144-ml volume in the occipitoparietal region of the human brain. The glucose signals were obtained at 8.4 mM plasma and at 4.8 mM. Both spectra are shown with the same vertical scale. Abscissa: chemical shift (resonance frequency) in ppm, ordinate: resonance amplitude in arbitrary units. Modified from Gruetter et al. (1992a).

The data obtained from the seven subjects are summarized in Table 1. The mean cerebral glucose concentration was 0.96 µmol/g at an average euglycemic plasma glucose concentration of 4.7 mM. The error of the NMR measurement of the glucose concentration was estimated to be between 0.1 and 0.4 µmol/g, depending on measurement time and fractional enrichment. During the localized ^{13}C NMR measurements of glucose, with the duration between 15 and 45 min, the standard deviation of the plasma glucose variation was less than 0.5 mM and that of the isotopic enrichment was less than 2.2%, indicating stable steady-state conditions.

Table 1. Measurement of brain glucose concentration at euglycemia and at hyperglycemia in seven normal subjects and corresponding Michaelis-Menten parameters of transport.

Study	Euglycemia		Hyperglycemia		Transport constants[1]	
	Plasma (mM)	Brain (μmol/g)	Plasma (mM)	Brain (μmol/g)	K_t (mM)	T_{max} (mmol g^{-1}min^{-1})
1	4.8	1.3	8.4	2.2	3.0	1.17
2	4.5	1.2	7.3	2.1	2.1	1.17
3	4.7	0.8	9.6	2.2	6.2	1.02
4	4.9	1.0	12.1	2.5	4.4	0.99
5	4.5	0.6	10.4	2.6	8.0	1.11
6	4.9	1.0	8.9	2.7	9.9	1.41
7	4.6	0.8	10.0	2.4	6.3	1.08
average	4.7	1.0	9.5	2.4	5.7	1.14
±SD	±0.2	±0.2	±1.5	±0.2	±2.8	±0.15

[1]Determined by fitting Equation 1 to the two measurements of brain glucose concentration in the corresponding study.

The relationship between brain and blood glucose can be analyzed for each study separately, which yielded highly consistent T_{max} values, as shown in Table 1. The kinetic constants derived by fitting Equation 1 to all fourteen data points simultaneously are consistent with the average values given in Table 1. The brain blood volume is ~5%, but the signal contribution of glucose in blood is difficult to determine. When subtracting 5% of the plasma glucose concentration from that measured in the brain (Table 1), K_t is generally lowered only by approximately 10% and T_{max}/V_{gly} is reduced by approximately 17%.

DISCUSSION

The brain glucose concentration of 1 μmol/g at euglycemia implies a rapid turnover time of a few minutes for labeled glucose, based on the glucose consumption rate V_{gly} of 0.3 μmol g^{-1}min^{-1}, reported from several other studies performed with auditory and visual deprivation (see 15).

Several positron emission tomography (PET) studies of glucose transport in the human brain have been reported. Using [^{11}C]3-O-methylglucose, K_t has been estimated by Brooks et al. (3) at 4.8 mM and by Feinendegen et al (5) at 3.8 mM, whereas Blomqvist et al. (2) obtained K_t=4.1 mM using D-[U-^{11}C]glucose. The K_t measured at steady state by ^{13}C NMR is in the range of these values. The NMR value for T_{max} of 1.1 μmol g^{-1}min^{-1} falls between the PET values of T_{max}=2.0 μmol g^{-1} min^{-1} reported by Feinendegen et al. (5) and the 0.6 μmol g^{-1} min^{-1} from Blomqvist et al. (2).

Using the steady-state form of the symmetric Michelis-Menten model (6,7,12-14) with V_{gly}=0.3 μmol/g min, a euglycemic (G_{plasma}=4.8mM) brain glucose concentration of 2.0 μmol/g is calculated from Equation 1 using the kinetic parameters given by Feinendegen et al. (5), whereas the constants given by Blomqvist et al. (2) imply a 0.2 μmol/g brain glucose concentration, both of which are different from the measured euglycemic brain glucose concentration (Table 1).

This study measured steady-state brain glucose concentrations in humans by ^{13}C NMR from 144-ml volumes, which has the advantage that glucose can be measured free from overlap with other signals. Its limitation, however, is the low sensitivity, although we have reduced the detection limit for localized ^{13}C NMR spectroscopy to 0.1 mM ^{13}C concentration for these large volumes (10). Even though the regional constancy of the distribution space of methylglucose in rats (6) and humans (3,5,8) suggest small variations in brain glucose content, a higher spatial resolution is potentially desirable.

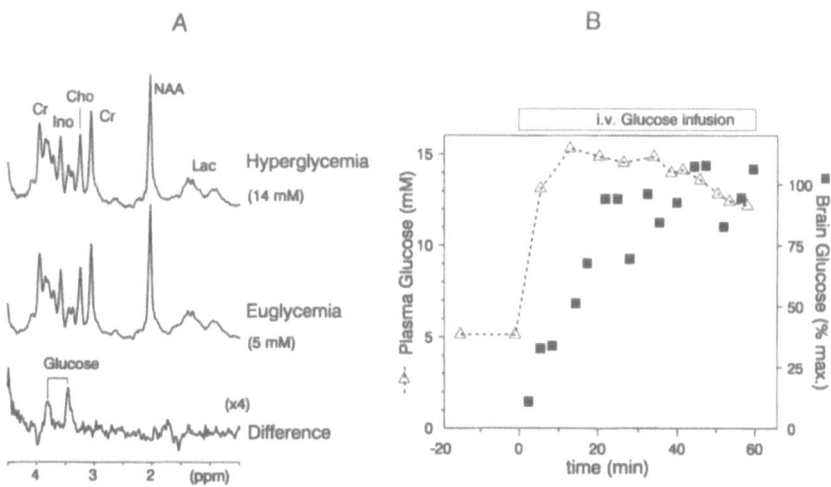

Figure 2. ^{1}H NMR observation of changes in brain glucose from a 36-ml volume in the occipitoparietal region of the human brain. (A) shows ^{1}H NMR spectra acquired at hyperglycemia (top) and euglycemia (middle). The small change is entirely attributed to glucose as shown by the four-fold enlarged difference spectrum (bottom). (B) shows the time course of the increment of brain glucose (solid squares) expressed in % of the maximum increase and the corresponding time course of plasma glucose (triangles). Modified from Gruetter et al. (11). Cr: (phospho)creatine; Ino: inositol; Cho:trimethylamine resonances, e.g., choline; NAA: N-acetyl-aspartate; Lac: lactate.

^{1}H NMR spectroscopy with its intrinsically higher sensitivity may provide an alternative means to study glucose transport in humans at a much higher spatial and temporal resolution. However, the glucose peaks are obscured by many other resonances in the ^{1}H NMR spectrum of human brain, as illustrated in Figure 2a. Nevertheless, we have shown that the change in brain glucose can be observed by ^{1}H NMR during intravenous glucose infusions (Figure 2b) with a 3-min resolution from 36-ml volumes (11), which may provide important insights into glucose transport and metabolism.

In conclusion, NMR spectroscopy provides a direct, non-invasive way to study the regulation of brain glucose transport in humans, which may be altered in diseases such as diabetes mellitus, seizures or hypoxic-ischemic encephalopathy. The method does not use ionizing radiation, which makes it particularly attractive to pediatric and longitudinal studies.

REFERENCES

1.   Beckmann, N., Turkalj, I., Seelig, J., and Keller, U., 13C NMR for the assessment of human brain
      glucose metabolism in vivo, *Biochem.* 30:6362, 1991.
2.   Blomqvist, G., Gjedde, A., Gutniak, M., Grill, V., Widén, L., Stone-Elander, S., and Hellstrand, E.,
      Facilitated transport of glucose from blood to brain in man and the effect of moderate hypoglycemia
      on cerebral glucose utilization, *Eur. J. Nucl. Med.* 18:834, 1991.

3. Brooks, D.J., Gibbs, J.S.R., Sharp, P., Herold, S., Turton, D.R., Luthra, S.K., Kohner, E.M., Bloom, S.R., and Jones, T., Regional cerebral glucose transport in insulin-dependent diabetic patients studied using [^{11}C]3-O-methyl-D-glucose and positron emission tomography, *J. Cereb. Blood Flow Metab.* 6:240, 1986.

4. Crone, C., Facilitated transfer of glucose from blood into brain tissue, *J. Physiol. (London)* 181:103, 1965.

5. Feinendegen, L.E., Herzog, H., Wieler, H., Patton, D.D., and Schmid, A., Glucose transport and utilization in the human brain: Model using carbon-11 methylglucose and positron emission tomography, *J. Nucl. Med.* 27:1867, 1986.

6. Gjedde, A., and Diemer, N.H., Autoradiographic determination of regional brain glucose content, *J. Cereb. Blood Flow Metab.* 3:303, 1984.

7. Gjedde, A., and Christensen, O., Estimates of Michaelis-Menten constants for the two membranes of the brain endothelium, *J. Cereb. Blood Flow Metab.* 4:241, 1984.

8. Gjedde, A., Wienhard, K., Heiss, W.D., Kloster, G., Diemer, N.H., Herholz, K., and Pawlik, G., Comparative regional analysis of 2-fluorodeoxyglucose and methylglucose uptake in brain of four stroke patients. With special reference to the regional estimation of the lumped constant, *J. Cereb. Blood Flow Metab.* 5:163, 1985.

9. Gruetter, R., Novotny, E.J., Boulware, S.D., Rothman, D.L., Shulman, G.I., Mason, G.F., Shulman, R.G., and Tamborlane, W.V., Direct measurement of brain glucose concentrations in humans by ^{13}C NMR spectroscopy, *Proc. Natl. Acad. Sci. USA* 89:1109, 1992.

10. Gruetter, R., Rothman, D.L., Novotny, E.J., and Shulman, R.G., Localized ^{13}C NMR spectroscopy of myo-inositol in the human brain in vivo, *Magn. Reson. Med.* 25:304, 1992.

11. Gruetter, R., Rothman, D.L., Novotny, E.J., Shulman, G.I., Prichard, J.W., and Shulman, R.G., Detection and assignment of the glucose signal in ^{1}H NMR difference spectra of the human brain, *Magn. Reson. Med.* 27:183, 1992.

12. Holden, J.E., Mori, K., Dienel, G.A., Cruz, N.F., Nelson, T., and Sokoloff, L., Modeling the dependence of hexose distribution volumes in brain on plasma glucose concentration: Implications for estimation of the local 2-deoxyglucose lumped constant, *J. Cereb. Blood Flow Metab.* 11:171, 1991.

13. Lund-Andersen, H., Transport of glucose from blood to brain, *Physiol. Rev.* 59:305, 1979.

14. Mason, G.F., Behar, K.L., Rothman, D.L., and Shulman, R.G., NMR determination of intracerebral glucose concentration and transport kinetics in rat brain, *J. Cereb. Blood Flow Metab.* 12:448, 1992.

15. Tyler, J.L., Strother, S.C., Zatorre, R.J., Alivisatos, B., Worsley, K.J., Diksic, M., and Yamamoto, Y.L., Stability of regional cerebral glucose metabolism in the normal brain measured by positron emission tomography, *J. Nucl. Med.* 29:631, 1988.

Electrolyte Transport

POTASSIUM TRANSPORT AT THE BLOOD-BRAIN
AND BLOOD-CSF BARRIERS

Richard F. Keep[1], Jianming Xiang[1], and A. Lorris Betz[1,2]

[1]Department of Surgery (Neurosurgery)
[2]Departments of Pediatrics and Neurology
University of Michigan
Ann Arbor, MI 48109-0532

INTRODUCTION

One of the earliest discoveries attributable to the blood-brain and blood-cerebrospinal fluid (CSF) barriers was the finding that the brain could maintain constant the concentrations of a number of ions (e.g., K, Ca, Mg) in CSF during fluctuations in plasma composition (reviewed in Bradbury, 8). These findings have since been extended to whole brain and brain interstitial fluid (e.g., 10,33). Although glial cells play a role in short-term buffering of potassium, long-term regulation must depend on control of the movement of ions between blood and brain across the blood-brain barrier (BBB), situated at the cerebral capillaries, and the blood-CSF barrier, situated at the choroid plexuses and the meninges. BBB ion transport may also be involved in the secretion of brain interstitial fluid, which accounts for about 30% of the total fluid secreted by the brain (42). This relationship between ion and fluid transport also means that blood-brain and blood-CSF barrier ion transport is important in brain volume regulation (19) and its modulation may have therapeutic advantages in the treatment of brain edema associated with diseases such as stroke.

Despite its potential importance, progress toward understanding the mechanisms of BBB ion transport is still at an early stage compared to many tissues. In vivo examination of BBB ion transport is complicated by a number of factors, and the in vitro models available, isolated cerebral microvessels and cultured endothelial cells, both have limitations. The purpose of this paper is to discuss some of the problems associated with using these techniques to study BBB ion transport, with particular reference to our own work on potassium, and to summarize our present knowledge of the transporters involved and their control. We will also compare results with those obtained from choroid plexus where better in vitro models are available and our knowledge is greater.

Frontiers in Cerebral Vascular Biology: Transport and Its Regulation
Edited by L.R. Drewes and A.L. Betz, Plenum Press, New York, 1993

IN VIVO MEASUREMENT OF BBB POTASSIUM TRANSPORT

Radiotracers, either ^{42}K or the longer lived ^{86}Rb, have been commonly used to study BBB potassium transport in vivo. Such studies have shown a higher BBB permeability to potassium than to sodium (reviewed in 8) suggesting that potassium is indeed transported across the BBB rather than moving passively through a paracellular route. This hypothesis is supported by the relative insensitivity of the rate of brain potassium influx to changes in plasma potassium (10) and the saturation kinetics of potassium loss from the brain (11). The hypothesis is further supported by the work of Cserr et al. (19), which showed that hyperosmotic stress induced a threefold increase in the potassium permeability surface area (PS) product of the BBB. This increase was specific because neither Na nor mannitol showed a similar change.

However, most attempts to modulate BBB ion transport in vivo have found only small effects, which are difficult to interpret. Some of the problems involved in such studies are demonstrated by our own experiments designed to examine whether a 5-hydroxytryptamine agonist, 1-(4-iodo-2,5-dimethoxyphenyl)-2-aminopropane hydrochloride (DOI), inhibits BBB K transport in the rat. DOI, 5 mg/kg, was given iv 10 min prior to the determination of a 10-min PS product for ^{86}Rb. ^{3}H-α-aminoisobutyric acid (AIB) was also given as a passive permeability marker. The PS products for the two isotopes were determined as the concentration of isotope in the brain divided by the integral of the plasma curve (45). Corrections for the vascular compartment of the brain (plasma and red blood cell) were determined in a separate series of experiments.

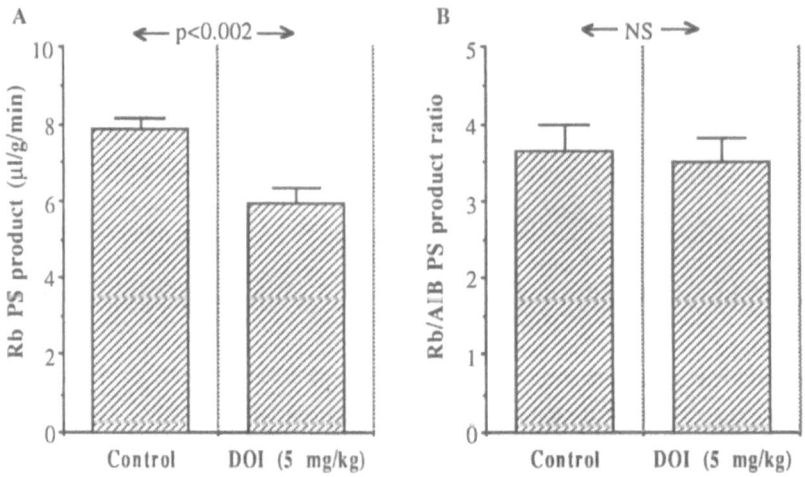

Figure 1. The effect of a 5HT agonist, DOI (5 mg/kg; iv), on (A) the ^{86}Rb PS product and (B) the ^{86}Rb/^{3}H-AIB PS product ratio in the anterior cortex of the rat. All values are means ± SEM; n=6.

DOI (a 5HT$_2$ / 5HT$_{1C}$ agonist) did cause a significant (p<0.002) reduction in the Rb (a marker for K) PS product (Figure 1). However, the interpretation of this result is complicated by a number of factors. The reduction was not specific because the Rb/AIB PS product ratio did not change. This suggests that the change in Rb PS product could reflect a change in the capillary surface area (S) available for transport. The action of DOI on the BBB could also have been secondary to central or systemic effects rather than a direct effect.

DOI caused a transient, 60 mmHg, increase in blood pressure and Ling et al. (39) have reported a similar decline in the PS product for water in response to hypertension. Although it is generally thought that the CSF has little influence on a 10-min Rb PS product measurement, distinguishing between BBB and blood-CSF barrier effects can be a problem for other ions (54). With other drugs, degradation and access to the abluminal membrane of the cerebral capillaries pose problems. In the case of K, the large intracellular pool of K in the brain also means that equilibration between blood and brain isotope concentrations requires days. This poses a difficulty in estimating the efflux rate constant, and it may be this factor rather than influx that is actively controlled.

Some of the problems associated with the typical in vivo isotopic experiments can be circumvented by using an in situ brain perfusion technique developed by Takasato et al. (57). This model allows for greater control of perfusate composition and avoids systemic effects. The low permeability of ions at the BBB does, however, mean that relatively long perfusion times are required. Another technique used to study BBB ion transport in vivo is the measurement of the transendothelial resistance and potential, which has been used to examine opening of the BBB (13,46), the development of the BBB (14) and the K transport mechanisms present on cerebral capillaries (1); see also Jones et al. in this volume. All of these different techniques have limitations and this has led to the use of in vitro models of BBB ion transport. However, to confirm and understand the physiological significance of the results of such in vitro experiments, it is ultimately necessary to return to in vivo models.

IN VITRO BBB POTASSIUM TRANSPORT: METHODOLOGY

Isolated Microvessels

A number of different methods are available to isolate cerebral microvessels. The work of Sussman et al. (55) would suggest there is a trade-off between preparation purity and microvessel viability as measured by tissue ATP levels. The method used in our laboratory is described in detail elsewhere (52) but, briefly, involves homogenizing brains from 4- to 6-week-old rats, centrifuging through dextran, sieving and passing through a glass bead column to trap the microvessels.

As a model to study BBB ion transport, isolated microvessels have a number of advantages and disadvantages. These will be discussed in general here, and additional examples will be given with the detailed results in the next section. As with all in vitro preparations, the main advantages over in vivo experiments are the control of the tissue environment and the removal of input from other tissues. In the case of the microvessels, other advantages are that the cells have not dedifferentiated and they permit ready access to the abluminal membrane. Disadvantages include the fact that cellular ATP levels are considerably depleted (55). This limits long-term experimentation and may mean that ion transport mechanisms have been altered. The lumen of the microvessels is also collapsed, preventing ready access to the transporters at that membrane and probably meaning that the transport measured is only across one, the abluminal, membrane. Ideally, one would like to study transcellular transport. Finally, even though the enzymes associated with the BBB are considerably enriched while those associated with glial and neuronal cells are depleted, the preparation is not totally pure. The major contaminants that could affect transport studies are pericytes, smooth muscle cells, vascular elements, free nuclei and glial end feet.

The importance of these problems varies with the techniques being used. For example, with patch clamp it is possible to visualize the cells being studied, thus avoiding cellular contaminants, but it is not feasible to patch the luminal membrane in intact microvessels. Because ^{86}Rb will not be concentrated in dead cells, isotopic efflux studies avoid some of the problems associated with viability, but some channels may be closed if the cells are depolarized. Whole cell homogenates that can be used for enzyme analysis and molecular biology avoid the luminal access problems associated with the intact microvessel, but these studies may be affected by contaminating cells and free nuclei.

Cultured Endothelial Cells

Described elsewhere in his volume, a plethora of methodologies have evolved for culturing cerebral endothelial cells. We have primarily limited our recent experimentation to freshly isolated microvessels because the advantages of using cultured cells, which include the purity of the preparation, its viability, the access to the whole cell and the hope that it is possible to recreate the BBB in vitro, are outweighed by a major disadvantage, the loss of some of the characteristics of the BBB. Progress has been made toward maintaining some of the characteristics of the BBB in culture, by co-culture, additions to the medium or by immortalizing the cells. However, even the highest of the electrical resistances recorded in such preparations, about 700 Ω/cm^2 (20), is still a factor of 10 less than the estimated value for the BBB in vivo (54). For studies of compounds with a high lipid solubility or that are rapidly transported, the leakiness of these monolayers is not important (although the number of transporters maintained in culture will be). For ion transport, however, the brain influx rate constants are very low (those for potassium and sodium being 10 and 2 μl/g/min, respectively) so that even small leaks would have an appreciable effect on the apparent transport of these ions.

IN VITRO BBB POTASSIUM TRANSPORT: UPTAKE

Overall Uptake Rate

In influx experiments, ^{86}Rb was used as a marker for potassium because of its longer half-life. Uptake into isolated rat brain microvessels was measured over 5 min, during which time the uptake was unidirectional. Measurements on rat choroid plexus were made using a method similar to that of Parmelee et al. (48). With that tissue, uptake measurements were limited to 30 sec.

Under control conditions, the rate of ^{86}Rb uptake into isolated rat cerebral microvessels was 2.0 ± 0.4 μl/mg protein/min (n=8). Measurements of choroid plexus uptake, yielded much greater values, 46 ± 4 μl/mg protein/min (n=5). This difference in uptake rate is not matched by a difference in the apparent number of Na/K ATPase units because a similar number of ouabain binding sites have been reported in both tissues (26). It may, however, reflect a depression in Na/K ATPase activity in the microvessels as a result of damage during the isolation procedure. Such concerns make exact quantification of the different K transport mechanisms difficult. The following section will concentrate on the types of mechanisms present and their control.

Potassium Uptake Mechanisms: Na/K ATPase

Sodium/potassium ATPase has been the most extensively studied of the potassium transport mechanisms at the BBB. It plays a central role in ion transport, both directly and by providing the electrochemical gradients used by other transporters. This enzyme has an asymmetric, abluminal (brain-facing) distribution (7,21,60) similar to that found at the choroid plexus (41), and it may also provide the driving force for brain interstitial fluid secretion (9). In common with other reports (e.g., 16), we find that the rate of ^{86}Rb uptake into isolated rat brain microvessels is predominantly via the Na/K ATPase being reduced by about 80% in the presence of 5 mM ouabain (Figure 2).

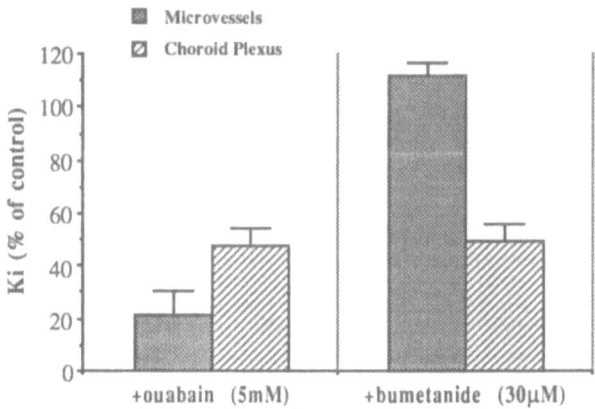

Figure 2. The effect of ouabain (5 mM) and bumetanide (30μM) on the influx rate constant for ^{86}Rb in isolated rat microvessels and choroid plexuses. All values are means ± SEM; n=4-5 microvessel preparations with triplicates performed on each microvessel preparation.

In most tissues, K transport rate is dependent on intracellular [Na] (22) suggesting that sodium entry mechanisms at the BBB may be an important regulator of K uptake. Experiments on isolated microvessels have, however, shown that the rates of potassium uptake and sodium efflux are also dependent on the extracellular [K] with a K_m of about 3 mM (52). This concentration is similar to the normal extracellular [K] of the brain and suggests that Na/K ATPase may be modulated by physiological changes in extracellular [K] and may participate in brain K homeostasis. Measurements of K efflux from brain in vivo support this hypothesis (11). The K_m of 3 mM K is similar to that found in the choroid plexus (47), and the ouabain binding properties of both tissues are similar. This suggests that the same isoforms of the Na/K ATPase are present in both tissues (52), probably the α-1 form found in the choroid plexus (56). However, work presented by Zlokovic et al. in this volume suggests that a number of isoforms are present in isolated microvessels, although the extent to which this could reflect contamination with other cell types is unclear.

The dependence of Na/K ATPase activity on extracellular [K] may explain an elevated rate of BBB Na transport in the early phase of cerebral ischemia (53), because this condition is associated with K loss to the extracellular space from neurons and glia. This stimulation may be an important factor in determining the rate of brain edema formation (6) since it has been suggested that the sodium flux generated by the Na/K ATPase may be responsible for the secretion of brain interstitial fluid, in analogy with the choroid plexus (9).

A ouabain-like factor, which probably is ouabain itself (24), may be a physiological modulator of the Na/K ATPase because it is present in CSF (23) and is blood. The concentration of ouabain required for 50% inhibition in isolated microvessels (50 μM; ref. 16) is much greater than that present in CSF, but the relationship between the dose required for inhibition and the time of exposure needed is unknown. Data from frog choroid plexus show that low doses of ouabain may require hours to reach maximal binding (61). Similar problems arise in assessing the potential modulation, both physiological and therapeutic, of Na/K ATPase activity by steroids where in vitro inhibition occurs only at very high concentrations (16). With isolated microvessels, preparation deterioration precludes the ability to perform long-term uptake studies, and the potential physiological role of these factors would probably be better performed with cultured cells.

Cerebral capillaries receive noradrenergic innervation from the locus ceruleus (34) and have both α– and β–adrenergic receptors (49). Harik et al. (25) have shown that locu ceruleus lesions reduce ouabain binding to cerebral microvessels, suggesting that there is adrenergic control of Na/K ATPase activity, and possibly BBB ion transport and interstitial fluid secretion. At the choroid plexus, noradrenergic innervation has been implicated in the control of CSF secretion (37,44).

BBB Na/K ATPase activity is not only affected by physiological modulators but can be reduced by oxygen free radicals (40). Such free radicals are produced during partial ischemia or ischemia/reperfusion. The ATPase can also be inhibited by environmental toxins such as mercury (3) and aluminum (15). Although mercury and aluminum have neurological effects, the extent to which these could be the result of BBB dysfunction is uncertain.

Potassium Uptake Mechanisms: Potassium Co-Transport

In isolated rat brain microvessels, we find no evidence that bumetanide or furosemide, two inhibitors of K/Cl co-transport, block ^{86}Rb uptake (Figure 2). This is in contrast to the choroid plexus where a major part of ^{86}Rb influx is via a bumetanide-sensitive mechanism (Figure 2; ref. 4). The choroid plexus transporter is probably of the Na/K/2Cl form since bumetanide acts via a Cl site and the electrochemical gradient for both Cl and K is out of the cell. Thus, only the Na electrochemical gradient could drive inward transport.

In contrast to our findings in isolated cerebral microvessels, Lin (35,36) found a furosemide-sensitive ^{86}Rb uptake that was Na and Cl dependent and that was stimulated by hyperosmotic stress. Ibaragi et al. (30) have also demonstrated a furosemide-sensitive Na influx into cerebral microvessels that we do not see in our preparation. An interpretation of these findings is that there may be a Na/K/2Cl transporter present at the BBB but that the access to the transporter varies in different preparations; i.e., it may be on the luminal membrane. In favor of this interpretation is preliminary evidence for the presence of such a co-transporter in cultured endothelial cells where the whole cell is accessible. Against this interpretation are in vivo experiments that failed to show furosemide-sensitive ^{86}Rb uptake into brain at the BBB (5). It should be noted, however, that many in vivo experiments have also failed to find furosemide or bumetanide inhibition of ^{22}Na or ^{86}Rb uptake at the choroid plexus (reviewed in 31), although in vitro evidence for such co-transport is much better in this tissue (4,32,51). This discrepancy may reflect difficulties in reaching the correct inhibitor concentrations in the cerebral circulation in vivo.

Other Potassium Uptake Mechanisms

Nothing is known about other possible potassium transport mechanisms in isolated microvessels. It has, however, been shown that omeprazole can significantly reduce the rate of fluid production in the brain, an effect that is not apparently a result of inhibition of choroid plexus Na/K ATPase (38). Omeprazole blocks gastric acid secretion by inhibiting a H$^+$/K$^+$ ATPase (28) and may have a similar effect at either the blood-CSF or BBB.

IN VITRO BBB POTASSIUM TRANSPORT: EFFLUX

Overall Efflux Rate

In efflux experiments, microvessels were allowed to accumulate ^{86}Rb for 30 min, then washed to remove extracellular Rb and sampled at 0 and 20 min. Initial experiments showed that a plot of the natural logarithm of the fraction of Rb remaining against time for this period was linear; i.e., the loss from the microvessels follows a single exponential with an efflux rate constant (K_0) given by the slope of that line. The control efflux rate constant from cerebral microvessels was 0.047 ± 0.002 min^{-1}. Data from Lin (35) and Chaplin et al. (16)

give a similar value. This figure is considerably less than that found at the choroid plexus (0.367 ± 0.016 min[-1]) that was measured over 2 min using a method similar to that described by Parmelee et al. (48).

Potassium Efflux Mechanisms (Isotopic Experiments)

A series of experiments were performed with and without 5 mM ouabain in the efflux buffer to examine whether reuptake might have reduced the apparent efflux rate of ^{86}Rb from the isolated rat brain microvessels. There was indeed a stimulation of efflux in the presence of ouabain (Figure 3), but this was not a result of an effect on reuptake because the effect could be blocked by 0.5 mM quinidine, a K-channel blocker. Possibly, this putative K-channel could have been opened by depolarization induced by ouabain. There is no quinidine effect on efflux in the absence of ouabain, suggesting the channel is not open under normal conditions. It is possible that the ouabain-like factor present in CSF (23) could modulate the opening of this channel.

Developmentally, the ouabain stimulation of efflux and the quinidine block of this stimulation both appear around 20 days postnatal in the rat (Keep et al., this volume) a time of major maturation in the rat brain. The developmental appearance of the quinidine-sensitive efflux coincides with the ability of the rat brain to accumulate potassium during hyperosmotic stress, and it is possible that the putative channel could be involved in the transport of potassium from blood to brain under these conditions.

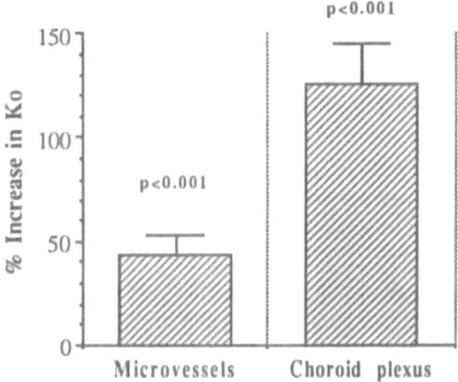

Figure 3. Effect of ouabain (5 mM) on the efflux rate constant for ^{86}Rb in isolated microvessels and isolated choroid plexuses. The ouabain stimulation of efflux was blocked by quinidine (0.5 mM) in microvessels and by bumetanide (30 μM) in the choroid plexus. All values are means ± SEM; n= 5-6 microvessel preparations with duplicates or triplicates performed.

Two other factors that affect Rb efflux from isolated microvessels are pH and intracellular Ca. Extracellular acidification reduced efflux and alkalinization increased it (Figure 4). Whether this action is through a transporter or an ion channel is as yet uncertain, although a pH-sensitive K channel has been described in the choroid plexus (12,18). Increasing intracellular Ca via a Ca-ionophore, A23187, enhanced Rb efflux (Figure 4) possibly via a Ca-sensitive K channel. Verapamil, a Ca-channel blocker was also used in an attempt to reduce intracellular Ca and reduce efflux. However, 0.5 mM verapamil failed to produce a significant reduction.

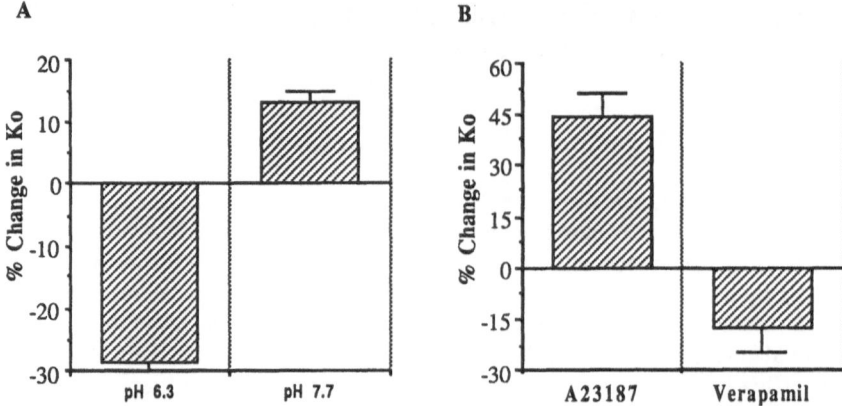

Figure 4. (A) Effect of pH on the efflux rate constant (K_o) for ^{86}Rb in isolated rat microvessels and choroid plexuses. (B) The effect of A23187 (5 µg/ml), a calcium ionophore, and verapamil (0.5 mM), a calcium-channel blocker, on ^{86}Rb efflux. All values are means ± SEM; n=6-13 microvessel preparations with triplicates performed on each. Measurements were made in the presence of 5 mM ouabain.

As with measurements of influx, bumetanide (30 µM) had no significant effect on the rate of ^{86}Rb efflux from isolated cerebral microvessels. This was in marked contrast to choroid plexus where this loop diuretic reduced the efflux rate constant by 60%. At low concentrations, bumetanide is a fairly selective blocker of either K/Cl or Na/K/2Cl co-transport. The presence of ouabain stimulates ^{86}Rb efflux from the plexus through a bumetanide-sensitive mechanism (Figure 3). This suggests that the transporter is of the Na co-transport form because ouabain would decrease the Na gradient opposing efflux through such a mechanism. The presence of K co-transport into and out of the plexus may reflect the same transporter shuttling across the membrane with the net fluxes dependent on the ion gradients across the membrane. As with the uptake mechanism, it is uncertain whether the lack of a bumetanide effect in microvessels reflects a true absence or a difficulty in accessing the luminal membrane in this preparation.

In common with the results from isolated microvessels, part of the Rb efflux from the choroid plexus was through a mechanism sensitive to potassium channel blockers. Barium (20 mM) reduced the rate of efflux by about 25%. The relative effectiveness of different channel blockers, though, does vary between the two tissues.

Potassium Efflux Mechanisms (Patch Clamp)

Patch clamp studies are now providing greater insight into the different types of K channels present at the cerebral microvessels and the choroid plexus. Vigne et al. (59) have demonstrated the existence of a non-selective cation channel in cultured cells with a conductance of about 23 pS. This channel has similar permeabilities for potassium and sodium (1 : 1.5) and is amiloride-sensitive. The exact location of this channel is uncertain but, at least under some conditions, systemic amiloride can reduce the rate of sodium entry into brain (5,43), suggesting a luminal distribution.

Hoyer et al. (29) have demonstrated the occurrence of two inwardly-rectifying potassium channels with conductances of 7 and 30 pS. Only the latter was seen in intact microvessels where the abluminal membrane was sampled, suggesting that these two channels have different distributions, with the 7 pS channel being present only on the luminal membrane. The term inwardly rectifying means that the channel closes on depolarization.

The degree to which they are normally open and their precise function are not known, but they could act as leak pathways for the Na/K ATPase since they are activated by hyperpolarization. Because of the inactivation of these channels by depolarization, these channels are difficult to study with tracers in isolated microvessels, some of which may be metabolically impaired and thus depolarized. There is some evidence that the larger rectifying channel is modulated by angiotensin II and vasopressin and that its activity is controlled by a G-protein.

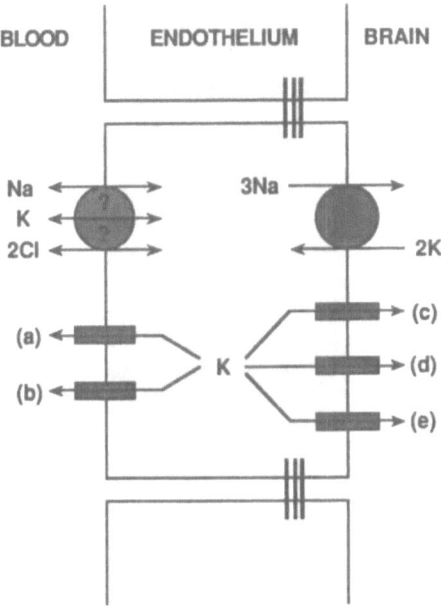

Figure 5. Model of K transport at the BBB. Available evidence points to the presence of abluminal Na/K ATPase and possibly luminal Na/K/2Cl co-transport. There is also evidence for a number of K channels. On the luminal membrane: (a) an amiloride-sensitive non-selective cation channel and (b) an inwardly-rectifying K channel. On the abluminal membrane: (c) a Ba-sensitive inwardly-rectifying K channel, (d) a quinidine sensitive, ouabain-stimulated K channel and (e) a Ca-sensitive, non-selective (?) cation channel. There is also a pH-sensitive efflux mechanism. Whether this is through a K channel (perhaps one of the above) or a transporter is unknown. The ionic permeability of the tight junctions that link the endothelial cells is the subject of debate.

Abbott and Revest (2) have reported in abstract form a Ca-sensitive non-selective cation channel. This might be the mechanism for the A23187 stimulation of Rb efflux described above. A similar channel has been described in choroid plexus (17) where it is activated by osmotic stress via the opening of a Ca channel. Cerebral endothelial intracellular Ca is modulated by a number of hormones (27,50) and it is possible that some of these could modulate endothelial potassium transport. Such an effect occurs with bradykinin in aortic endothelial cells (58).

SUMMARY

Figure 5 gives a summary of K transporters at the BBB based on the available evidence. It appears that the cerebral endothelial cells have an array of potassium channels, although the degree to which each is open under physiological conditions is uncertain. Different channels are present on the luminal and abluminal membranes, and the opening and closing of these channels may allow modulation of the brain K influx and efflux rates and play a role in brain K homeostasis. These channels may also play a role in hyperosmotic brain volume regulation by increasing the entry rate of potassium into brain and may be involved in volume regulation of the endothelial cell itself. The nature of fluid transport at the BBB remains to be fully elucidated, with the presence of a Na/K/2Cl co-transporter being uncertain. The abluminal inwardly-rectifying channel may act as a leak pathway to allow modulation of fluid secretion by the Na/K ATPase without altering the K concentration of that fluid. Finally, there is some evidence that K transport at the BBB is under hormonal and neuronal control. The cerebral capillaries possess receptors for many of the hormones present in blood and brain.

ACKNOWLEDGMENTS

This work was aided by Basic Research Grant No 1-FY91-0-131 from the March of Dimes Birth Defects Foundation and by grants NS-23870 and HL-18575 from the National Institutes of Health. We would also like to thank Dr. Steven R. Ennis, Dr. Gerald P. Schielke, Dr. Mathew Whitico, Ms. Mary Beer and Mr. Marc Anthony Vellila for their contributions to this work.

REFERENCES

1. Abbott, N.J., Butt, A.M., and Wallis, W., The Na-K ATPase of the blood-brain barrier: a microelectrode study, *Ann New York Acad Sci* 481:390-391, 1986.
2. Abbott, N.J., and Revest, P.A., Single-channel currents recorded from rat brain endothelial cells in culture, *J Physiol* 423:105P, 1990.
3. Albrecht, J., Durable inhibition of rat cerebral capillary Na^+/K^+-ATPase after in vivo administration of mercuric chloride, *Toxicol Lett* 59:133-138, 1991.
4. Bairamian, D., Johanson, C.E., Parmelee, J.T., and Epstein, M.H., Potassium cotransport with sodium and chloride in the choroid plexus, *J Neurochem* 56:1623-1629, 1991.
5. Betz, A.L., Sodium transport from blood to brain: inhibition by furosemide and amiloride, *J Neurochem* 41:1158-1164, 1983.
6. Betz, A.L., Ennis, S.R., and Schielke, G.P., Blood-brain barrier sodium transport limits development of brain edema during partial ischemia in gerbils, *Stroke* 20:1253-1259, 1989.
7. Betz, A.L., Firth, J.A., and Goldstein, G.W., Polarity of the blood-brain barrier: distribution of enzymes between the luminal and antiluminal membranes of brain capillary endothelial cells, *Brain Res* 192:17-28, 1980.
8. Bradbury, M.W.B. "The Concept of a Blood-Brain Barrier," John Wiley & Sons, Chichester, p. 465, 1979.
9. Bradbury, M.W.B., The blood-brain barrier. Transport across the cerebral endothelium, *Circ Res* 57:213-222, 1985.
10. Bradbury, M.W.B., and Kleeman, C.R., Stability of the potassium content of cerebrospinal fluid and brain, *Am J Physiol* 213:519-528, 1967.
11. Bradbury, M.W.B., and Stulcova, B., Efflux mechanism contributing to the stability of the potassium concentration in cerebrospinal fluid., *J Physiol* 208:415-430, 1970.
12. Brown, P.D., Loo, D.D.F., and Wright, E.M., Ca^{2+}-activated K^+ channels in the apical membrane of *Necturus* choroid plexus, *J Membrane Biol* 105:207-219, 1988.

13. Butt, A.M., and Jones, H.C., Effect of histamine and antagonists on electrical resistance across the blood-brain barrier in rat brain-surface microvessels, *Brain Res* 569:100-105, 1992.

14. Butt, A.M., Jones, H.C., and Abbott, N.J., Electrical resistance across the blood-brain barrier in anaesthetised rats: a developmental study, *J Physiol* 429:47-62, 1990.

15. Caspers, M.L., Kwaiser, T.M., and Grammas, P., Control of [^{3}H]ouabain binding to cerebromicrovascular (Na$^+$ + K$^+$)-ATPase by metal ions and proteins, *Biochem Pharmacol* 39:1891-1895, 1990.

16. Chaplin, E.R., Free, R.G., and Goldstein, G.W., Inhibition by steroids of the uptake of potassium by capillaries isolated from rat brain, *Biochem Pharmacol* 30:241-245, 1981.

17. Christensen, O., Mediation of cell volume regulation by Ca^{2+} influx through stretch-activated channels, *Nature* 330:66-68, 1987.

18. Christensen, O., and Zeuthen, T., Maxi K$^+$ channels in leaky epithelia are regulated by intracellular Ca^{2+}, pH and membrane potential, *Pflügers Arch* 408:249-259, 1987.

19. Cserr, H.F., DePasquale, M., and Patlak, C.S., Volume regulatory influx of electrolytes from plasma to brain during acute hyperosmolality, *Am J Physiol* 253:F530-F537, 1987.

20. Dehouck, M.-P., Méresse, S., Delorme, P., Fruchart, J.-C., and Cecchelli, R., An easier, reproducible, and mass-production method to study the blood-brain barrier in vitro, *J Neurochem* 54:1798-1801, 1990.

21. Firth, J.A., Cytochemical localization of the K$^+$ regulation interface between blood and brain, *Experientia* 33:1093-1094, 1976.

22. Gadsby, D.C., The Na/K pump of cardiac cells., *Ann Rev Biophys Bioeng* 13:373-398, 1984.

23. Halperin, J.A., Shaeffer, R., Galvez, L., and Malave, S., Ouabain-like activity in human cerebrospinal fluid, *Proc Natl Acad Sci USA* 80:6101-6104, 1983.

24. Hamlyn, J.M., Blaustein, M.P., Bova, S., DuCharme, D.W., Harris, D.W., Mandel, F., Mathews, W.R., and Ludens, J.H., Identification and characterization of a ouabain-like compound from human plasma, *Proc Natl Acad Sci USA* 88:6259-6263, 1991.

25. Harik, S.I., Blood-brain barrier sodium/potassium pump: modulation by central noradrenergic innervation, *Proc Natl Acad Sci USA* 83:4067-4070, 1986.

26. Harik, S.I., Doull, G.H., and Dick, A.P.K., Specific ouabain binding to brain microvessels and choroid plexus, *J Cereb Blood Flow Metab* 5:156-160, 1985.

27. Hess, J., Jensen, C.V., and Diemer, N.H., The vasopressin receptor of the blood-brain barrier in the rat hippocampus is linked to calcium signalling, *Neurosci Lett* 132:8-10, 1991.

28. Holt, S., and Howden, C.W., Omeprazole. Overview and opinion, *Digest Dis Sci* 36:385-393, 1991.

29. Hoyer, J., Popp, R., Meyer, J., Galla, H.-J., and Gögelein, H., Angiotensin II, vasopressin and GTP[γ-S] inhibit inward-rectifying K$^+$ channels in porcine cerebral capillary endothelial cells, *J Membrane Biol* 123:55-62, 1991.

30. Ibaragi, M.-A., Niwa, M., and Ozaki, M., Atrial natriuretic peptide modulates amiloride-sensitive Na$^+$ transport across the blood-brain barrier, *J Neurochem* 53:1802-1806, 1989.

31. Javaheri, S., Role of NaCl cotransport in cerebrospinal fluid production: Effects of loop diuretics, *J Appl Physiol* 71:795-800, 1991.

32. Johanson, C.E., Sweeney, S.M., Parmelee, J.T., and Epstein, M.H., Cotransport of sodium and chloride by adult mammalian choroid plexus, *Am J Physiol* 258:C211-C216, 1990.

33. Jones, H.C., and Keep, R.F., The control of potassium concentration in the cerebrospinal fluid and brain interstitial fluid of developing rats, *J Physiol* 383:441-453, 1987.

34. Kalaria, R.N., Stockmeier, C.A., and Harik, S.I., Brain microvessels are innervated by locus ceruleus noradrenergic neurons, *Neurosci Lett* 97:203-208, 1989.

35. Lin, J.D., Potassium transport in isolated cerebral microvessels from rat, *Japan J Physiol* 35:817-830, 1985.

36. Lin, J.D., Effect of osmolarity on potassium transport in isolated cerebral microvessels, *Life Sci* 43:325-333, 1988.

37. Lindvall, M., Edvinsson, L., and Owman, Ch., Sympathetic nervous control of cerebrospinal fluid production from the choroid plexus, *Science* 201:176-178, 1978.

38. Lindvall-Axelsson, M., Nilsson, C., Owman, Ch., and Winbladh, B., Inhibition of cerebrospinal fluid formation by omeprazole, *Exp Neurol* 115:394-399, 1992.

39. Ling, R.T.K., Hartman, B.K., and Clark, H.B., Hypertension induced changes in cerebral capillary permeability to water are mediated by afferent neuronal connections from the carotid sinus to the brain, *Brain Res* 308:301-308, 1984.

40. Lo, W.D., and Betz, A.L., Oxygen free-radical reduction of brain capillary rubidium uptake, *J Neurochem* 46:394-398, 1986.

41. Masuzawa, T., Saito, T., and Sato, F., Cytochemical study on enzyme activity associated with cerebrospinal fluid secretion in the choroid plexus and ventricular ependyma, *Brain Res* 222:309-322, 1981.

42. Milhorat, T.H., Cerebrospinal fluid and the brain edemas, *in*: "Cerebrospinal Fluid and the Brain Edemas," ed., Neuroscience Society of New York, New York, pp. 1-168, 1987.

43. Murphy, V.A., and Johanson, C.E., Acidosis, acetazolamide, and amiloride: effects on ^{22}Na transfer across the blood-brain and blood-CSF barriers., *J Neurochem* 52:1058-1063, 1989.

44. Nathanson, J.A., Adrenergic regulation of cerebrospinal fluid and aqueous humor, *Trends Pharmacol Sci* 3:452-454, 1982.

45. Ohno, K., Pettigrew, K.D., and Rapoport, S.I., Lower limits of cerebrovascular permeability to nonelectrolytes in the conscious rat, *Am J Physiol* 235:H299-H307, 1978.

46. Olesen, S.-P., and Crone, C., Substances that rapidly augment ionic conductance of endothelium in cerebral venules, *Acta Physiol Scand* 127:233-241, 1986.

47. Parmelee, J.T., Bairamian, D., and Johanson, C.E., Response of infant and adult rat choroid plexus potassium transporters to increased extracellular potassium, *Dev Brain Res* 60:229-233, 1991.

48. Parmelee, J.T., and Johanson, C.E., Development of potassium transport capability by choroid plexus of infant rats, *Am J Physiol* 256:R786-R791, 1989.

49. Peroutka, S.J., Moskowitz, M.A., Reinhard, J.F., and Snyder, S.H., Neurotransmitter receptor binding in bovine cerebral microvessels, *Science* 208:610-612, 1980.

50. Revest, P.A., Abbott, N.J., and Gillespie, J.I., Receptor-mediated changes in intracellular [Ca^{2+}] in cultured rat brain capillary endothelial cells., *Brain Res* 549:159-161, 1991.

51. Saito, Y., and Wright, E.M., Regulation of intracellular chloride in bullfrog choroid plexus, *Brain Res* 417:267-272, 1987.

52. Schielke, G.P., Moises, H.C., and Betz, A.L., Potassium activation of the Na,K-pump in isolated brain microvessels and synaptosomes, *Brain Res* 524:291-296, 1990.

53. Schielke, G.P., Moises, H.C., and Betz, A.L., Blood to brain sodium transport and interstitial fluid potassium concentration during early focal ischemia in the rat, *J Cereb Blood Flow Metab* 11:466-471, 1991.

54. Smith, Q.R., and Rapoport, S.I., Cerebrovascular permeability coefficients to sodium, potassium, and chloride, *J Neurochem* 46:1732-1742, 1986.

55. Sussman, I., Carson, M.P., McCall, A.L., Schultz, V., Ruderman, N.B., and Tornheim, K., Energy state of bovine cerebral microvessels: comparison of isolation methods, *Microvasc Res* 35:167-178, 1988.

56. Sweadner, K.J., Isozymes of the Na$^+$/K$^+$-ATPase, *Biochim Biophys Acta* 988:185-220, 1989.

57. Takasato, Y., Rapoport, S.I., and Smith, Q.R., An in situ brain perfusion technique to study cerebrovascular transport in the rat, *Am J Physiol* 247:H484-H493, 1984.

58. Thuringer, D., Diarra, A., and Sauve, R., Modulation by extracellular pH of bradykinin-evoked activation of Ca^{2+}-activated K$^+$ channels in endothelial cells., *Am J Physiol* 261:H656-H666, 1991.

59. Vigne, P., Champigny, G., Marsault, R., Barbry, P., Frelin, C., and Lazdunski, M., A new type of amiloride-sensitive cationic channel in endothelial cells of brain microvessels, *J Biol Chem* 264:7663-7668, 1989.

60. Vorbrodt, A.W., Lossinsky, A.S., and Wisniewski, H.M., Cytochemical localization of ouabain-sensitive, K$^+$-dependent p-nitro-phenylphosphatase (transport ATPase) in the mouse central and peripheral nervous systems, *Brain Res* 243:225-234, 1982.

61. Zeuthen, T., and Wright, E.M., Epithelial potassium transport: tracer and electrophysiological studies in choroid plexus, *J Memb Biol* 60:105-128, 1981.

EXPRESSION OF Na,K-ATPase AT

THE BLOOD-BRAIN INTERFACE

Berislav V. Zlokovic, Liang Wang, Jasmina B. Mackic, Asma J. Saraj,
J. Gordon McComb and Alicia McDonough

Departments of Neurosurgery, Physiology and Biophysics, and Division of
Neurosurgery, Childrens Hospital, Los Angeles, USC School of Medicine,
Los Angeles, CA 90033

INTRODUCTION

A closed regulation of ionic composition in the brain is a prerequisite for normal
neuronal functions and for the maintenance of physiological brain volume and intracranial
pressure. The blood-brain barrier (BBB) and choroid plexus play the chief roles in
preserving relatively constant K^+ and Na^+ concentrations in extracellular fluid (ECF) and
cerebrospinal fluid (CSF) of the brain despite variations in plasma levels of these ions (1).
Na,K-ATPase is a key enzyme that catalyzes active transport of K^+ and Na^+ between blood,
ECF and CSF.

Na,K-ATPase is a ubiquitous plasma membrane-bound enzyme that consists of two
protein subunits in a 1:1 ratio. The α subunit (112 kDa) is the catalytic site for ATP
hydrolysis and serves as the receptor for cardiac glycosides, such as ouabain, which
specifically inhibit the pump activity (2). The exact function of the β glycoprotein subunit (35
kDa) has not been established, but the current hypothesis states that the β subunit may
regulate proper folding of Na,K-ATPase and/or delivery of the enzyme to the plasma
membrane (3). Three α (α_1, α_2, α_3) (2) and two β (β_1, β_2)(4) isoforms are detected in the
brain. The α isoforms are found to be distinct with respect to their mobility in SDS-PAGE,
antigenic determinants and antibody specificity (2). The α_1 form is found in non-neural cells
and neuronal soma, α_2 alone in myelinated axons, and the α_3 predominantly in neural cells.
The α_2 and α_3 subunits differ from α_1 by having a higher affinity for ouabain (2,5). In situ
hybridization studies have suggested that six structurally different $\alpha\beta$ heterodimers may be
expressed by different cells in the brain (6).

Regulation of the expression of Na,K-ATPase α and β subunits in the brain may be
affected by changes in the rates of synthesis and/or degradation of the α and/or β isoforms at
the level of mRNA and/or protein (3,7), as summarized in Figure 1. The α_1, α_2, α_3, β_1 and
β_2 subunits are encoded by separate genes, independently transcribed into RNA, processed
into their respective mRNAs and cotranslationally inserted into endoplasmic reticulum (ER)
membranes. The α and β subunits assemble very rapidly after synthesis, and that assembly
is essential for export from the ER, and for expression of functional activity (3,7).

Frontiers in Cerebral Vascular Biology: Transport and Its Regulation
Edited by L.R. Drewes and A.L. Betz, Plenum Press, New York, 1993

55

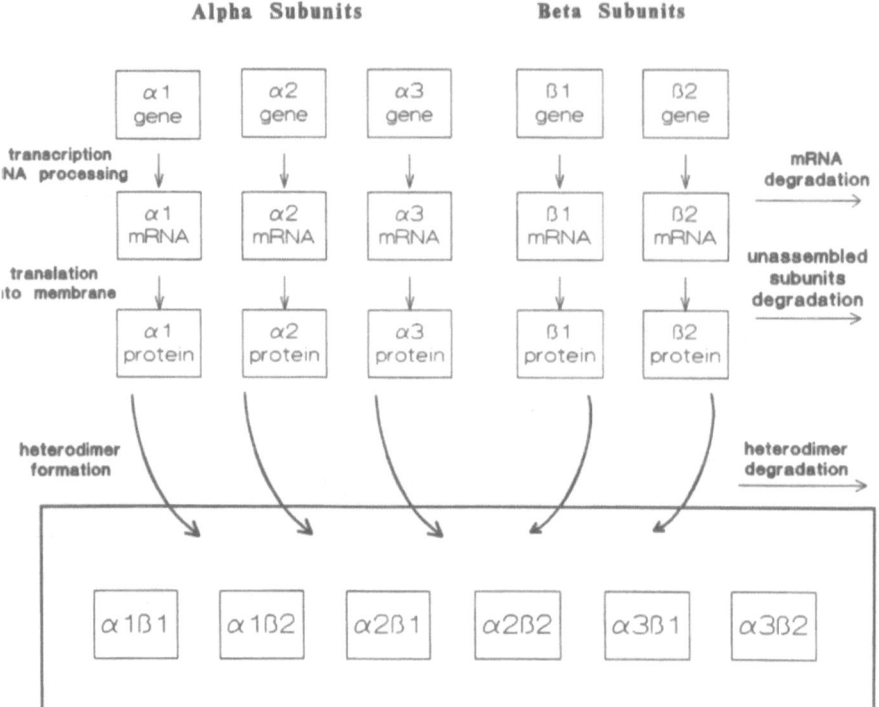

Figure 1. Compartmental model for synthesis and assembly of sodium pump isoforms in the brain. The boxes enclose α_1, α_2, α_3, β_1 and β_2 subunits: DNA, mRNA, unassembled protein, and heterodimers. The levels in each compartment are determined by the rates of synthesis into and degradation from each compartment. These rates are indicated by arrows. Each arrow represents a potential site of regulation.

Previous cytochemical and ouabain-binding studies have demonstrated localization and relative enrichment of Na,K-ATPase in cerebrovascular endothelium (8-10) and choroid plexus epithelium (11). Expression of subunits at the blood-brain interface has been recently studied in our laboratory (12-14). The results described below are derived from our reported (12-14) and on-going studies.

MATERIALS AND METHODS

Tissue Preparations

Cerebral microvessels, capillary-depleted brain tissue and choroid plexuses were obtained from guinea pig (Hartley), rat (Sprague-Dawley) and bovine brains. Cerebral microvessels were isolated by a mechanical homogenization technique (15) with modifications as described (16). The choroid plexuses were dissected from both lateral ventricles. Microvessels, capillary-depleted brain tissue, and choroid plexuses were stored at -70°C until assayed. Endothelial cells from rat cerebral resistance microvessels were provided by Dr. Diglio. Crude cellular homogenates were prepared and samples assayed for protein content (17). Na,K-ATPase activity was estimated in tissue homogenates by K-dependent p-nitrophenyl phosphatase (K-pNPPase) reaction (18), and gamma-glutamyl transpeptidase (GGT) activity was determined by a quantitative colorimetric Sigma Assay (19).

Immunoblotting

The presence of α and β subunits in tissue samples was determined by Western blot analysis with isoform specific antibodies. Previously described monoclonal antibodies McK1 (anti-rat α_1, cross reacts weakly with guinea pig, but not with bovine) and McB2 (anti-rat α_2, cross reacts with guinea pig and bovine) were provided by Dr. Sweadner; polyclonal antibodies against guinea pig holoenzyme were from the McDonough laboratory; polyclonal anti-rat α_3 (does not cross react with guinea pig and bovine) antibodies were provided by Dr. Mercer; polyclonal anti-rat β_1 and anti-rat β_2 antibodies were provided by Dr. Levenson; monoclonal anti-chicken α_1 (cross reacts with bovine) was provided by Dr. Fambrough.

To verify linearity of the detection system, different amounts (0.5 to 10 µg) of sample tissue homogenate proteins were resolved by SDS-PAGE (20). Gels were blotted onto diazotized paper and incubated overnight for rat samples with either anti-α_1 (McK1, 1:200 dilution), anti-α_2 (McB2, 1:100 dilution), anti-α_3 (1:500 dilution), anti-β_1 (1:400 dilution) and anti-β_2 (1:500 dilution); for guinea pig samples with McK1 (1:200), McB2 (1:200) and anti-holoenzyme guinea pig antibodies (1:500); for bovine samples with anti-chicken α_1 (1:100) and McB2 (1:200). The enhanced chemiluminescence (ECL) and/or [^{125}I]-protein A method were used to detect subunits. Autoradiograms of Western blots were scanned to determine levels of α and β subunits relative to a constant amount of proteins in different samples.

RESULTS AND DISCUSSION

We have initially reported differential expression of α_1 and α_2 protein subunits in guinea-pig cerebral microvessels and choroid plexus based on studies with McK1 and McB2 antibodies (12). Since cross-reaction of McK1 antibodies with guinea pig α_1 isoform was relatively weak, the study has been repeated with more specific anti-guinea pig antibodies. These results indicated that both low and high affinity α isoforms were present in microvessels, but only α_1 was expressed in choroid plexus while the α_2 signal was negligible. Relative regional levels of α_1 and α_2 in guinea pig cerebral cortex, cerebellum, caudate nucleus, hippocampus and midbrain were as previously reported in rat brain (2). The expression of α_1 was the highest in guinea pig choroid plexus. Both α_1 and α_2 were somewhat higher at the BBB than in capillary-depleted guinea pig brain tissues, and β_1 was found in all tissues studied.

Expression of α_1 and α_2 isoforms in bovine microvessels, choroid plexus, and capillary-depleted bovine brain tissue have been studied with anti-chicken α_1 and McB2 antibodies. Results obtained with bovine tissues paralleled findings obtained with guinea pig tissues.

Subsequent studies were performed with rat tissues (13), and expression of α_1 and α_2 at rat BBB and choroid plexus were comparable with guinea pig (12) and bovine data. Recently we have simultaneously examined the distribution of all five Na,K-ATPase subunit isoforms in rat cerebral microvessels, choroid plexus and capillary-depleted brain, and correlated the levels of subunit expression with functional enzymatic Na,K-ATPase activity (14).

As shown in Figure 2 six structurally distinct Na,K-ATPase isoenzymes (αβ heterodimers) may be expressed in cerebral microvessels and two in the choroid plexus (14). Previous studies have demonstrated homogeneity of ouabain binding sites in rat cerebral microvessels (9), suggesting the presence of only high-affinity α isoforms. However, it is unlikely that the low concentrations of ouabain used in these studies were able

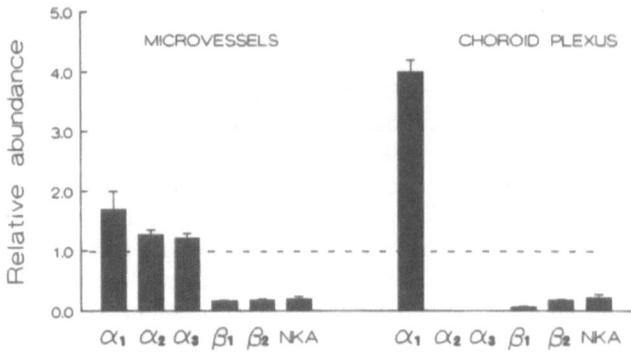

Figure 2. Distribution of sodium pump isoform subunits and Na,K-ATPase activity (NKA) in rat cerebral microvessels and choroid plexus relative to capillary-depleted brain homogenates determined by quantitation of Western blots. All values are normalized to a constant amount of protein in studied samples. The dashed line is included at a value of 1 and represents either subunit abundance or enzyme activity in the brain defined arbitrarily as 1.

to detect the low affinity α_1 isoform that was found to have a 200-fold lower sensitivity to ouabain than the α_2 and α_3 isoforms, as reported in other studies (2). As indicated by the present quantitative immunoblot analysis, the α_1 subunit was most abundant in the choroid plexus, but whether this represents functional $\alpha\beta$ heterodimers remains to be determined. Expression of the α_1 isoform in microvessels and choroid plexus is consistent with the concept that the α_1 form is expressed in all tissues (2), as well as in virtually all cell types and structures in the brain (6). Although the mRNA encoding for the α_2 isoform was shown to be expressed in choroidal tissue (6), the α_2 protein isoform was undetectable and may not be translated in the choroid plexus.

Several observations indicated that separation of the β subunit from the $\alpha\beta$ heterodimers results in irreversible inactivation of both enzymatic and transport activities of the Na,K-ATPase (3). We reported recently that β subunits may be limiting in microvessels and choroid plexus for the $\alpha\beta$ assembly (Figures 1 and 2), despite the relative overexpression of α_1 isoform (14). This in turn may limit the expression of functionally active Na,K-ATPases at the BBB and plexus relative to the brain, as suggested by the high correlation between lower expression of β_1 and β_2 isoforms in microvessels and plexus and lower enzymatic activity. Na,K-ATPase activity in rat microvessels and choroid plexus was only about one fifth of the activity in capillary-depleted brain.

The present studies do not address cell-specific expression of Na,K-ATPase in microvessels and choroid plexus. Although the microvascular preparation is primarily regarded as a model of capillary endothelial cells, the presence of small amounts of the astrocytic end feet, pericytes and some axon terminals cannot be excluded and may, therefore, influence the interpretation of results. Thus, present data may indicate that regulation of ion transport across the BBB requires a complex interaction of endothelial cells, astrocytes and pericytes, which in turn may express different Na,K-ATPase isoenzymes. For example, the increased extracellular K^+ generated during neuronal activity may be buffered and re-distributed into plasma and CSF by an active $\alpha_2\beta_2$ system expressed in astrocytes (Shyjan and Levenson, unpublished observation) that may be more efficient at higher K^+ concentrations. Also, there may be a possibility that cerebral endothelial cells and/or choroidal epithelial cells express more than one type of $\alpha\beta$ heterodimer. In an attempt to answer this question we recently studied isoform expression in cultured rat cerebral endothelial cells (Zlokovic et al., unpublished observation). This study demonstrated expression of α_1 and β_1, but the signals from α_2 and α_3 were almost negligible. Since cultures studied were maintained for 16 passages, there may be a possibility that the signal from both high affinity isoforms is extremely weak because of loss of a differentiated phenotype.

At present we have little insight into the functional significance of Na,K-ATPase isoenzyme diversity in cerebral microvessels and choroid plexus. The presence of all three α isoforms in microvessels suggests the possibility for differential function or regulation and, in particular, α_2 has been shown to be highly regulated in other tissues (7). Our recent studies indicated that α_2 is also down-regulated at the BBB in hypokalemic rats (Zlokovic et al., unpublished observations), as it is in brain, muscle and heart (21). Subsequent immunocytochemical and in situ hybridization studies should establish the cellular distribution pattern and structural differences among members of Na,K-ATPase multigene family at the blood-brain interface.

ACKNOWLEDGMENTS

This work was supported by funds provided by the Cigarette and Tobacco Surtax Fund of the State of California through the Tobacco-Related Disease Research Program of the University of California, grant 2RT0071.

REFERENCES

1. Betz, A.L., and Goldstein, G.W., Specialized properties and solute transport in brain capillaries, *Annu. Rev. Physiol.* 48:241-250, 1986.

2. Sweadner, K.J., Isozymes of the Na,K-ATPase, *Biochim. Biophys. Acta* 988:185-220, 1989.

3. McDonough, A.A., Geering, K., and Farley, R.A., The sodium pump needs its β subunit, *FASEB J.* 4:1598-1605, 1990.

4. Shyjan, A.W., Gottardi, C., and Levenson, R., The Na,K-ATPase beta 2 subunit is expressed in rat brain and copurifies with Na,K-ATPase activity, *J. Biol. Chem.* 265:5166-5169, 1990.

5. Urayama, O., and Sweadner, K.J., Ouabain sensitivity of alpha 3 isozyme of rat Na,K-ATPase, *Biochem. Biophys. Res. Commun.* 156:796-800, 1988.

6. Watts, A.G., Sanchez-Watts, G., Emanuel, J.R., and Levenson, R., Cell-specific expression of mRNAs encoding Na,K-ATPase α and β subunit isoforms within the rat central nervous system, *Proc. Natl. Acad. Sci.* 88:7425-7429, 1991.

7. McDonough A.A., Hensley, C.B., and Azuma, K.K., Differential regulation of sodium pump isoforms in heart, *Sem. Nephrol.* 12:49-55, 1992.

8. Betz, A.L., Firth, J.A., and Goldstein, G.W., Polarity of the blood-brain barrier: distribution of enzymes between the luminal and the antiluminal membranes of brain capillary endothelial cells, *Brain Res.* 192:17-28, 1980.

9. Harik, S.I., Doull, G.H., and Dick, P.K.A., Specific ouabain binding to brain microvessels and choroid plexus, *J. Cereb. Blood Flow Metab.* 5:156-160, 1985.

10. Nag, S., Ultracytochemical localization of Na^+,K^+-ATPase in cerebral endothelium in acute hypertension, *Acta Neuropathol.* 80:7-11, 1990.

11. Miwa, S., Inagaki, C., Fujiwara, M., and Takaori, S., Na,K-, Mg- and HCO_3-adenosine triphosphatases in the rabbit brain choroid plexus, *Japan. J. Pharmacol.* 30:337-345, 1980.

12. Saraj, A.J., McDonough, A.A., McComb, J.G., and Zlokovic, B.V., Differential expression of Na,K-ATPase isoforms in the choroid plexus and cerebral microvessels of the guinea-pig brain, *Soc. Neurosci. Abstr.*, 17:240, 1991.

13. Zlokovic, B.V., Mackic, J.B., Wang, L., Magyar, C., McComb, J.G., and McDonough, A.A., Expression of Na,K-ATPase at the blood-brain interface in rats, *Soc. Neurosci. Abstr.*, in press, 1992.

14. Zlokovic, B.V., Mackic, J.B., Wang, L., McComb, J.G., and McDonough, A., Differential expression of Na,K-ATPase α and β subunit isoforms at the blood-brain barrier and the choroid plexus, *J. Biol. Chem.*, submitted, 1992.

15. Goldstein, G.W., Wolinsky, J.C., Csejtey, J., and Diamond, I., Isolation of metabolically active capillaries from rat brain, *J. Neurochem.* 25:715-717, 1975.

16. Pardridge, W.M., Eisenberg, J., and Yamada, T., Rapid sequestration and degradation of somatostatin analogues by isolated brain microvessels, *J. Neurochem.* 44:1178-1184, 1985.

17. Lowry, O.H., Rosebrough, N.J., Farr A.L. and Randall, R.J., Protein measurement with the Folin phenol reagent, *J. Biol. Chem.* 193:265-275, 1951.

18. Murer, H., Ammann, E., Biber, J., and Hopfer, U., The surface membrane of the small intestinal epithelial cell. I. Localization of adenyl cyclase, *Biochim. Biophys. Acta* 433:509-519, 1976.

19. Naftalin, L., Sexton, M., Whitaker, J.F., and Tracey, D., A routine procedure for estimating gamma-glutamyl transpeptidase activity, *Clin. Chim. Acta* 26:293-296, 1969.

20. McDonough, A.A., and Schmitt, C., Comparison of subunits of cardiac, brain and kidney Na^+,K^+-ATPase., *Am. J. Physiol.* 248:C247-C251, 1985.

21. Azuma, K.K., Hensley, C.B., Putnam, D. S., and McDonough, A.A., Hypokalemia decreased Na,K-ATPase α_2 but not $\alpha 1$ isoform abundance in heart, muscle and brain, *Am. J. Physiol.* 260:C958-C964, 1991.

EFFECTS OF PROTEIN KINASE C AGONISTS ON

Na,K-ATPase ACTIVITY IN RAT BRAIN MICROVESSELS

Hiroo Johshita[1], Takao Asano[2], Tohru Matsui[2], Tohru Koide[3]

[1]Saitama Cardiovascular Center (Project Office), Saitama Prefectural Government, 3-15-1, Takasago, Urawa, Saitama 336 Japan
[2]Saitama Medical Center, Department of Neurosurgery, Saitama Medical Center, 1981, Kamoda, Kawagoe, 350 Japan
[3]Chugai Pharmaceutical Co. Ltd. 2-1-9, Kyobashi, Chuo-ku, 104, Tokyo, Japan

INTRODUCTION

Growing evidence for the role of protein kinase C (PKC) as an intracellular second messenger system (1) raises the possibility that this enzyme might also modulate membrane function such as the ion conductance mediated by Na,K-ATPase. We examined the effects of an sn-diacylglycerol (sn-1,2-dioctanoylglycerol, diC$_8$), a cell-permeable, endogenous PKC activator, and phorbol esters (phorbol 12,13-diacetate, PDA; phorbol 12-myristate 13-acetate, PMA; and 4-alpha phorbol, 4α-P) as exogenous activators of PKC on the sodium pump in rat brain microvessels (MV) in vitro.

MATERIALS AND METHODS

Rat brain microvessels were prepared (2) from the hemispheres of saline perfused SD rats (n=3-4), minced with scissors in 0.25 M sucrose, 5 mM EDTA and 50 mM Tris·HCl (pH 7.3). The suspension was passed serially through nylon mesh having pore sizes of 670 μm and 335 μm and ultracentrifuged at 58,000 g for 60 min in a discontinuous sucrose density gradient. The Na,K-ATPase activity was obtained by subtracting the 1 mM ouabain-insensitive activity from the total phosphatase activity. Inorganic phosphate was determined using the phospho-molybdate method (2). Stocks of diC$_8$, PDA, PMA and 4α-P were made in dimethylsulfoxide (DMSO). The vehicle concentration was under 5% at 10^{-3} M of each agent per tube and was less at lower concentrations of agents examined. The protein concentration was determined by the Bio-Rad protein assay (Richmond, CA). The effects of the PKC inhibitors, H-7 (3) and staurosporine (4) on the ATPase activation induced by diC$_8$ at 10^{-4} M were further studied.

Frontiers in Cerebral Vascular Biology: Transport and Its Regulation
Edited by L.R. Drewes and A.L. Betz, Plenum Press, New York, 1993

61

RESULTS

The control Na,K-ATPase activity in mV was 6.7 ± 0.9 µmol P_i/mg protein/hr (n=9). The Na,K-ATPase activity with the DMSO vehicle present at a concentration equivalent to 10^{-3} M agent level was 7.8 ± 1.3 µmol P_i/mg protein/hr (n=9) and was not statistically different (P=0.31, Wilcoxon rank sum) with the control. DiC_8 significantly enhanced Na,K-ATPase activity at 10^{-4} M ($139\%\pm6$, n=9, mean $\pm$ SE, P< 0.01, Wilcoxon rank sum), and at 10^{-6} M ($136\%\pm18$, n=9, P<0.05) compared with the control value, and PDA ($109\%\pm6$, n=8) or PMA ($75\%\pm7$, n=7) or 4α-P ($92\%\pm9$) at 10^{-6} M in any range examined showed no remarkable activation of Na,K-ATPase in mV (Figure 1). In a separate experiment, the PKC inhibitors H-7 and staurosporine suppressed the Na,K-ATPase activation caused by diC_8 at 10^{-4} M (Table 1).

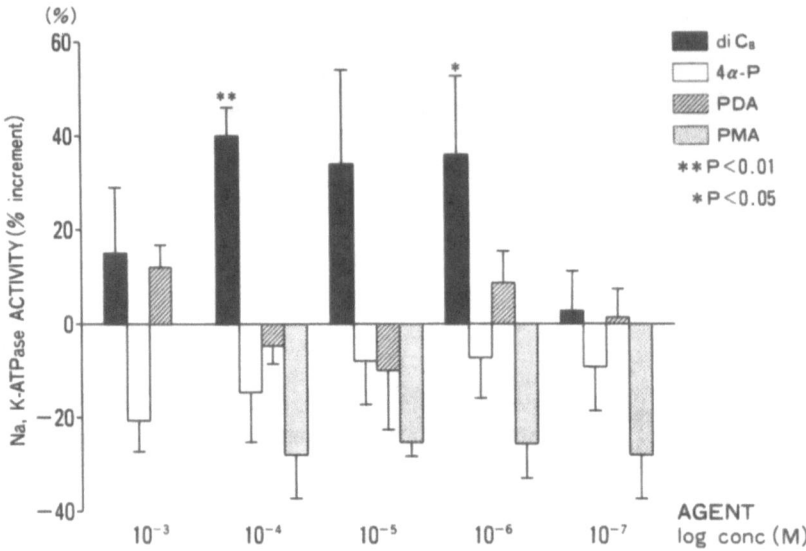

Figure 1. Effects of PKC agonists on Na,K-ATPase activity in rat brain microvessels. Each value represents the mean $\pm$ SE of the percent change from the control activity.

Table 1. Effects of PKC inhibitors on Na,K-ATP activity in microvessels with diC_8 at 10^{-4} M.

Agents	Concentration of PKC inhibitor (M)					
	0	10^{-6}	10^{-7}	10^{-8}	3×10^{-8}	10^{-9}
DiC$_8$ (10^{-4} M)	149±20					
DiC$_8$ (10^{-4} M) + H-7		98±06	103±16			
DiC$_8$ (10^{-4} M) + Staurosporine				47±11	67±8	69±8

Percentage of control Na,K-ATPase activity (mean $\pm$ SE, n=3)

DISCUSSION

Our results show that, under normal conditions, the intrinsic PKC activator, diC_8 significantly enhanced Na,K-ATPase activity in rat brain microvessels, but all phorbol esters examined failed to activate it. The effects of the specific PKC inhibitors H-7 and staurosporine against this activation indicate the possibility that this intracellular lipid messenger system might be involved in the activity of the sodium pump. However, the reason for the lack of effect of active phorbols is unclear at present. In previous studies that focused on the effect of PKC on Na,K-ATPase, the results were diverse, i.e., stimulation of Na,K-ATPase was shown in rat hepatocytes (5) and in rabbit diabetic nerve (6) directly by both diC_8 and PMA, and in rat cerebral cortex (7) synaptosomes, indirectly using a PKC inhibitor such as sphingosine. In contrast, suppression of Na,K-ATPase was reported in rat kidney proximal tubule cells (8) by OAG (1-oleoyl-2-acetyl-sn-glycerol) or PDBu (phorbol 12,13-dibutyrate). This might indicate that PKC activation is cell specific. Alternatively, taking into account that the mode of PKC activation by phorbols and diacylglycerol may be different (9), the difference in the effects of PKC activators used in the present study may be somewhat different from the results presented by other reports. This might be the result of a technical difficulty inherent in the technique employed in the present study. However, it is of interest that the lipid requirement of Na,K-ATPase (10), which has been recognized for some time, might have some relation to our findings. Furthermore, according to the report by Greene et al. (6), exogenous and endogenous PKC activators have not been shown to have any effects under normal condition. Therefore, further studies are necessary to examine the effects of the above agents in pathological conditions, i.e., ischemic situations.

ACKNOWLEDGMENTS

The authors gratefully acknowledge the expert technical assistance of Tohru Kametani and Kenbu Ishido and helpful suggestions of Dr. Shinichi Yoshida.

REFERENCES

1. Nishizuka, Y., Studies and perspectives of protein kinase C, *Science* 233:305, 1986.
2. Koide, T., Asano, T., Matsushita, H., and Takakura, K., Enhancement of ATPase activity by a lipid peroxide of arachidonic acid in rat brain microvessels, *J. Neurochem.* 46:235, 1986.
3. Hidaka, H., Inagaki, M., Kawamoto, S., and Sasaki, Y., Isoquinolinesulfonamides, novel and potent inhibitors of cyclic nucleotide dependent protein kinase and protein kinase C, *Biochemistry* 23:503, 1984.
4. Tamaoki, T., Nomoto, H., Isami, T., Kato, Y., Morimoto, M., and Tomita, F., Staurosporine, a potent inhibitor of phospholipid/Ca^{++} dependent protein kinase, *Biochem. Biophys. Res. Commun.* 135:397, 1986.
5. Lynch, C.J., Wilson, P.W., Blackmore, P.F., and Exton, J.H., The hormone-sensitive hepatic Na^+-pump, *J. Biol. Chem.* 261:14551, 1986.
6. Greene, D.A., and Lattimer, S.A., Protein kinase C agonists acutely normalize decreased ouabain-inhabitable respiration in diabetic rabbit nerve. Implications for (Na,K)-ATPase regulation and diabetic complications, *Diabetes* 35:242, 1986.
7. Oishi, K., Zheng, B., and Kuo, J.F., Inhibition of Na,K-ATPase and sodium pump by protein kinase C regulators sphingosine, lysophosphatidylcholine, and oleic acid, *J. Biol. Chem.* 265:70, 1990.
8. Bertorello, A., and Aperia, A., Na^+,K^+-ATPase is an effector protein for protein kinase C in renal proximal tubule cells, *Am. J. Physiol.* 254:F370, 1988.

9. Kolesnik, R.N., and Clegg, A., 1,2-Diacylglycerols, but not phorbol esters, activate a potential inhibitory pathway for protein kinase C in GHP$_3$ pituitary cells, *J. Biol. Chem.* 263:6534, 1988.

10. Roelofsen, B., The (non)specificity in the lipid-requirement of calcium- and (sodium plus potassium)-transporting adenosine triphosphatases, *Life Sci.* 29:2235, 1981.

BRAIN VOLUME REGULATION DURING DEVELOPMENT: THE ROLE OF BLOOD-BRAIN BARRIER POTASSIUM TRANSPORT

Richard F. Keep[1], Jianming Xiang[1], Steven R. Ennis[1], Mary E. Beer[1] and A. Lorris Betz[1,2]

[1]Department of Surgery
[2]Departments of Pediatrics and Neurology
University of Michigan, Ann Arbor, MI 48109-0532

INTRODUCTION

In adult mammals, acute plasma hyperosmolality causes brain shrinkage but because of the uptake of solute into brain from either blood, across the blood-brain barrier (BBB), or CSF, (4,5,16) or the generation of idiogenic osmoles, there is some degree of brain volume regulation. In addition, brain cells are themselves capable of volume regulation. This again involves the uptake and loss of solute, in particular Na^+, K^+, Cl^-, and amino acids (13,15).

Although brain volume regulation has been extensively studied in the adult (reviewed in (6,7)), such regulation during development has received sparse attention despite the importance of a number of conditions associated with brain water imbalance, for example, hydrocephalus, water intoxication, ischemia and perhaps intracerebral hemorrhage. In this study, we examined the degree of brain volume regulation in response to acute hyperosmotic stress in the rat during development. In two previous studies, it has been shown that potassium and calcium homeostasis improve markedly at birth in the rat (9,10), and we examined whether brain volume regulation improves during the same period. We also investigated the role of BBB potassium transport in the development of brain volume regulation as, in the adult, hyper-osmotic stress induces a pronounced increase in the permeability of the blood-brain barrier (BBB) to potassium (5).

METHODS

Volume Regulation

Initial experiments were performed on fetal (21 days gestation; 1-2 days prior to birth), neonatal (1-2 days after birth) and adult Sprague Dawley rats. Animals were anesthetized with sodium pentobarbital (50 mg/kg) and maintained at 37°C. Acute hyperosmotic stress

Frontiers in Cerebral Vascular Biology: Transport and Its Regulation
Edited by L.R. Drewes and A.L. Betz, Plenum Press, New York, 1993

65

was induced by intraperitoneal injection of 2 M NaCl, 2 ml/100g. Controls received either isotonic saline or no injection; there were no significant differences between these two groups. The sacrifice of animals between 5 and 90 min after NaCl injection showed that plasma osmolality was fairly constant between 10 and 90 min, and brain water content stabilized after 60 min. Therefore, the age comparison presented is limited to animals sacrificed between 60 and 90 min. As well as determining brain water content (by wet/dry weight) and plasma osmolality, brain Na and K contents were determined by flame photometry. In further experiments, the study was expanded to include 10-, 20- and 30-day postnatal rats.

Potassium Transport

For transport studies in isolated cerebral microvessels, ^{86}Rb was used as a marker for potassium. Both in vivo (5) and in vitro (14) experiments have shown that ^{86}Rb is a good marker for potassium transport at the BBB. Microvessels were isolated from the brains of rats aged between 21 days gestation and adult using density gradient centrifugation, sieving and collection on a glass bead column (17). These vessels were then incubated for 30 min in buffer containing ^{86}Rb. After this accumulation phase, the microvessels were washed and transferred to fresh buffer, and the rate of ^{86}Rb efflux was examined during the following 20 min. A plot of the natural log of the fraction of isotope remaining against time was linear, demonstrating that the efflux follows a single exponential. Where drugs were added, they were given at the start of the efflux measurement.

RESULTS

In Vivo Volume Regulation

There was a marked improvement in brain volume regulation around birth in the rat. Despite a similar increase in plasma osmolality (Δ osmolality was 63±3 and 67±6 mOsm/kg in fetuses and neonates, respectively), fetuses lost more brain water (p<0.01) during the hyperosmotic stress. In fetuses, brain water content fell from 7.39±0.06 to 6.27±0.05 g/g dry wt (n= 6 and 12), and in neonates it fell from 7.46±0.04 to 6.74±0.04 g/g dry wt (n= 13 and 15). Postnatal comparisons of the degree of volume regulation are hampered by

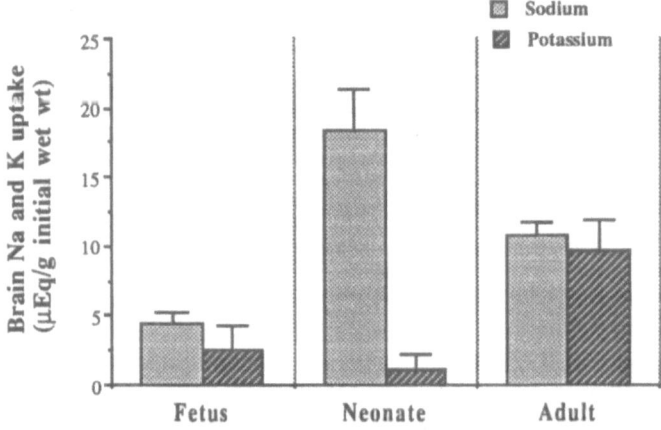

Figure 1. Net sodium and potassium uptake into brain 60-90 min after the induction of plasma hyperosmolality in fetal, neonatal and adult rats. Values are means ± SE; n=10-12.

marked developmental changes in control brain water content. However, if compared in terms of a perfect osmometer, where osmotic equilibrium is achieved by movements of water and not solute (i.e., no volume regulation), neonatal and adult rats volume regulate equally. Neonatal water loss was $56\pm4\%$ (n=15) and adult loss was $54\pm4\%$ of that expected from an osmometer.

Brain volume regulation during hyperosmotic stress is brought about by the uptake of solutes from blood to brain. The improvement in brain volume regulation around birth is associated with a marked increase in net sodium uptake (Figure 1). There is, however, also a postnatal difference in ion uptake. Neonates take up sodium ($p<0.001$), and adults take up both sodium and potassium ($p<0.05$; Figure 1). Further results indicate that this change occurs between 10 and 20 days after birth.

In Vitro Potassium Transport

In the adult rat, brain volume regulation during acute hyperosmotic stress is associated with a pronounced increase in the permeability of the BBB to potassium (as measured with [42]K or [86]Rb [5]) and we have examined whether developmental change in net potassium uptake in response to hyperosmotic stress could reflect developmental changes in BBB potassium transport by examining potassium (^{86}Rb) efflux from cerebral microvessels isolated from rats of different ages. This efflux is probably from the abluminal (brain-facing) membrane of the microvessel and thus forms one step of the transport of potassium from blood to brain.

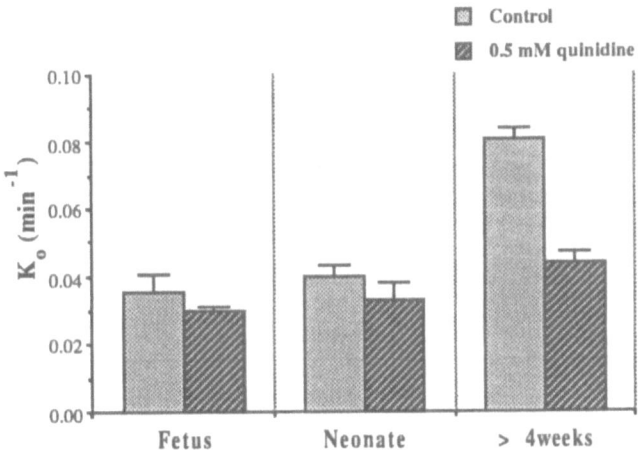

Figure 2. Developmental changes in the efflux rate constant (K_o) for ^{86}Rb in cerebral microvessels in the presence or absence of 0.5 mM quinidine. All measurements were made in the presence of 5 mM ouabain. Values are means $\pm$ SE, n = 4-12 preparations with triplicates performed on each.

Initially, measurements of the Rb efflux rate constant were made in the presence of 5 mM ouabain to prevent reuptake. Under these conditions, the efflux rate showed no significant change between fetus and neonate, but doubled in rats aged 4- to 6-weeks old (Figure 2; $p<0.001$). Quinidine (0.5 mM), a K-channel blocker, had no significant effect on Rb efflux in the fetus and neonate, but blocked the developmental increase found in the 4- to 6-week-old animals (Figure 2). In 4- to 6-week-old rats, quinidine blocked $45\pm3\%$ of the total efflux. Further experiments demonstrated that developmental increase in the efflux rate constant for Rb occurred between 10 and 20 days postnatal and that this coincided with the appearance of the quinidine-sensitive pathway. The developmental appearance of this pathway also coincides with the first net uptake of K during hyperosmotic stress, suggesting that this pathway may have a role in brain volume regulation.

Measurements were next made of ^{86}Rb efflux in the absence of ouabain. Under these conditions, in 4- to 6-week-old rats, there was no quinidine-sensitive efflux. The K_0 was 0.047 ± 0.002 and 0.043 ± 0.003 min^{-1} in the absence and presence of quinidine respectively. Thus, it appears that ouabain acts to open the quinidine-sensitive pathway and that this pathway is not, at least in isolated microvessels, normally open.

DISCUSSION

The marked improvement in brain volume regulation that occurs around birth coincides with a number of other changes related to ion transport in the brain. Transendothelial resistance increases four-fold just prior to birth in the rat (3), an event paralleled by a marked decrease in the BBB permeability to rubidium (used as a marker for potassium) and urea (unpublished observations). There is also little or no regulation of brain extracellular [K] and [Ca] in the rat fetus, but regulation improves greatly by the neonate (9,10).

The improvement in brain volume regulation at birth is linked to an increased net uptake of Na into the brain in response to the hyperosmotic stress. The mechanism behind the increased sodium uptake has yet to be elucidated. In the adult rat, however, the main route of such Na entry is from the CSF (16) and the results presented above suggest that this mechanism is much reduced in the fetus. This could reflect a reduced fetal CSF secretion rate that would limit the rate that CSF can move into the brain. Structurally, there are a number of changes in the choroid plexus over this perinatal period (12).

The improvement in net potassium uptake occurred later than that seen for sodium. Thus, it appears that not all BBB functions mature simultaneously. The period between 10 and 20 days postnatal coincides with a number of cerebral changes such as myelination and an increase in vascularization (1,11). It is also coincident with a number of changes in BBB enzymes. For example, the activity of γ-glutamyl transpeptidase increases 3- or 4-fold over this period (2).

In adult rats, the net uptake of potassium during hyperosmotic stress is associated with an increased blood-brain barrier permeability to potassium (5). During development, it coincides with the appearance of a quinidine-sensitive efflux mechanism at the BBB. This is probably a K-channel situated on the abluminal membrane of the endothelial cell as there is little access to the luminal membrane in the isolated microvessel preparation. In vitro this efflux mechanism was activated by the presence of ouabain, possibly through cell depolarization, although a direct effect cannot be ruled out at present. The mechanism(s) of activation in vivo is unknown. A ouabain-like factor is present in CSF (8) but at lower concentrations than used in these experiments. Alternatively, the channel might be opened by another depolarizing or hormonal signal.

ACKNOWLEDGMENTS

This work was aided by Basic Research Grant No 1-FY91-0-131 from the March of Dimes Birth Defects Foundation and by grants NS-23870 and HL-18575 from the National Institutes of Health.

REFERENCES

1. Bär, T., The vascular system of the cerebral cortex, *Adv Anat Cell Biol* 59:1-62, 1980.
2. Betz, A.L., and Goldstein, G.W., Developmental changes in metabolism and transport properties of capillaries isolated from rat brain, *J Physiol* 312:365-376, 1981.
3. Butt, A.M., Jones, H.C., and Abbott, N.J., Electrical resistance across the blood-brain barrier in anaesthetized rats: a developmental study, *J Physiol* 429:47-62, 1990.
4. Cserr, H.F., DePasquale, M., and Patlak, C.S., Regulation of brain water and electrolytes during acute hyperosmolality in rats, *Am J Physiol* 253:F522-F529, 1987.
5. Cserr, H.F., DePasquale, M., and Patlak, C.S., Volume regulatory influx of electrolytes from plasma to brain during acute hyperosmolality, *Am J Physiol* 253:F530-F537, 1987.
6. Cserr, H.F., and Patlak, C.S., Regulation of brain volume under isosmotic and anisosmotic conditions, *in*: "Advances in Comparative and Environmental Physiology," Vol. 9, Gilles R., et al. eds., Springer-Verlag, Berlin Heidelberg, pp. 61-80, 1991.
7. Fenstermacher, J.D., Volume regulation of the central nervous system, *in*: "Edema," Staub N.C., and Taylor A.E., ed., Raven Press, New York, pp. 383-404, 1984.
8. Halperín, J., Shaeffer, R., Galvez, L., and Malavé, S., Ouabain-like activity in human cerebrospinal fluid., *Proc Natl Acad Sci USA* 80:6101-6104, 1983.
9. Jones, H.C., and Keep, R.F., The control of potassium concentration in the cerebrospinal fluid and brain interstitial fluid of developing rats, *J Physiol* 383:441-453, 1987.
10. Jones, H.C., and Keep, R.F., Brain fluid calcium concentration and response to acute hypercalcaemia during development in the rat, *J Physiol* 402:579-593, 1988.
11. Keep, R.F., and Jones, H.C., Cortical microvessels during brain development. A morphometric study in the rat, *Microvasc Res* 40:412-426, 1990.
12. Keep, R.F., and Jones, H.C., A morphometric study on the development of the lateral ventricle choroid plexus, choroid plexus capillaries and ventricular ependyma in the rat., *Dev. Brain Res.* 56:47-53, 1990.
13. Kimelberg, H.K., and Ransom, B.R., Physiological and pathological aspects of astrocytic swelling, *in*: "Astrocytes," Vol. 3, S. Federoff, and A. Vernadakis, eds., Academic Press, Orlando, pp. 129-166, 1986.
14. Lin, J.D., Potassium transport in isolated cerebral microvessels from rat., *Japan. J. Physiol* 35:817-830, 1985.
15. Pollock, A.S., and Arieff, A.I., Abnormalities of cell volume regulation and their functional consequences, *Am J Physiol* 239:F195-F205, 1980.
16. Pullen, R.G.L., DePasquale, M., and Cserr, H.F., Bulk flow of cerebrospinal fluid into brain in response to acute hyperosmolality, *Am J Physiol* 253:F538-F545, 1987.
17. Schielke, G.P., Moises, H.C., and Betz, A.L., Potassium activation of the Na,K-pump in isolated brain microvessels and synaptosomes, *Brain Res* 524:291-296, 1990.

THE TRANSENDOTHELIAL DC POTENTIAL OF
RAT BLOOD-BRAIN BARRIER VESSELS IN SITU

Patricia A. Revest,[1] Hazel C. Jones,[2]* and N. Joan Abbott[1]

[1]Physiology Group, Division of Biomedical Sciences,
King's College London, London WC2R 2LS, UK
[2]Department of Pharmacology, University of Florida
Health Science Center, Gainesville, FL 32610, USA

INTRODUCTION

A small DC potential between cerebrospinal fluid (CSF) and blood has been recorded in many species (11,12). Possible sources of the potential are the choroid plexus, the brain capillaries and the arachnoid membrane, all of which contribute to the control of the brain microenvironment. The choroid plexus, however, is a low resistance epithelium (14) with a transepithelial potential difference (PD) of <1.0 mV (13) and, hence, is unlikely to contribute significantly to the overall PD. The ion regulatory properties of the blood-brain barrier (BBB) are thought to be the most likely source of the potential (3) and the properties of single BBB vessels have been studied in pial surface vessels that are readily accessible for experimentation in situ in anesthetized animals. Pial vessels have BBB structural properties but, unlike parenchymal vessels, do not have an astrocytic covering (4,6). They have been shown to have a high electrical resistance of 2000 Ωcm^2 in frog (8) and 500-6000 Ωcm^2 in rat (6). The transendothelial PD in single vessels in frogs has been studied and is 3-4 mV lumen negative (1,8). It has an active component that is inhibited by ouabain that can be attributed to Na^+-K^+-ATPase situated on the abluminal endothelial membrane (2), the remaining PD being attributed to passive diffusion (1).

There are three possible routes for ion movement across the brain capillary endothelium: a passive non-selective paracellular route through the interendothelial tight junctions, a transendothelial route via carrier-mediated transport, and a passive transendothelial route via ion channels. In these experiments we have measured the transendothelial potential in exposed rat pial vessels and used it as a tool to investigate the contributions of the different routes to ion movement across the BBB.

---

* To whom correspondence should be addressed.

Frontiers in Cerebral Vascular Biology: Transport and Its Regulation
Edited by L.R. Drewes and A.L. Betz, Plenum Press, New York, 1993

71

MATERIALS AND METHODS

Rats (24-30 days old) were anesthetized with pentobarbitone (60 mg/kg, i.p.). A piece of the parietal bone about 5 mm^2 was thinned with a dental drill and a plastic ring about 4 mm in diameter and 2 mm deep was fixed over the thinned area with cyanoacrylate adhesive. The ring was firmly clamped between two horizontal bars to reduce respiratory movements and a continuous flow of warmed artificial CSF was started (composition in mM: NaCl 138, KCl 3.0, NaHCO$_3$ 4.2, MgSO$_4$ 1.2, CaCl$_2$ 1.26, Hepes 10, pH 7.4, and sterilized through a 0.22 μm filter). The thinned bone was removed and the dura and arachnoid reflected to expose the cortical surface vessels, which consisted of small veins and venules and fewer arteries and arterioles. A glass microelectrode (2-10 MΩ) filled with 3M KCl was inserted into arterial or venous microvessels, 20-60 μm diameter, and connected to an amplifier (WPI duo773) via a Ag/AgCl pellet. The circuit was completed with a reference electrode consisting of a Ag/AgCl pellet in a 3M KCl reservoir attached to a CSF-agar bridge that was placed in the CSF bath. The abluminal conditions were varied by changing the composition of the superfusate.

RESULTS AND DISCUSSION

Control Potentials

The PD recorded on penetration into venous vessels was 3.2 ±0.16 mV, n=133, lumen negative, and in arteries it was significantly higher (p<0.01), at 4.5 ±0.48 mV, n=23. The change in PD on withdrawing the electrode from the vessel was 2.8 ±0.18 mV, n=102 for veins and 3.8 ±0.48 mV, n=19, for arteries. These were not significantly different from the penetration values, suggesting that vessels were not damaged by electrode insertion.

Effect of Changing Superfusate [K$^+$] and [Na$^+$]

Changing the abluminal K$^+$ concentration from 3 to 100 mM for 90 sec made the lumen less negative in both arteries and veins by 3.1 ±0.2mV, n=165. The PD was restored on returning to control CSF. High K$^+$ solutions, however, caused arterial vessels to constrict, probably through contraction of the surrounding smooth muscle layers, and therefore, most subsequent experiments were performed on venous vessels, which were more numerous in any case. The 100 mM K$^+$ CSF was prepared by substituting Na$^+$ for K$^+$ and thus contained only 41 mM Na$^+$. Reducing the abluminal Na$^+$ from 141 to 41 mM using Tris instead of K$^+$, made the lumen more negative than control by 2.0 ±0.21 mV, n=7, giving a total PD of 5 mV lumen negative. Increasing [K$^+$] to 100 mM at this point caused a PD change of +5 mV showing that the high K$^+$ and low Na$^+$ have opposite effects but that the effect on PD is less for Na$^+$ than for K$^+$.

Effect of Changes in Paracellular Permeability

Superfusion with histamine (10^{-3} M) which decreases the electrical resistance (5) and increases BBB permeability in rats (10), did not alter the control PD, nor did it affect the response to 100 mM K$^+$. There is evidence that the increase in BBB permeability caused by histamine is mediated via histamine H$_2$ receptors since the effects of histamine can be inhibited by H$_2$-receptor antagonists (9,10), and the antagonist cimetidine increased the electrical resistance of pial vessels in the rat (5). In these experiments, pretreatment of rats with cimetidine (20 mg/kg i.p.) did not affect the control PD or the response to 100 mM K$^+$. This suggests that the PD is insensitive to the permeability of the paracellular pathway through the interendothelial tight junctions.

Effect of Ouabain and K+-Channel Blockers

The application of CSF containing ouabain (10^{-3} - 10^{-6} M) for 60 sec had no significant effect on the control PD in veins and also had no effect on the response to 100 mM K^+. This suggests that the abluminal Na-K pump makes little contribution to the PD. A range of agents that block K^+ channels (7): 4-aminopyridine, CsCl, $BaCl_2$, and tetraethylammonium (TEA), were tested by adding them at a concentration of 10 mM to CSF containing 3 mM K^+ and 41 mM Na^+ and superfusing for 60 sec during a control recording with the same CSF. The response to 100 mM K^+ was tested immediately after the 60 sec application. Of the four drugs, the only significant effect was with TEA, which significantly reduced the response to high K^+ by 20%, from 6.2 ±0.82 to 5.1 ±0.94 mV. This suggests that the PD may be at least partly related to passive permeability via ion channels.

SUMMARY AND CONCLUSIONS

The recorded potential between the abluminal CSF and the vessel lumen, 3-4 mV, blood negative is similar to that recorded in frog and can account for most or all of the previously-reported PD between CSF and blood. It is not affected by substances that alter paracellular permeability and hence is mainly generated by the properties of endothelial cell membranes. Unlike the PD recorded in frog brain vessels, the PD in rat is not sensitive to the Na^+-K^+ATPase inhibitor, ouabain, which suggests that although electrogenic Na^+-K^+ transport is known to be present, the contribution it makes to the PD is not detectable. This suggests that the changes in PD recorded when abluminal $[K^+]$ or $[Na^+]$ are altered, are a result of the passive permeability properties of the endothelial cell membranes.

ACKNOWLEDGMENTS

This work was supported by the Medical Research Council and the Wellcome Trust.

REFERENCES

1. Abbott, N.J., Butt, A.M., Joels, S., Pitchford, S., and Wallis, W., Na^+-K^+ transport at the blood-brain barrier - a micro-electrode study in the frog, *J. Physiol.* 372:68P, 1986.
2. Betz, A.L., Firth, J.A., and Goldstein, G.W., Polarity of the blood-brain barrier: distribution of enzymes between the luminal and antiluminal membranes of brain capillary endothelial cells, *Brain Res.* 192:17, 1980.
3. Bradbury, M.W.B., "The Concept of a Blood-Brain Barrier," Wiley, Chichester, 1979.
4. Bundgaard, M., Ultrastructure of frog cerebral and pial microvessels and their impermeability to lanthanum ions, *Brain Res.* 241:57, 1982.
5. Butt, A.M., and Jones, H.C., Effect of histamine and antagonists on electrical resistance across the blood-brain barrier in rat brain-surface microvessels, *Brain Res.* 569:100, 1992.
6. Butt, A.M., Jones, H.C., and Abbott, N.J., Electrical resistance across the blood-brain barrier in anaesthetised rats, *J. Physiol.* 429:47, 1990.
7. Cook, N.S., The pharmacology of potassium channels and their therapeutic potential, *TIPS* 9:21, 1988.
8. Crone, C., and Olesen, S.P., Electrical resistance of brain microvascular endothelium, *Brain Res.* 241:49, 1982.
9. Dux, E., and Joo, F., Effects of histamine on brain capillaries, *Exp. Brain Res.* 47:252, 1982.
10. Gross, P.M., Teasdale, G.M., Graham, D.I., Angerson, W.J., and Harper, A.M., Intra-arterial histamine increases blood-brain barrier transport in rats, *Am. J. Physiol.* 243:H307, 1982.

11. Held, D., Fencl, V., and Pappenheimer, J.R., Electrical potential of cerebrospinal fluid, *J. Neurophysiol.* 27:942, 1964.

12. Loeschcke, H.H., DC potentials between CSF and blood, *in:* "Ion Homeostasis of the Brain." B.K. Siesjö, and S.C. Sorensen, eds., Academic Press, New York, 1971.

13. Wright, E.M., Mechanisms of ion transport across the choroid plexus, *J. Physiol.* 226:545, 1972.

14. Zeuthen, T., and Wright, E.M., Epithelial potassium transport: Tracer and electrophysiological studies in choroid plexus, *J. Memb. Biol.* 60:105, 1981.

PHOSPHATE TRANSPORT IN CAPILLARIES

OF THE BLOOD-BRAIN BARRIER

Richard Béliveau, Lise Dallaire and Sylvie Giroux

Laboratoire de Membranologie
Groupe de Recherche en Biothérapeutique Moléculaire
Département de biochimie, Université du Québec à Montréal
C.P. 8888, Succ. A, Montréal (Québec), Canada, H3C 3P8

INTRODUCTION

Endothelial cells of cerebral capillaries form a continuous barrier between blood and brain interstitium. These endothelial cells, sealed together by tight junctions, do not possess fenestrations or transendothelial channels (1,2). Brain capillaries also contain a large number of mitochondria and are able to metabolize a variety of substrates (3,4). This specialized endothelium regulates the movement of solutes between blood and brain. This is accomplished through carrier-mediated transport systems for hexoses, monocarboxylic acids, amino acids, nucleosides, purines and amines (5). Active transport pumps in the endothelial cells of brain capillaries appear to maintain the volume and composition of brain's interstitial fluid (6). This extracellular fluid differs from plasma in that it is almost free of protein and differs from an ultrafiltrate of plasma by maintaining the concentration of various ions at distinct levels. In fact, the concentration of inorganic phosphate in the interstitial fluid of the brain is held between 0.5 and 1.0 mM, and the plasma level is maintained between 1.5 and 1.8 mM (7). Maintenance of this transcellular phosphate concentration gradient could be a result of transport mechanisms at the luminal and anti-luminal sites of the membranes of cerebral capillaries. The relationship of inorganic phosphate to the phosphorylation processes could account for a regulation of the concentration of this ion by a membrane transport mechanism. Furthermore, capillary cells require phosphate for energy demands, as reflected by the large number of mitochondria, and for synthesis of nucleic acids and complex lipids (3). Phosphate transport in brain capillaries was studied with metabolically active capillaries isolated from bovine cortex (8).

MATERIALS AND METHODS

Isolation of Capillaries

Capillaries of the blood-brain barrier were obtained from bovine brain cortex by a

Frontiers in Cerebral Vascular Biology: Transport and Its Regulation
Edited by L.R. Drewes and A.L. Betz, Plenum Press, New York, 1993

75

procedure previously described by Dallaire et al. (8). The final pellet was resuspended in a cryoprotective medium containing 147 mM NaCl, 4 mM KCl, 3 mM $CaCl_2$, 5 mM glucose, 14% (w/v) glycerol, 1.4% (w/v) sorbitol and 15 mM Hepes/Tris, pH 7.4. Following isolation, the capillaries were stored in liquid nitrogen until use. γ-Glutamyl transpeptidase enrichment of the capillary preparations was routinely 20-fold greater than the homogenate.

Preparation of Capillaries for Transport Studies

Capillaries were washed (1:30 v/v) with a solution containing 147 mM NaCl, 4 mM KCl, 5 mM glucose and 15 mM Hepes/Tris, pH 7.5, and collected by centrifugation at 25,000 g for 10 min at 4°C. The pellet was resuspended in the same medium to a concentration of 7-10 µg of protein/µl.

Phosphate Transport

A rapid filtration technique was used to measure transport. Phosphate uptake, performed in triplicate at 25°C, was initiated by the addition of 50 µl of incubation medium to 70-100 µg of protein/10 µl, previously incubated 60 s at 25°C. The incubation medium contained [$^{32}P_i$]phosphate (2 µCi), 147 mM NaCl, 4 mM KCl, 5 mM glucose and 15 mM Hepes/Tris, pH 7.5. After incubation, the reaction was stopped by addition of 1 ml of ice-cold stop solution (147 mM NaCl, 4 mM KCl and 15 mM Hepes/Tris, pH 7.5). The suspension was filtered under vacuum through a 0.45 µm-pore-size Millipore filter. The filter was rinsed with 8 ml of stop solution and the radioactivity counted. Non-specific binding to the filter was determined by filtering the incubation medium without capillaries. Non-specific binding of the radioactivity to the capillaries was determined by the addition of the ice-cold stop solution to the capillaries before adding the incubation medium. This value was subtracted from uptake values at different incubation times. K_m and V_{max} were estimated by non-linear regression analysis using the statistical software "KaleidaGraph" and a Macintosh computer.

Chemicals

[$^{32}P_i$]Orthophosphate (carrier-free) was obtained from ICN Biomedicals. Other chemicals were of the highest purity commercially available.

RESULTS

The uptake of phosphate by capillaries as a function of time is illustrated in Figure 1. The phosphate influx in the presence and absence of sodium showed a similar time course which indicates that the uptake was not dependent on the concentration of sodium in the incubation medium. The uptake reached equilibrium after 30 min (results not shown). After 5 min, capillary cells had accumulated 2.09 ± 0.62 (SE) nmol phosphate/mg protein. On the basis of this experiment, incubation of 30 s was selected as an early time point to approximate initial phosphate uptake in all subsequent measurements. These were done in the presence of 147 mM external sodium. Variations in phosphate uptake were observed between different preparations of purified capillaries, and could reflect uncontrollable parameters such as origin, age, sex and diet of the animals. An appropriate control was thus performed with each experiment.

Phosphate uptake by the isolated bovine brain capillaries was studied over a phosphate concentration range of 10 µM to 800 µM (Figure 2). Over this concentration range, phosphate uptake can be described as a single saturable process. The K_m and V_{max}

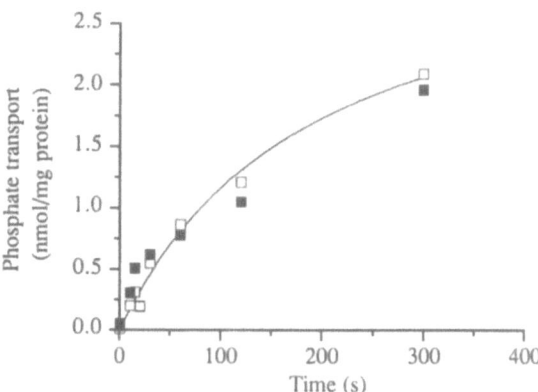

Figure 1. Time course of phosphate uptake by isolated bovine brain capillaries. Capillaries were treated as described in Materials and Methods and incubated at 25°C. The incubation medium contained 200 µM [^{32}P$_i$]orthophosphate (2 µCi), 5 mM glucose, 15 mM Hepes (pH 7.5) and (□) 147 mM NaCl with 4 mM KCl or (■) 151 mM KCl. A representative experiment is shown.

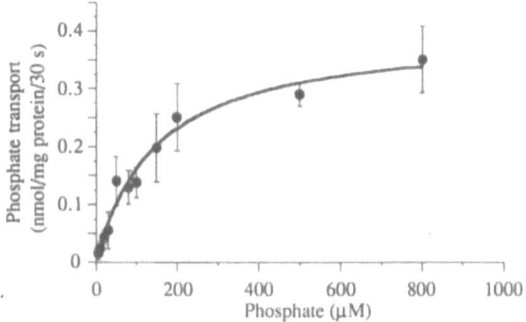

Figure 2. Influence of phosphate concentrations on phosphate uptake by the isolated brain capillaries. Capillaries were treated as described in Materials and Methods and incubated 30 s at 25°C with different concentrations of phosphate in the incubation medium. The incubation medium contained 147 mM NaCl, 4 mM KCl, 5 mM glucose, 15 mM Hepes (pH 7.5) and the concentration of phosphate indicated in the figure. [^{32}P$_i$]orthophosphate was constant at 2 µCi. Each value represents the mean ± SE of data derived from three different experiments each done in triplicate.

77

estimated by non-linear regression or by a Lineweaver-Burk plot were similar (160 μM and 0.37 nmol/mg protein/30 s, respectively). The saturability of the influx indicated that a specific transport system of high affinity is involved in the uptake of phosphate by the blood-brain barrier.

Db cAMP (0.2 mM), a permeant analog of cAMP, significantly inhibited the phosphate uptake by 19%. Db cAMP together with 10 μM IBMX, an inhibitor of the breakdown of cAMP by phosphodiesterases, reduced phosphate uptake by 40%. Phosphate influx, in the presence and absence of IBMX was not changed. The decrease of phosphate transport into capillaries was significantly different in the presence of IBMX and db cAMP than with db cAMP alone. On the basis of this experiment, all subsequent preincubations of the capillaries with db cAMP were done in the presence of 10 μM IBMX. Preincubation of capillary cells with different concentrations of db cAMP resulted in a dose-dependent inhibition of phosphate uptake (Figure 3). Half-maximal inhibition of phosphate influx was observed at 10^{-9}M db cAMP.

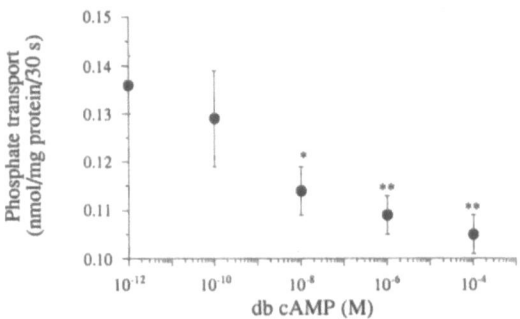

Figure 3. Dose-response of db cAMP on phosphate uptake in brain capillaries. Experiments were performed after a 20 min preincubation at 25°C, in the presence of various concentrations of db cAMP and 10 μM IBMX. Data are means ± SE of five experiments in quadruplicate. * p < 0.01 vs control. ** p < 0.005 vs control.

DISCUSSION

The uptake of phosphate by bovine brain capillaries is time-dependent and is not stimulated by the presence of sodium in the incubation medium. This influx is not coupled to cations for the translocation of the phosphate across the capillary membranes, as it is for the phosphate transport in the renal proximal tubule and intestine (9,10). The phosphate uptake in the capillaries would represent a passive facilitated mode of transport (11).

Such transport processes occur through a limited number of sites and exhibit saturation kinetics. Kinetic analysis of the uptake process showed that phosphate influx occurred by a saturable mechanism. The full displacement of $[^{32}P_i]$orthophosphate with different concentrations of cold phosphate solutions indicated kinetics without diffusion constant (K_D) and a one site component. The apparent K_m value for phosphate is close to those reported in the brush-border membrane of mammalian kidney (80 μM) (9,12,13) and intestine (250 μM) (10). The uptake of phosphate in the capillaries of the blood-brain barrier reflect transport via a high affinity system.

Formation of intracellular cAMP from ATP is catalyzed by a membrane-bound adenylate cyclase (14). cAMP regulates the activity of cAMP-dependent protein kinases (protein kinase A). The rate of pinocytosis was increased by db cAMP in brain capillaries, suggesting a role of protein kinase A in the permeability of the blood-brain barrier (15). The inhibition of phosphate uptake by db cAMP suggests a role for protein kinase A in the regulatory mechanisms of blood-brain phosphate transport. Our studies performed with db cAMP in the presence of IBMX led to a stronger inhibition of phosphate transport in isolated bovine capillaries. Capillaries isolated from bovine cortex were shown to exhibit significant cAMP phosphodiesterase activity and this activity was best inhibited by compounds of the xanthine family (16).

In summary, the findings reported here demonstrate the presence of a facilitated phosphate transport mechanism in the blood-brain barrier, which is regulated by cAMP. Further investigations will be required to determine the nature of this phosphate carrier and to assess a possible modulation of anion exchange activities by intracellular messengers.

ACKNOWLEDGMENTS

This work was supported by grants from the Natural Sciences and Engineering Research Council of Canada.

REFERENCES

1. Reese, T.S., Fine structural localization of a blood brain barrier to exogenous peroxidase, *J. Cell Biol.* 34:207-217, 1967.

2. Brightman, M.W., Morphology of blood brain barrier interfaces, *Exp. Eye Res.* 25, Suppl. 1:1-25, 1977.

3. Oldendorf, W.H., Marcia, E., and Brown, W.J., The large apparent work capability of the blood-brain barrier: A study of the mitochondrial content of capillary endothelial cells in brain and other tissues of the rat, *Ann. Neurol.* 1:409-417, 1977.

4. Djuricic, B.M., Rogac, Lj., Spatz, M., Rakic, Lj. M., and Mrsulja, B.B., Brain microvessels, I. Enzymatic activities, *Adv. Neurol.* 20:197-205, 1978.

5. Pardridge, W.M., Recent advances in blood-brain barrier transport, *Ann. Rev. Pharmacol. Toxicol.* 28:25-39, 1988.

6. Bradbury, M.W.B., The structure and function of the blood brain barrier, *Federation Proc.* 43:186-190, 1984.

7. Ganong, W.F., Circulation throught special region (chapter 32), in *Review of Medicinal Physiology*, 14th Ed., Appleton & Lange, CA, pp. 514-546, 1989.

8. Dallaire, L., Tremblay, L., and Béliveau, R., Purification and characterization of metabolically active capillaries of the blood-brain barrier, *Biochem. J.* 276:745-752, 1991.

9. Hoffmann, N., Thees, M., and Kinne, R., Phosphate transport by isolated renal brush border vesicles, *Pflugers Archiv.* 362:147-156, 1976.

10. Berner, W., Kinne, R., and Murer, H., Phosphate transport into brush-border membrane vesicles isolated from rat small intestine, *Biochem. J.* 160:467-474, 1976.

11. Béliveau, R., Vésicules membranaires purifiées: un outil pour l'étude de la réabsorption rénale, *Medecine/Sciences*, 3:589-598, 1987.

12. Takuwa, Y., and Ogata, E., Characterization of Na^+-dependent phosphate uptake in cultured kidney cells (JTC-12) from monkey, *Biochem. J.* 230:715-721, 1985.

13. Chen, M.L., King, R.S., and Armbrecht, H.J., Sodium-dependent phosphate transport in primary cultures of renal tubule cells from young and adult rats, *J. Cell. Physiol.* 143:488-493, 1990.

14. Jain, M.K., Transbilayer response of signals, *in*: "Introduction to Biological Membranes," Wiley, J. & Sons, eds., Wiley-Interscience, New York, pp. 356-373, 1988.

15. Joó, F., Effect of N^6O^2-dibutyryl cyclic 3', 5'-adenosine monophosphate on the pinocytosis of brain capillaries of mice, *Experientia* 28:1470-1471, 1972.

16. Stefanovich, V., Cyclic 3', 5'-adenosine monophosphate phosphodiesterase (cAMP PDE) and cyclic 3', 5'-guanosine monophosphate phosphodiesterase (cGMP PDE) in microvessels isolated from bovine cortex, *Neurochem. Res.* 4:681-687, 1979.

Drug and Amino Acid
Transport

DRUG DELIVERY TO BRAIN AND THE ROLE

OF CARRIER-MEDIATED TRANSPORT

Quentin R. Smith

Laboratory of Neurosciences, National Institute on Aging
National Institutes of Health, Bethesda, MD 20892

INTRODUCTION

The blood-brain barrier restricts the brain uptake of many valuable hydrophilic drugs and limits their efficacy in the treatment of brain diseases. This has been shown for numerous polar antibiotic (32,56), anticancer (17,19,20) and antiviral drugs (9,60) and is likely to gain increased importance with the development of new protein, peptide and genetic pharmaceuticals.

Various strategies have been developed to overcome this restriction, as are summarized in Table 1. This paper reviews the current status of each of these strategies and then presents new data on the design and development of high-affinity drugs that are taken into brain by carrier-mediated mechanisms. More than 10 separate transport carriers have been identified at the blood-brain barrier that mediate the brain uptake of essential nutrients, vitamins and hormones. With increased knowledge of the selectivity and capacity of these carriers, it may be possible to design a wide variety of agents that gain ready access to brain via facilitated diffusion.

STRATEGIES TO ENHANCE DRUG DELIVERY TO BRAIN

Six strategies have been developed to enhance brain uptake and accumulation of water-soluble drugs (Table 1). Of these, the predominant method is to prepare lipid-soluble derivatives or analogs that cross the barrier more readily by simple passive diffusion. Brain capillaries (i.e., the blood-brain barrier), unlike capillaries in most other organs of the body, contain few paracellular aqueous pores and exhibit minimal pinocytotic activity. As a result, for a drug to cross the barrier, it must either dissolve in and diffuse across the lipoid endothelial cell membranes or be shuttled across by carrier-mediated or receptor-mediated mechanisms (37). Passive diffusion through cell membranes depends highly upon solute lipid solubility as well as solute pK_a and molecular weight (55). Many studies have demonstrated a linear relation between brain capillary permeability and solute lipid solubility,

Frontiers in Cerebral Vascular Biology: Transport and Its Regulation
Edited by L.R. Drewes and A.L. Betz, Plenum Press, New York, 1993

83

Table 1. Strategies to augment brain uptake and accumulation of water-soluble drugs

---

1. Lipid soluble analogues or prodrugs

2. Intracarotid infusion to maximize brain arterial concentration

3. Direct injection into brain or CSF

4. Blood-brain barrier modification
   a. Osmotic Opening
   b. Specific Solutes

5. Carrier- or receptor-mediated transport

6. Inhibition of active removal or breakdown

---

as measured by the octanol/water partition coefficient, for compounds with molecular weights of <400 daltons (14,27,44). A partition coefficient on the order of 1-10 is often sought to optimize passive lipid-soluble diffusion. However, the benefits achieved in barrier permeability are, in some instances, offset by changes in plasma protein binding, peripheral distribution or elimination, that reduce the concentration of free drug available for transport. Numerous examples are known of lipophilic drugs that show minimal brain uptake because of tight plasma protein binding (40,66). Furthermore, lipophilic modification can reduce or abolish drug activity by disrupting receptor binding or action. Thus, both the potential effects of each chemical modification must be thoroughly considered when designing new lipophilic agents.

A related, but more difficult approach is to prepare lipophilic "prodrugs" that are taken up into tissues and metabolized to release active, parent compound. The prototypical prodrug is heroin, which is a di-acetylated morphine. The acetyl groups increase the lipid solubility of morphine and enhance brain uptake by >10-fold (37). Once within brain, the acetyl groups are cleaved to generate active morphine. This same approach has been employed with different substituents to enhance brain delivery of chlorambucil (21), dopamine (6) and GABA (24). Side groups can be used to target and retain drug within brain, as in the case of dihydropyridine, which is selectively metabolized within the central nervous system to the polar quaternary pyridinium salt, thereby trapping the complex for future release (6,7). Successful implementation of the prodrug approach requires careful balancing of the various rates of 1) prodrug uptake, 2) efflux, 3) biotransformation to polar intermediate (when applicable) and 4) release of active parent compound. Full control of all these rates is not easy to achieve (for example, see 17,21,22). Relative to lipophilic analogues, prodrugs offer the theoretical advantage of selective sequestration and release if the enzymes that mediate breakdown are expressed preferentially at the disease site or tissue. As with other lipophilic compounds, prodrugs can be prevented from entering brain by plasma protein binding or rapid peripheral distribution and elimination.

Intracarotid infusion is beneficial for compounds that exhibit rapid peripheral distribution and elimination (15). Delivery of such solutes can be enhanced severalfold or more, especially when combined with techniques, such as osmotic modification, that enhance barrier permeability (38). The relative advantage of intra-arterial infusion (R_d) for a compound can be calculated as $R_d = 1 + Cl_{tb}/F_a$ where Cl_{tb} equals the total body clearance of the drug and F_a equals the rate of blood flow in the infused artery (17). Though the method allows greater drug exposure to brain, it is limited by the invasiveness of the procedure and by the fact that poor intraarterial mixing and streaming can cause marked variability in regional delivery (3).

Direct administration of drug to brain or CSF also sounds attractive, but has many limitations because of problems in achieving effective drug concentrations over wide areas. Diffusion within brain is slow, requiring hours to days to obtain 50% equilibration over distances of ~1 cm (28). Similarly, CSF administration delivers significant drug concentrations only to the outer few millimeters of brain tissue, i.e., to tissue immediately adjacent to the ventricular and subarachnoid spaces (4,5). Thus, such methods are of interest only for delivery of drugs for treatment of CSF or meningeal diseases or for drug delivery to very localized areas (e.g., brain tumors).

Barrier modification can be used to enhance brain uptake of water-soluble compounds directly from the circulation. Rapoport and colleagues have demonstrated that intracarotid infusion of hypertonic arabinose solutions increases barrier permeability to both large and small solutes by at least 20-fold, with a return to normal permeability within several hours (45,67). Barrier opening is thought to occur through disruption of the interendothelial junctional complexes to form transient aqueous pores (46). The osmotic method, though invasive, has been shown to augment drug delivery to brain tumors and to lead to increased survival and therapeutic effect (34,35). New chemical means to open the barrier are currently being tested and include infusion of leukotrienes (2), calcium-dependent adhesion molecules (47), polyamines (25) and pharmacologic manipulation of endothelial cyclic nucleotides (47).

Carrier- or receptor-mediated transport provides an alternative method to deliver drugs at active concentrations to the central nervous system. In this volume, Friden describes progress in the use of transferrin receptor antibodies to deliver neuropeptides across the blood-brain barrier (see also 42,43). In addition, Zlokovic (this volume) reports on the further characterization of blood-brain barrier peptide transport systems. The facilitative nutrient carriers of the blood-brain barrier may also be of use, and this is described in detail below.

Finally, drug accumulation in brain can be limited by active efflux or metabolism. Inhibition of such processes could lead to enhanced brain drug concentration and improved therapeutic effect.

CARRIER-MEDIATED DRUG TRANSPORT INTO BRAIN

The brain capillary endothelium contains more than 20 separate transport systems for various nutrients, ions, metals, peptides and hormones. A partial list of some of the better-characterized systems is presented in Table 2. Most of the transport carriers are passive and act simply to enhance the bidirectional transfer of solutes between plasma and brain. The transporter with the greatest capacity (V_{max}) is the cerebrovascular hexose carrier, which facilitates D-glucose uptake into brain for cerebral oxidative metabolism. Other carriers are present in lesser quantities for monocarboxylic acids, amino acids, amines and various vitamins, cofactors, nucleic acid precursors and peptides.

Many drugs are now known to be taken up into brain via cerebrovascular nutrient carriers. The large neutral amino acid transporter mediates the brain uptake of L-DOPA, alpha-methyl DOPA, melphalan, acivicin, 6-diazo-5-oxo-norleucine (DON), azaserine, buthionine sulfoximine, and baclofen (10,13,18,30,58,61-63). Similarly, the carboxylic acid carrier transports benzylpenicillin, ceftriaxone, cefodizime (31,53,57), and the purine carrier mediates the brain uptake of the antidepressant, 2-amino-2-oxazoline (11).

L-DOPA is the prototypical drug of this class and was one of the first drugs to be shown to be taken up into brain by a carrier mechanism. Melphalan, a related compound, is the nitrogen mustard derivative of the amino acid L-phenylalanine (Figure 1). Both melphalan

Table 2. Carrier-mediated transport systems at the brain capillary endothelium

Transport Carrier	Representative Substrate	V_{max} (nmol/min/g)	$PA_{in\ vivo}$ (ml/s/g $\times 10^4$)
Hexose	Glucose	4400	40
Monocarboxylic Acid	Pyruvate	420	23
Neutral Amino Acid	Leucine	64	17
Basic Amino Acid	Lysine	25	5.7
Amine	Choline	10	3.8
Nucleoside	Adenosine	0.75	5.0
Purine Base	Adenine	0.50	3.8
Pantothenic Acid	Pantothenate	0.21	1.8
Thyroid Hormone	T3	0.19	1.9
Enkephalin Peptide	Leu-Enk	0.16	<0.7
Acidic Amino Acid	Glutamate	0.028	<0.5
Thiamine	Thiamine	0.018	4.0

and phenylalanine are transported into brain via the large neutral amino acid carrier, showing self-saturation, cross inhibition and sensitivity to BCH (2-aminobicyclo[2.2.1]-heptane2-carboxylic acid), the defining substrate of the large neutral amino acid carrier system (10,18,33,51,58) (Figure 2). Brain uptake of amphetamine, lidocaine and propranolol is also saturable, but transfer may not be mediated by a carrier mechanism because V_{max} and K_m values are high (39,41).

Maximal transport capacity varies among brain capillary transport systems by >5 orders of magnitude. This, however, does not indicate that only the high capacity systems are of interest for drug uptake, as brain capillary permeability-surface area (PA) products, which give a better indication of the overall rate of equilibration of a solute between plasma and brain ($t_{1/2} \sim \ln2/[PA/V_d]$), vary to a far lesser degree (Table 2). In fact, for 9 of the 12 carriers listed in Table 2, PA values fall within one order of magnitude.

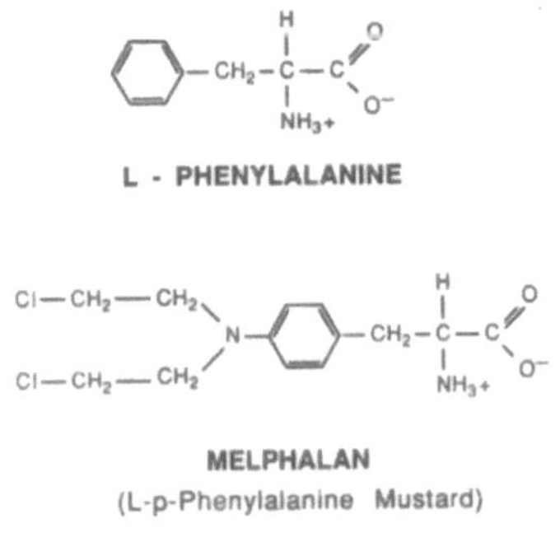

L - PHENYLALANINE

MELPHALAN
(L-p-Phenylalanine Mustard)

Figure 1. Chemical structures of L-phenylalanine and its para-substituted nitrogen mustard analogue, melphalan.

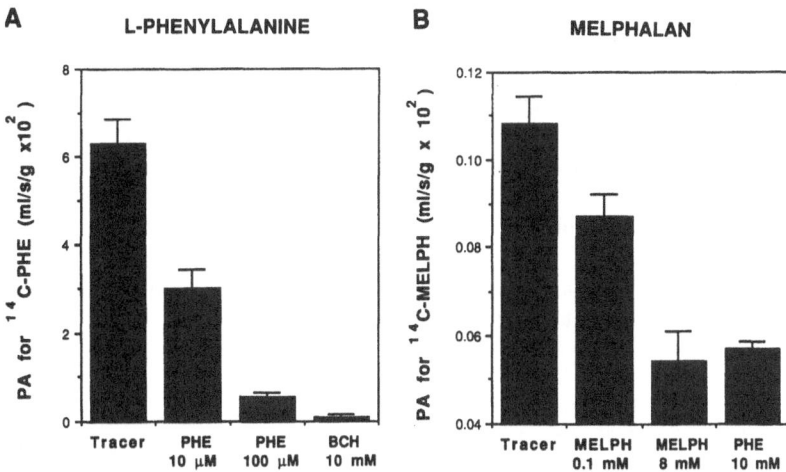

Figure 2. Cerebrovascular permeability-surface area (PA) products for brain uptake of L-[^{14}C]phenylalanine (A) and L-[^{14}C]melphalan (B) during rat brain perfusion with physiologic saline (T=15-60 s). Values are means ± SEM for n=4-8. Uptake was inhibited by addition to the perfusate of unlabeled substrate or BCH (2-aminobicyclo[2.2.1]heptane-2-carboxylic acid), the defining substrate of the sodium-independent, large neutral amino acid carrier system. Modified from Momma et al., (33) and Greig et al., (18).

This difference can be understood with reference to the following equations that describe the concentration-dependence of saturable transport across the blood-brain barrier (50,51),

$$\text{Influx} \approx \text{PA} \times C_{plasma} \qquad (1)$$

$$\text{PA} = V_{max}/[K_m(app) + C_{plasma}] \qquad (2)$$

where $K_m(app)$ is the apparent in vivo half-saturation concentration, and C_{plasma} is the plasma concentration of solute. For most of the transport systems listed in Table 2, $K_m(app)$ and C_{plasma} vary directly with V_{max} so that changes in cerebrovascular PA are minimized. Thus, it should be possible to obtain good brain transfer rates on most carriers as long as 1) plasma drug concentration is maintained at or below the apparent K_m, and 2) the K_m of the drug is comparable or less than that for endogenous substrates. Of course, the small K_m values of the low capacity systems will restrict the plasma drug concentration range over which the low capacity systems can be efficiently used.

One potential problem with carrier-mediated drug transport is that brain delivery may be subject to competition effects from endogenous substrates. This has been clearly shown for modulation of L-DOPA uptake into brain by plasma large neutral amino acids (i.e., phenylalanine, tryptophan, leucine, isoleucine, methionine and valine) (26,63). Raising plasma concentrations of large neutral amino acids decreases L-DOPA uptake, whereas lowering plasma large neutral amino acid concentrations has the opposite effect. Competition can also arise from drug metabolites that share the same transport system, such as 3-O-methyl-DOPA, a metabolite of L-DOPA (64). However, for L-DOPA, the effect of 3-O-methyl-DOPA is not significant under most in vivo situations (29,36).

Care must be taken in the design and administration of carrier-mediated drugs to ensure that 1) plasma fluctuations of endogenous substrates do not severely disrupt drug delivery to the central nervous system, and 2) the drug does not reduce brain nutrient delivery below the minimum tolerable level. In regard to the latter, the brain uptake of D-glucose is of special concern because the brain requires a continuous and stable supply of glucose to sustain cerebral oxidative metabolism. Severe hypoglycemia can result in seizures and death (48). Competition, in some instances, can be beneficial, as shown by Williams et al. (65) who used infusions of neutral amino acids to prevent brain uptake and neurotoxicity of acivicin in cats. As noted previously, acivicin is taken up into brain by the cerebrovascular neutral amino acid transporter (8,58). Thus, plasma competitor concentration becomes an additional factor that can be modulated to control drug delivery to brain.

IMPROVEMENT IN BRAIN DELIVERY THROUGH THE DESIGN OF HIGH-AFFINITY LIGANDS

Over the past five years, a number of drugs have been identified that are transported into brain via carrier-mediated mechanisms. However, for most, affinities ($1/K_m$) are low and limit uptake into brain. A good example of this is found in the study by Greig et al. (18) of melphalan transport across the blood-brain barrier. Although melphalan was shown to be transported into brain by the large neutral amino acid carrier, transport affinity for the drug was one-tenth that of L-phenylalanine, and the measured uptake PA was only a fraction of that of the endogenous substrate. Subsequent work demonstrated that melphalan, unlike phenylalanine, bound heavily to plasma albumin and that maximal brain concentrations following systemic injection were less than 10% of those of plasma (19). A follow-up study demonstrated that transport affinity was also low for a number of other anticancer amino acid drugs, including acivicin, azaserine, diazo-oxo-norleucine and buthionine sulfoximine (58).

The results of the above studies suggested that high-affinity, rapidly transported drugs might best be developed by tailoring the compounds to the requirements of the carriers. Surveys of the binding requirements of the cerebrovascular large neutral amino acid carrier indicated a strong preference for amino acids with large hydrophobic side groups (16,33,50-52). The critical influence of side-chain hydrophobicity on transport affinity is shown in Figure 3 for a series of neutral amino acids that are transported into brain via the facilitated diffusion. The relation between apparent affinity ($1/K_m$) and hydrophobicity, as measured by the octanol/water partition coefficient, is linear over 3-4 orders of magnitude. From this we hypothesized that transport affinity and uptake might best be improved by design of drugs with increased side-chain hydrophobicity.

An analog of melphalan with increased side-chain hydrophobicity was identified and examined for transport into brain using the in situ rat brain perfusion technique (50). This compound, D,L-2-amino-7-bis[(2-chloroethyl)amino]-1,2,3,4-tetrahydro-2-naphthoic acid (D,L-NAM)(Figure 4), had been prepared previously by Haines et al. (23) and shown to have increased affinity for the large neutral amino acid transporter of L1210 leukemia cells. D,L-[3H]NAM was found to be taken up into brain at a 20- to 50-fold greater rate than L-[14C]melphalan (59), consistent with NAM's greater side-chain hydrophobicity. Influx was saturable, sodium independent and inhibitable by BCH, indicating mediation by the cerebrovascular large neutral amino acid transporter. D,L-NAM exhibited minimal plasma protein binding and was transported into brain at a rate consistent with a $t_{1/2}$ for brain equilibration of ≤5 min. Several isomers of NAM were examined (59), but none exhibited greater affinity than the parent compound.

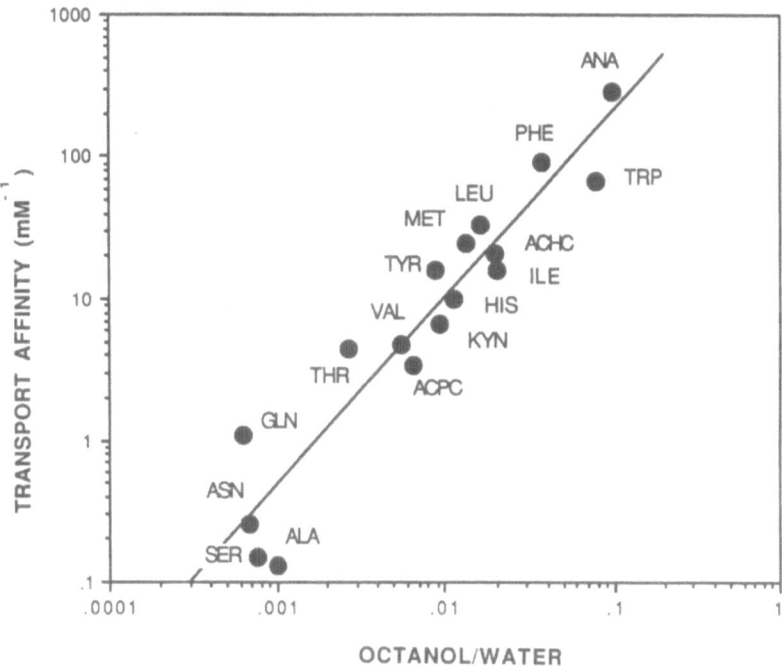

Figure 3. Relation of amino acid transport affinity ($1/K_m$) to octanol/water partition coefficient for 18 amino acids that are taken up into brain by the cerebrovascular large neutral amino acid transporter. Data compiled from Smith et al. (50-52), Momma et al. (33), Fukui et al. (16), Aoyagi et al. (1) and Takada et al. (59).

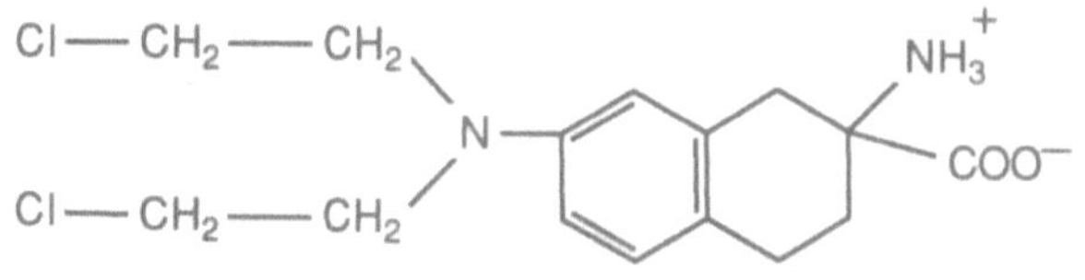

$$Cl-CH_2-CH_2$$
$$Cl-CH_2-CH_2$$

D,L-2-NAM-7

(D,L-2-Amino-7-bis(2-chloroethyl)amino-
1,2,3,4-tetrahydro-2-naphthoic acid)

Figure 4. Chemical structure of D,L-2-amino-7-bis[(2-chloroethyl)amino]-1,2,3,4-tetrahydro-2-naphthoic acid (D,L-NAM).

Preliminary analysis of the pharmacokinetics of D,L-NAM in rats following systemic administration (Figure 5) indicates that the compound is rapidly taken up into brain and reaches levels comparable to those in plasma within 10-15 min after injection. Primary deposition is to kidney and liver. The central nervous system appears to receive ~1% of the injected dose. NAM exhibits good alkylating activity and reduced myelosuppressive activity, as compared to melphalan (23). Together, the data suggest that NAM may show improved in vivo activity against brain tumors and thus merits further investigation.

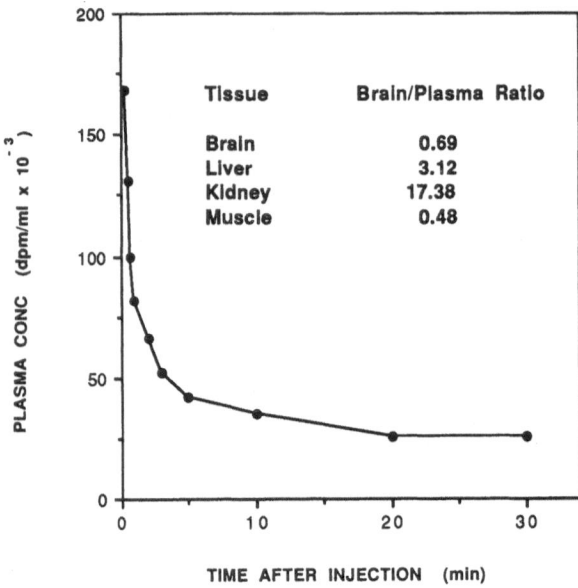

Figure 5. Time course of plasma ^{3}H-NAM concentration and 30-min tissue/plasma concentration ratios following i.v. injection in the rat.

SUMMARY

In summary, the results suggest that carrier-mediated transport can be used to augment the brain delivery of a wide variety of hydrophilic therapeutic drugs. A large number of carriers are now known to be present at the brain capillary endothelium, and in many instances these carriers have been shown to mediate the brain uptake of exogenous drugs. The findings with D,L-NAM demonstrate that brain delivery can be improved through design of selective, high affinity agents. Although NAM was developed for the large neutral amino acid carrier, high affinity drugs could be produced for other systems, as shown by the work of Schein et al. (49) with nitrogen mustard monosaccharides and by the work of Deves and Krupka (12) on choline derivatives. Lastly, the method may allow some selectivity of delivery because of differential expression of transport carriers between tissues and in various disease states.

REFERENCES

1. Aoyagi, M., Agranoff, B.W., Washburn, L.C., and Smith, Q.R., Blood-brain barrier transport of 1-aminocyclohexanecarboxylic acid, a nonmetabolizable amino acid for in vivo studies of brain transport, *J. Neurochem.* 50:1220, 1988.

2. Black, K., King, W.A., and Ikezaki, K., Selective opening of the blood-tumor barrier by intracarotid infusion of leukotriene C4, *J. Neurosurg.* 72:912, 1990.

3. Blacklock, J.B., Wright, D.C., Dedrick, R.L., Blasberg, R.G., Lutz, R.J., Doppman, J.L., and Oldfield, E.H., Drug streaming during intra-arterial chemotherapy, *J. Neurosurg.* 64:284, 1986.

4. Blasberg, R.G., Patlak, C.S., and Fenstermacher, J.D., Intrathecal chemotherapy: brain tissue profiles after ventriculocisternal perfusion, *J. Pharmacol. Exp. Ther.* 195:73, 1975.

5. Blasberg, R.G., Patlak, C.S., and Shapiro, W.R., Distribution of methotrexate in the cerebrospinal fluid and brain after intraventricular administration, *Cancer Treat. Rep.* 61:633, 1977.

6. Bodor, N., and Simpkins, J.W., Redox delivery system for brain-specific, sustained release of dopamine, *Science* 221:65, 1983.

7. Bodor, N., and Kaminski, J.J., Prodrugs and site-specific chemical delivery systems, *Ann. Rep. Med. Chem.* 22:303, 1987.

8. Chastain, J.E., and Borchardt, R.T., Acivicin transport across brain microvessel endothelial cell monolayers: a model of the blood-brain barrier, *Neurosci. Res. Commun.* 6:51, 1990.

9. Chou, S., and Dix, R.D., Viral infections and the blood-brain barrier, *in:* "Implications of the Blood-Brain Barrier and Its Manipulation, Vol. 2," E.A. Neuwelt, ed., Plenum, New York, 1989.

10. Cornford, E.M., Young, D., Paxton, J.W., Finlay, G.J., Wilson, W.R., and Pardridge, W.M., Melphalan penetration of the blood-brain barrier via the neutral amino acid transporter in tumor-bearing brain, *Cancer Res.* 52:138, 1992.

11. Damaj, M.I., Urien, S., Trouvin, J.H., Chanut, E., Lambrey, B., and Jacquot, C., In vivo evidence for carrier-mediated brain uptake of a new 2-amino-2-oxazoline (COR3224) via the purine transport system in rat, *Brain Res.* 554:333, 1991.

12. Deves, R., and Krupka, R.M., The binding and translocation steps in transport as related to substrate structure: a study of the choline carrier of erythrocytes, *Biochim. Biophys. Acta* 557:469, 1979.

13. Fekete, I., Griffith, O.W., Schlageter, K.E., Bigner, D.D., Friedman, H.S., and Groothuis, D.R., Rate of buthionine sulfoximine entry into brain and xenotransplanted human gliomas, *Cancer Res.* 50:1251, 1990.

14. Fenstermacher, J.D., Pharmacology of the blood-brain barrier, *in:* "Implications of the Blood-Brain Barrier and Its Manipulation, Vol. 1," E.A. Neuwelt, ed., Plenum, New York, 1989.

15. Fenstermacher, J.D., and Cowles, A.L., Theoretical limitations of intracarotid infusions in brain tumor chemotherapy, *Cancer Treat. Rep.* 61:519, 1977.

16. Fukui, S., Schwarcz, R., Rapoport, S.I., Takada, Y., and Smith, Q.R., Blood-brain barrier transport of kynurenines: implications for brain synthesis and metabolism, *J. Neurochem.* 56:2007, 1991.

17. Greig, N.H., Drug delivery to the brain by blood-brain barrier circumvention and drug modification, *in:* "Implications of the Blood-Brain Barrier and Its Manipulation, Vol. 1," E.A. Neuwelt, ed., Plenum, New York, 1989.

18. Greig, N.H., Momma, S., Sweeney, D.J., Smith, Q.R., and Rapoport, S.I., Facilitated transport of melphalan at the rat blood-brain barrier by the large neutral amino acid carrier system, *Cancer Res.* 47:1571, 1987.

19. Greig, N.H., Sweeney, D.J., and Rapoport, S.I., Comparative brain and plasma pharmacokinetics and anticancer activities of chlorambucil and melphalan in the rat, *Cancer Chemother. Pharmacol.* 21:1, 1988.

20. Greig, N.H., Soncrant, T.T., Shetty, H.U., Momma, S., Smith, Q.R., and Rapoport, S.I., Brain uptake and anticancer activities of vincristine and vinblastine are restricted by their low cerebrovascular permeability and binding to plasma constituents in rat, *Cancer Chemother. Pharmacol.* 26:263, 1990a.

21. Greig, N.H., Daly, E., Sweeney, D.J., and Rapoport, S.I., Pharmacokinetics of chlorambucil-tertiary butyl ester, a lipophilic chlorambucil derivative that achieves and maintains high concentrations in brain, *Cancer Chemother. Pharmacol.* 25:320, 1990b.

22. Greig, N.H., Genka, S., Daly, E.M., Sweeney, D.J., and Rapoport, S.I., Physiolchemical and pharmacokinetic parameters of seven lipophilic chlorambucil esters designed for brain penetration, *Cancer Chemother. Pharmacol.* 25:311, 1990c.

23. Haines, D.R., Fuller, R.W., Ahmad, S., Vistaca, D.T., and Marquez, V.E., Selective cytotoxicity of a system L specific amino acid nitrogen mustard, *J. Med. Chem.* 30:542, 1987.

24. Jacob, J.N., Shashoua, V.E., Campbell, A., and Baldessarini, R.J., Gamma-aminobutyric acid esters. 2. Synthesis, brain uptake, and pharmacological properties of lipid esters of gamma-aminobutyric acid, *J. Med. Chem.* 28:106, 1985.

25. Koenig, H., Goldstone A.D., and Chung, Y.L., Polyamines mediate the reversible opening of the blood-brain barrier by the intracarotid infusion of hyperosmolal mannitol, *Brain Res.* 483:110, 1989.

26. Leenders, K.L., Poewe, W.H., and Palmer, A.J., Inhibition of L-[18F]fluorodopa uptake into human brain by amino acids demonstrated by positron emission tomography, *Ann. Neurol.* 20:258, 1986.

27. Levin, V.A., Relationship of octanol/water partition coefficient and molecular weight to rat brain capillary permeability. *J. Med. Chem.* 23:682, 1980.

28. Levin, V.A., Patlak, C.S., and Landahl, H.L., Heuristic modeling of drug delivery to malignant brain tumors. *J. Pharmacokin. Biopharm.* 8:257, 1980.

29. Luquin, M.R., Vaamonde, J., and Obeso, J.A., Levodopa and 3-O-methyldopa plasma levels in parkinsonian patients with stable and fluctuating motor response, *Clin. Neuropharmacol.* 12:46, 1989.

30. Markovitz, D.C., and Fernstrom, J.D., Diet and uptake of Aldomet into brain: competition with natural large neutral amino acids, *Science* 197:1014, 1977.

31. Matsushita, H., Suzuki, H., Sugiyama, Y., Sawada, Y., Iga, T., Kawaguchi, Y., and Hanano, M., Facilitated transport of cefodizime into the rat central nervous system, *J. Pharmacol. Exp. Ther.* 259:620, 1991.

32. Meulemans, A., Vicart, P., Mohler, J., and Vulpillat, M., Determination of antibiotic lipophilicity with a micromethod: application to brain permeability in man and rats, *Chemotherapy* 34:90, 1988.

33. Momma, S., Aoyagi, M., Rapoport, S.I., and Smith, Q.R., Phenylalanine transport across the blood-brain barrier as studied with the in situ brain perfusion technique, *J. Neurochem.* 48:1291, 1987.

34. Neuwelt, E.A., and Dahlborg, S.A., Blood-brain barrier disruption in the treatment of brain tumors: clinical implications, *in:* "Implications of the Blood-Brain Barrier and Its Manipulation, Vol. 2," E.A. Neuwelt, ed., Plenum, New York, 1989.

35. Neuwelt, E.A., Goldman, D.L., Dahlborg, S.A., Crossen, J., Ramsey, F., Roman-Goldstein, S., Braziel, R., and Dana, B., Primary CNS lymphoma treated with osmotic blood-brain barrier disruption: prolonged survival with preservation of cognitive function, *J. Clin. Oncol.* 9:1580, 1991.

36. Nutt, J.G., Woodward, W.R., Gancher, S.T., and Merrick, D., Methyldopa and the response to levodopa in Parkinson's disease, *Ann. Neurol.* 21:584, 1987.

37. Oldendorf, W.H., Drug penetration of the blood-brain barrier, *in:* "Narcotics and the Hypothalamus," E. Zimmermann and R. George, eds., Raven Press, New York, 1974.

38. Ohno, K., Pettigrew, K.D., and Rapoport, S.I., Lower limits of cerebrovascular permeability to nonelectrolytes in the conscious rat, *Am. J. Physiol.* 235:H299, 1979.

39. Pardridge, W.M., and Connor, J.D., Saturable transport of amphetamine across the blood-brain barrier, *Experientia* 29:302, 1973.

40. Pardridge, W.M., and Mietus, L.J., Palmitate and cholesterol transport through the blood-brain barrier, *J. Neurochem.* 34:463, 1980.

41. Pardridge, W.M., Sakiyama, R., and Fierer, G., Blood-brain barrier transport and brain sequestration of propranolol and lidocaine, *Am. J. Physiol.* 247:R582, 1984.

42. Pardridge, W.M., Triguero, D., and Buciak, J.L., ß-Endorphin chimeric peptides: transport through the blood-brain barrier in vivo and cleavage of disulfide linkage by brain, *Endocrinology* 126:977, 1990.

43. Pardridge, W.M., Buciak, J.L., and Friden, P.M., Selective transport of an anti-transferrin receptor antibody through the blood-brain barrier, *J. Pharmacol. Exp. Ther.* 259:66, 1991.

44. Rapoport, S.I., Ohno, K., and Pettigrew, K.D., Drug entry into the brain, *Brain Res.* 172:354, 1979.

45. Rapoport, S.I., Fredericks, W.R., Ohno, K., and Pettigrew, K.D., Quantitative aspects of reversible osmotic opening of the blood-brain barrier, *Am. J. Physiol.* 238:R421, 1980.

46. Robinson, P.J., and Rapoport, S.I., Size selectivity of blood-brain barrier permeability at various times after osmotic opening, *Am. J. Physiol.* 253:R459, 1987.

47. Rubin, L.L., Barbu, K., Bard, F., Cannon, C., Hall, D.E., Horner, H., Janatpour, M., Liaw, C., Manning, K., Morales, J., Porter, S., Tanner, L., Tomaselli, K., and Yednock, T., *New York Acad. Sci.* 633:420, 1991.

48. Siesjö, B.K., "Brain Energy Metabolism," John Wiley & Sons, Chichester, 1978.

49. Schein, P.S., Green, D., Dean, S.W., and McPherson, E., 6-[Bis(2-chloroethyl)amino]-6-deoxygalactopyranose hydrochloride (C6-galactose mustard), a new alkylating agent with reduced bone marrow toxicity, *Cancer Res.* 47:696, 1987.

50. Smith, Q.R., Takasato, Y., Sweeney, D.J., and Rapoport, S.I., Regional cerebrovascular transport of leucine as measured by the in situ brain perfusion technique, *J. Cereb. Blood Flow Metabol.* 5:300, 1985.

51. Smith, Q.R., Momma, S., Aoyagi, M., and Rapoport, S.I., Kinetics of neutral amino acid transport across the blood-brain barrier, *J. Neurochem.* 49:1651, 1987.

52. Smith, Q.R., Aoyagi, M., and Rapoport, S.I., Structural specificity of the brain capillary neutral amino acid transporter, *Soc. Neurosci. Abstr.* 15:1025, 1989.

53. Spector, R., Ceftriaxone transport through the blood-brain barrier, *J. Infect. Dis.* 156:209, 1987.

54. Spector, R., and Lorenzo, A.V., Inhibition of penicillin transport from the cerebrospinal fluid after intracisternal inoculation of bacteria, *J. Clin. Invest.* 54:316, 1974.

55. Stein, W.D., "Transport and Diffusion across Cell Membranes," Academic Press, Orlando, 1986.

56. Strausbaugh, L.J., Meningitis, antimicrobial agents, and the blood-brain barrier, *in:* "Implications of the Blood-Brain Barrier and Its Manipulation, Vol. 2," E.A. Neuwelt, ed., Plenum, New York, 1989.

57. Suzuki, H., Sawada, Y., Sugiyama, Y., Iga, T., and Hanano, M., Facilitated transport of benzylpenicillin through the blood-brain barrier in rats, *J. Pharmacobio-Dyn.* 12:182, 1989.

58. Takada, Y., Greig, N.H., Vistica, D.T., Rapoport, S.I., and Smith, Q.R., Affinity of antineoplastic amino acid drugs for the large neutral amino acid transporter of the blood-brain barrier, *Cancer Chemother. Pharmacol.* 29:89, 1991.

59. Takada, Y., Vistica, D.T., Greig, N.H., Purdon, D., Rapoport, S.I., and Smith, Q.R., Rapid high-affinity transport of a chemotherapeutic amino acid across the blood-brain barrier, *Cancer Res.* 52:2191, 1992.

60. Terasaki, T., and Pardridge, W.M., Restricted transport of 3'-azido-3'-deoxythymidine and dideoxynucleosides through the blood-brain barrier, *J. Infect. Dis.* 158:630, 1988.

61. van Bree, J.B.M.M., Audus, K.L., and Borchardt, R.T., Carrier-mediated transport of baclofen across monolayers of bovine brain endothelial cells in primary culture, *Pharm. Res.* 5:369, 1988.

62. van Bree, J.B.M.M., Heijligers-Feijen, C.D., de Boer, A.G., Danhof, M., and Breimer, D.D., Stereoselective transport of baclofen across the blood-brain barrier in the rats as determined by the unit impulse response methodology, *Pharm. Res.* 8:259, 1991.

63. Wade, L.A., and Katzman, R., Synthetic amino acids and the nature of L-DOPA transport at the blood-brain barrier, *J. Neurochem.* 25:837, 1975a.

64. Wade, L.A., and Katzman, R., 3-O-Methyldopa uptake and inhibition of L-dopa at the blood-brain barrier, *Life Sci.* 17:131, 1975b.

65. Williams, M.G., Earhart, R.H., Bailey, H., and McGovern, J.P., Prevention of central nervous system toxicity of the antitumor antibiotic acivicin by concomitant infusion of an amino acid mixture, *Cancer Res.* 50:5475, 1990.

66. Levitan, H., Ziylan, Z., Smith, Q.R., Takasato, Y., and Rapoport, S.I., Brain uptake of a food dye, erythrosin B, prevented by plasma protein binding, *Brain Res.* 322:131, 1984.

67. Ziylan, Y.Z., Robinson, P.J., and Rapoport, S.I., Blood-brain barrier permeability to sucrose and dextran after osmotic opening, *Am. J. Physiol.* 247:R634, 1984.

VASOPRESSIN MODULATES THE BLOOD-BRAIN

TRANSFER OF AMINO ACIDS - STUDIES WITH

[11C]METHIONINE IN DOGS

Peter Brust[1], Dean F. Wong[2], Albert Gjedde[3],
and Armin Ermisch[1]

[1] Department of Biosciences, University of Leipzig
Germany
[2] Department of Radiology, Johns Hopkins Medical Institutions
Baltimore, MD USA
[3] PET Laboratories, Montreal Neurological Institute
Montreal, Quebec, Canada.

INTRODUCTION

Peptides are chemical signals among different cell populations (1,2). Brain endothelial cells are specific targets for peptide hormones (3,4,). Vasopressin (VP) binds specifically to brain microvessels (5,6), and there is strong evidence that blood-borne VP affects the transport of neutral amino acids (NAA) across the BBB of rats via a receptor-mediated mechanism. After intracarotid injection of VP, a decline of the maximum velocity, V_{max}, and an increase of the affinity, $1/K_t$, of the transporter for large NAA was observed (7,8). Accordingly, intravenous injection of VP decreased the PS-product of NAA (9,10).

If the changes of NAA transport are the result of a specific receptor-mediated effect of VP, changes can be predicted also for species other than rodents. In the present study, we used adult male mongrel dogs to measure the effects of VP on the blood-brain transfer of [11C]methionine by means of a simplified probe detection system (11) and positron emission tomography (PET). The methodological strength of these techniques in relation to other in vivo methods is the non-invasive type of measurement that allows for multiple studies to be performed. Therefore, we were able to measure the effects of VP in each animal studied.

METHODS

The positron detection system consisted of a pair of two 3-in x 3-in sodium iodide (NaI) scintillators positioned in cylindrical lead collimators for coincident detection of γ-photons, thereby measuring the tracer radioactivity in regions of the brain viewed by the detectors. The collimator-to-collimator separation was 19.6 cm, with the scintillator

Frontiers in Cerebral Vascular Biology: Transport and Its Regulation
Edited by L.R. Drewes and A.L. Betz, Plenum Press, New York, 1993

95

crystals being recessed 2.5 cm from the collimator edge. The faces of the crystals were collimated to a 5-cm diameter circle using 2.5-cm thick lead rings.

The theoretical response of the dual detector system was determined using a computer simulation that calculates the three-dimensional response through the head of a dog. In addition, a single study was performed using a Neuro-Ecat II PET scanner to verify the results obtained with the dual-probe system and to calculate the percentage of coincidences (23%) originating in the extracerebral tissue (ECT). The kinetics of [11C]methionine accumulation in brain and in plasma was recorded up to 30 min after i.v. injection of the tracer. The same procedure was performed before and after injection of 1 μg VP.

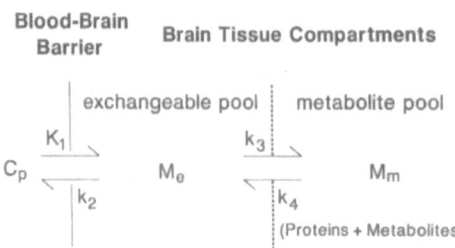

Figure 1. Three-compartment model to the kinetics of [11C]methionine in dog brain.

A three-compartment model (Figure 1) was applied to the kinetics of [11C]methionine in dog brain. It consisted of a vascular compartment with the plasma concentration of [11C]methionine (C_p) and the initial volume of distribution (V_0), a brain tissue compartment corresponding to [11C]methionine present in the brain extravascular space (M_e), and a brain tissue compartment (M_m) corresponding to methionine incorporated into proteins and methionine retained in the brain as different metabolic intermediates. Four kinetic parameters (K_1, k_2, k_3, k_4) describe the transfer of the tracer between the compartments. Two different analyses, a graphic (slope-intercept plot) and a kinetic approach were used for the data processing.

Graphic Analysis

The net accumulation rate of methionine (K) was obtained by using the graphic approach (12,13). In this analysis it was assumed that the tracer is practically irreversibly trapped in the brain. The relationship between the normalized plasma integral and the volume of distribution in this plot was linear after the initial tracer uptake in both cases, i.e., before and after injection of VP suggesting that $k_4 \ll k_3$ (Table 2).

The ratio between the concentrations of the radioactivity in the brain tissue and in the plasma was plotted against the normalized plasma integral, i.e., the ratio between the time-integrated plasma concentration and the plasma concentration. The slope of the linear fit to these data (assuming k_4 is negligible) represents the net accumulation rate of the tracer in the tissue:

$$K = \frac{K_1 k_3}{[k_2 + k_3]} \tag{1}$$

whereas the y-axis intercept is the volume V_g, the lower limit of which theoretically is:

$$V_g = \frac{K_1 k_2}{[k_2 + k_3]^2} \tag{2}$$

Kinetic Analysis

Since the slope-intercept plot assumes that k_4 is negligible, non-linear regression analysis of five parameters (K_1, k_2, k_3, k_4, V_0) was performed to account for the release of methionine during and after protein synthesis. With this kinetic approach, the transfer parameters were estimated from multilinear differential equations (14). The parameters were estimated based on the solution of the following equations. The total brain accumulation of $[^{11}C]$methionine was computed as:

$$M(t) = M_e(t) + M_m(t) + V_0 C_p(t) \tag{3}$$

The changes of the total tracer amounts within the compartments are described by:

$$\frac{dM_e(t)}{dt} = K_1 C_p(t) - k_2 M_e(t) - k_3 M_e(t) + k_4 M_m(t) \tag{4}$$

$$\frac{dM_m(t)}{dt} = k_3 M_e(t) - k_4 M_m(t) \tag{5}$$

The ratio $K_1/(k_2 + k_3)$ represents the distribution volume of the free tracer in the brain (V_f).

Table 1. Blood-brain clearance K_1, net accumulation rate K, and the initial distribution volume V_0 of $[^{11}C]$methionine obtained from probe and PET data.

	Probe	PET(brain)	PET (extracerebral)
K(ml ml-1 min-1)	0.040	0.054	0.011
K_1(ml ml-1 min-1)	0.075	0.070	0.021
V_0(ml ml-1)	0.029	0.067	0.013

RESULTS AND DISCUSSION

Table 1 compares estimates of the blood-brain clearance (K_1), the net accumulation rate (K) and the initial distribution volume of $[^{11}C]$methionine (V_0) obtained from probe and PET data, respectively. It shows that these parameters were much lower in the

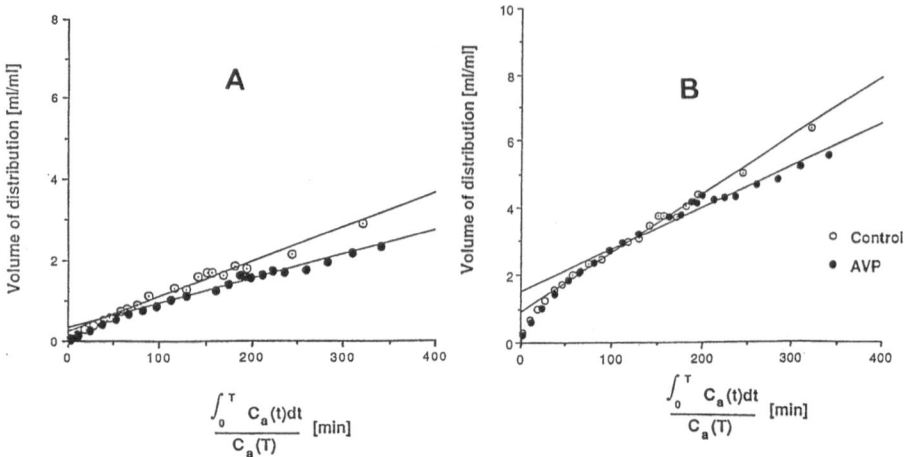

Figure 2. Plots of the normalized plasma integral versus the total volume of L-[^{11}C]methionine distribution in the extracerebral (A) and brain tissue (B) of a dog before and after i.v. injection of 1 μg of vasopressin. The lines represent a linear fit of the experimental PET data between 5 and 30 min after tracer injection. The slope of the lines are equal to K, the net accumulation of [^{11}C]methionine in the region of interest.

extracerebral tissue than in the brain. Because the probe detectors measured coincidences that originated in both tissues, the calculated parameters are different from those obtained from PET data.

However, the effects of VP were similar in brain and in the extracerebral tissue (Figure 2). Therefore, our conclusions drawn from the effect of VP in the probe studies should not be influenced by this fact. In all studies, K and K_1 were lower after injection of VP. K, estimated from the graphic plots (Figure 2), declined on average by 33%. The changes of K were similar (-38%) when this parameter was calculated from the kinetically determined transfer coefficients (Table 2). K_1 declined by 45% (Table 2). Since the value of K_1 is much smaller than the cerebral blood flow in dogs (15), it is expected to be close to the PS-product of [^{11}C]methionine. Also the rate constant of brain-blood transfer of [^{11}C]methionine (k_2) and the rate of protein incorporation of the tracer (k_3) tended to decline by 39% and 23%, respectively (Table 2). Notably, these changes occurred at VP levels within the physiological range. The VP concentrations measured in the arterial blood plasma by radioimmunoassay were 7±3 pg ml^{-1} before and 108±20 pg ml^{-1} at 5 min after injection of 1 μg of VP.

Table 2. Kinetic transfer coefficients (means ± S.E.M.) of L-[^{11}C]methionine accumulation in dog brain, volume of free methionine in brain (V_f) and initial volume of distribution (V_0) before and after i.v. injection of 1 μg of vasopressin.

	K_1	k_2	k_3	k_4	V_f	V_0
	(ml ml^{-1} min^{-1})	(min^{-1})	(min^{-1})	(min^{-1})	(ml ml^{-1})	(ml ml^{-1})
Before VP	0.075±0.011	0.204±0.032	0.271±0.029	0.039±0.009	0.172±0.023	0.029±0.004
After VP	0.041±0.009[1]	0.125±0.051	0.209±0.050	0.039±0.012	0.147±0.027	0.028±0.005

[1]$p \leq 0.05$ (after VP)

Together with the present data, a total of four different NAA (PHE, LEU, MET, VAL) have been used to study the effects of VP (7-10,16). The BBB transport of all of them was influenced by the peptide. The effect probably results from interaction with receptors localized at the luminal membrane of the brain endothelial cells, perhaps by changing the conformation of the BBB transporter of NAA. On isolated microvessels from rat brain, binding of vasopressin to the V_1 receptor type (17) caused an increase of intracellular calcium most likely by stimulation of calcium influx (18). The phosphoinositol metabolism seems to be less involved (19). From studies demonstrating inhibition of inward-rectifying K^+ channels in porcine cerebral capillary endothelial cells, it was suggested that the effect of vasopressin is mediated via G-protein(s) (20) similar to the V_1 vascular smooth muscle receptor (21,22). Concomitant activation of protein kinase C is expected to result in the phosphorylation of transporter molecules similar to mechanisms regulating protein phosphorylation in muscle cells after V_1 receptor occupation (21). The phenomenon seems to be only a special case of a general principle concerning tight epithelial cell layers that separate compartments containing fluids of different composition. Recently, we have demonstrated that β-casomorphins, peptides originating from enzymatic cleavage of milk proteins and perhaps of other sources, alter the intestinal transport in rats (23) and earthworms (24).

ACKNOWLEDGMENTS

The authors wish to thank Christine Steinert, Elias K. Shaya, and Keith J. Jeffries for their help during these studies as well as Robert F. Dannals, Hayden T. Ravert, and Alan A. Wilson for providing us with [^{11}C]methionine. The study was supported in part by grants PO1 HD24061, USPHS NS15080, and USPHS CA32845. Peter Brust was supported by the MRC of Canada (SP-30).

REFERENCES

1. Scharrer, B., Neurosecretion: beginnings and new direction in neuropeptide research, *Annu. Rev. Neurosci.* 10:1, 1987.
2. Scharrer, B., The neuropeptide saga, *Amer. Zool.* 30:887, 1990.
3. Ermisch, A., Ruhle, H.J., Landgraf, R., and Hess, J., Blood-brain barrier and peptides, *J. Cereb. Blood Flow Metab.* 5:350, 1985.
4. Ermisch, A., Peptide receptors of the blood-brain barrier and substrate transport into the brain, *in:* "Progress in Brain Research: Circumventricular Organs and Brain Fluid Environment-Molecular and Functional Aspects," A. Ermisch, R. Landgraf, and H.-J. Rühle, eds., Amsterdam-New York-Oxford, Elsevier, in press, 1992.
5. Kretzschmar, R., Landgraf, R., Gjedde, A., and Ermisch, A., Vasopressin binds to microvessels from rat hippocampus, *Brain Res.* 380:325, 1986.
6. Kretzschmar, R., and Ermisch, A., Arginine-vasopressin binding to isolated hippocampal microvessels of rats with different endogenous concentrations of the neuropeptide, *Exp. Clin. Endocrinol.* 94:151, 1989.
7. Brust, P., Changes in regional blood-brain transfer of L-leucine elicited by arginine-vasopressin, *J. Neurochem.* 46:534, 1986.
8. Ermisch, A., Reichel, A., and Brust, P., Changes in the blood-brain transfer of L-phenylalanine elicited by arginine-vasopressin, *Endocrine Regul.*, in press, 1992.
9. Reith, J., Ermisch, A., Diemer, N.H., and Gjedde, A., Saturable retention of vasopressin by hippocampus vessels in vivo, associated with inhibition of blood-brain transfer of large neutral amino acids, *J. Neurochem.* 49:1471, 1987.

10. Brust, P., and Diemer, N.H., Decrease of regional blood-brain transfer of L-phenylalanine after peripheral but not central application of vasopressin, *J. Neurochem.* 55:2098, 1990.

11. Bice, A.N., Wagner, H.N., Jr., Frost, J.J., Natarajan, T.K., Lee, M.C., Wong, D.F., Dannals, R.F., Ravert, H.T., Wilson, A.A., and Links, J.J., Simplified detector system for neuroreceptor studies in the human brain, *J. Nucl. Med.* 27:184, 1986.

12. Gjedde, A., High- and low-affinity transport of D-glucose from blood to brain, *J. Neurochem.* 36:1463, 1981.

13. Patlak, C.S., Blasberg, R.G., and Fenstermacher, J.D., Graphical evaluation of blood-to-brain transfer constants from multiple-time uptake data, *J. Cereb. Blood Flow Metab.* 3:1, 1983.

14. Gjedde, A., and Wong, D.F., Modeling of neuroreceptor binding of radioligands in vivo, *in:* "Quantitative Imaging: Neuroreceptors, Neurotransmitters, and Enzymes," J.J. Frost and H.N. Wagner, eds., New York, Raven Press, p. 51, 1990.

15. Moursi, M.M., van Wylen, D.G.L., and D'Alecy, L.G., Regional blood flow changes in response to mildly pressor doses of triglycyl desamino lysine and arginine vasopressin in the conscious dog, *J. Pharmacol. Exp. Ther.* 232:360, 1985.

16. Reichel, A., Brust, P., and Ermisch, A., Effects of arginine-vasopressin on the blood-brain transfer of certain large neutral amino acids, *in:* "Third IBRO World Congress of Neurosciences" (Abstracts), Montreal, P 59.38, 1991.

17. Jard, S., Vasopressin isoreceptors in mammals: relation to cyclic AMP-dependent and cyclic AMP-independent transduction mechanisms, *Curr. Top. Membr. Transp.* 18:255, 1983.

18. Hess, J., Jensen, C.V., and Diemer, N.H., The vasopressin receptor of the blood-brain barrier in the rat hippocampus is linked to calcium signalling, *Neurosci. Lett.* 132:8, 1991.

19. Xu, J., Qu, Z.-X. , Moore, S.A., Hsu, C.Y., and Hogan, E.L., Receptor-linked hydrolysis of phosphoinositides and production of prostacyclin in cerebral endothelial cells, *J. Neurochem.* 58:1930, 1992.

20. Hoyer, J., Popp, R., Meyer, J., Galla, H.J., and Gogelein, H., Angiotensin-II,vasopressin and GTP[γ-S] inhibit inward-rectifying-K^+ channels in porcine cerebral capillary endothelial cells, *J. Membrane Biol.* 123:55, 1991.

21. Thibonnier, M., Signal transduction of V $_1$-vascular vasopressin receptors, *Reg. Peptides* 38:1, 1992.

22. Thibonnier, M., Bayer, A.L., Simonson, M.S., and Kester, M., Multiple signaling pathways of V_1-vascular vasopressin receptors of A7r5 cells, *Endocrin.* 129:2845, 1991.

23. Ermisch, A., Brust, P., and Brandsch, M., β-Casomorphins alter the intestinal accumulation of L-leucine, *Biochim. Biophys. Acta* 982:79, 1989.

24. Brandsch, M., Brust, P., Neubert, K., and Ermisch, A., β–Casomorphins - chemical signals for intestinal transport systems, *in:* "Proc. 2nd Int. Symp. on β-Casomorphins and Related Peptides," Titisee, Germany, in press, 1992.

AMINO ACID TRANSPORT AT THE BLOOD-NERVE
BARRIER OF THE RAT PERIPHERAL NERVE

Kishena C. Wadhwani, Quentin R. Smith, and Stanley I. Rapoport

Laboratory of Neurosciences, NIA
NIH, Bethesda, MD 20892

INTRODUCTION

The nerve fibers and associated supporting cells (e.g., Schwann cells) of vertebrate peripheral nerve are protected from circulating ions and toxins by a blood-nerve barrier (BNB), which exhibits many similarities to the blood-brain barrier (BBB) of the central nervous system. The BNB consists of endoneurial capillaries and a perineurial sheath (6,9). Tight junctions (zonulae occludens) between capillary endothelial cells and between cell layers of the perineurium restrict water soluble substances from entering the nerve from the blood (6,9). Although no evidence exists for a saturable transport of essential ions, such as Na, K (20-22) and Ca (17), at the BNB, D-glucose, a primary substrate of nerve oxidative metabolism (5), has been shown to cross the BNB by a carrier system, having characteristics like those at the BBB (10,11).

Peripheral nerve, like the brain, requires amino acids for protein and lipid synthesis, CO_2 production, and for post-translational protein modification (1,2,14,23). Saturable carriers exist at the BBB to shuttle essential amino acids into the brain (7,13). Whether the BNB also possesses amino acid transport systems similar to those found at the BBB is not known. The aim of this study was to determine if there was any evidence for saturable uptake of neutral, basic, or acidic amino acids at the BNB of the rat peripheral nerve and, if so, to investigate the properties of the transport systems.

MATERIALS AND METHODS

Unidirectional uptakes of L-[^{14}C]leucine (a neutral amino acid), L-[^{14}C]arginine (a basic amino acid), and L-[^{14}C]glutamic acid (an acidic amino acid), from blood into rat sciatic nerve across the BNB, were quantitatively determined, using an in situ perfusion technique (16), which is illustrated in Figure 1.

After heating the perfusion fluid to 37°C, the Ringer's solution was delivered through an inlet (I) into the left common iliac artery (CIA) of an anesthetized rat. An outlet (O) was inserted into the left common iliac vein (CIV) to drain fluid from the system. Immediately

Frontiers in Cerebral Vascular Biology: Transport and Its Regulation
Edited by L.R. Drewes and A.L. Betz, Plenum Press, New York, 1993

101

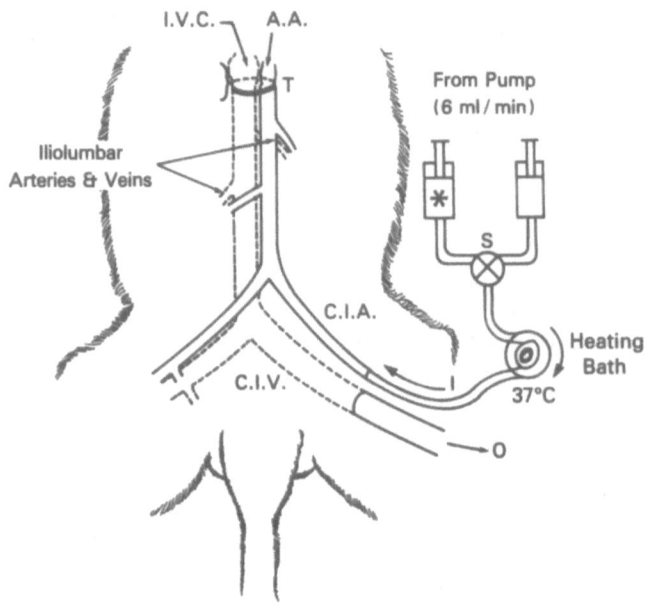

Figure 1. In situ arrangement to study transport at the blood-nerve barrier (BNB) in rat sciatic nerve. Detailed explanations are in the text.

after the perfusion pump was turned on, the abdominal aorta (AA) and inferior vena cava (IVC) were ligated caudal to renal arteries and veins (T). The vasculature of right hindlimb was perfused for 2 min with nonradioactive rat's Ringer. Perfusion medium was then switched to Ringer solution containing radiotracers (*), by means of a 3-way stopcock (S). [^{3}H]dextran (M.W. ~70,000) was included in all experiments as a vascular marker.

At the end of perfusion with radioactive Ringer (e.g., 1 min), the sciatic nerve was removed, frozen in liquid isopentane prechilled to -70°C. The epi-perineurial sheath of the frozen nerve was removed from the endoneurium, as previously described (18), and the desheathed nerve segments were dried and dried weights were recorded. The samples were then transferred to scintillation vials, digested overnight at 65°C in 10% piperidine, and processed for radioactivity counting using a liquid scintillation spectrometer equipped with a dual label counting program For comparison, a portion of biceps femoris muscle, adjacent to the removed nerve, also was collected and processed for radioactivity counting.

Permeability-surface area products (PAs, ml.s^{-1}.g wet wt^{-1}) for unidirectional tracer uptake from blood into sciatic nerve and into biceps femoris muscle were determined by the following equation (16):

$$PA_{tissue} = C^*_{tissue}/([C^*]_0 \cdot T) \tag{1}$$

where T is perfusion time (s), $[C^*]_0$ is the concentration of radiotracer in the perfusion medium (dpm/ml), and C^*_{tissue} (dpm/g wet wt) is the concentration of permeant tracer in the desheathed nerve or in the biceps femoris muscle, exclusive of intravascular tracer, and is given by (16)

$$C^*_{tissue} = (Q^*_{tissue}/W) - (I^* \cdot [C^*]_0/[I^*]_0 \cdot W) \tag{2}$$

where Q^*_{tissue} is total radioactivity of permeable tracer (dpm) in the tissue, I^* is radioactivity (dpm) of [³H]dextran in the tissue, $[I^*]_0$ is the concentration of [³H]dextran in the perfusion medium (dpm/ml) and W is tissue wet weight (g).

RESULTS

BNB Transport of Neutral Amino Acid L-[¹⁴C]leucine

Increasing the concentration of unlabeled L-leucine in the perfusion media up to 10 mM significantly decreased mean BNB PAs to L-[¹⁴C]leucine by more than 90% [from (61.0 ± 5.3) to (4.1 ± 0.3) x 10^{-5} ml/s.g (means±S.E; n=5); Figure 2]. Addition of 50 mM unlabeled L-leucine to the perfusion medium did not further decrease mean BNB PA (Figure 2). Addition of 10 mM [D-leucine] to the perfusion medium also decreased BNB PA, but to a degree less than that caused by an equimolar L-leucine (Figure 2). Ten millimolar BCH, a model substrate for the L-system (3,4), reduced mean BNB PA by more than 80% [from (61.0 ± 5.3) to (8.7 ± 0.5) x 10^{-5} ml/s.g; Figure 2].

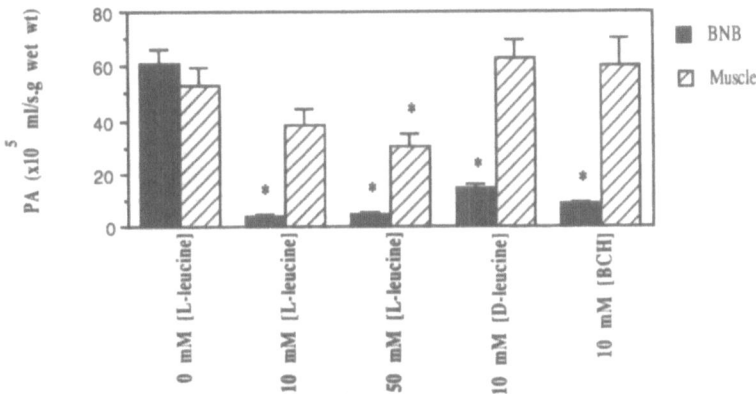

Figure 2. Means PAs for the BNB and for the biceps femoris muscle to L-[¹⁴C]leucine in the presence of 0, 10 or 50 mM unlabeled [L-leucine],10 mM [D-leucine], or 10 mM [BCH] in the perfusion media. Upper bars represent 1 SE.
* = significantly differs from 0 mM [L-leucine] group at p < 0.05 (n = 5 - 6 rats), Bonferroni t-statistics.

BNB Transport of Basic Amino Acid L-[¹⁴C]arginine

Unidirectional uptake of L-[¹⁴C]arginine into the rat sciatic nerve also exhibited facilitated and saturable characteristics. Addition of unlabeled [L-arginine] to the perfusion media significantly decreased mean BNB PA to L-[¹⁴C]arginine by more than 75% of its original value [from (20.0 ± 1.8) to (4.4 ± 0.6) x 10^{-5} ml/s.g (n=5); Figure 3]. As Figure 3 shows, mean BNB PA to L-[¹⁴C]arginine did not change in either the presence of [BCH], a model substrate for L-system, [MeAIB], a model substrate for A-system, or Na-free, Tris Ringer. However, addition of 50 mM [L-homoarginine], a model substrate for the basic amino acid transporter (3,4), to the perfusion medium decreased mean BNB PA by about 80% (Figure 3). Estimated kinetic constants for the saturable components (K_m and V_{max}) and nonsaturable diffusion component (K_d) of L-arginine, from the Michaelis-Menten model (12), are 6.2 mM, 1.5 nmol/g.min, and 4.5 x 10^{-5} ml/s.g, respectively.

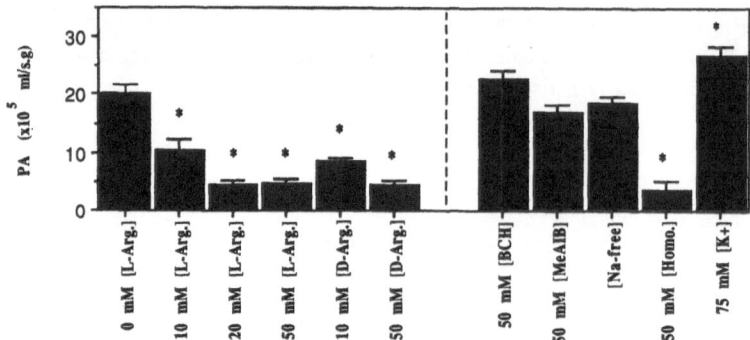

Figure 3. Means PAs for the BNB to L-[^{14}C]arginine in the presence of various concentrations of unlabeled [L-arginine], [D-arginine], [BCH], [MeAIB], Na-free [165 mM-Tris] Ringer, [L-homoarginine], and [K$^+$]-Ringer in the perfusion media Upper bars represent 1 SE.

* = significantly differs from 0 mM [L-arginine] group at p < 0.05 (n = 5 - 6 rats), Bonferroni t-statistics.

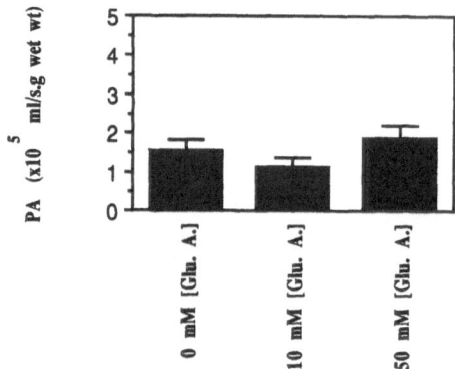

Figure 4. Means PAs for the BNB to L-[^{14}C]glutamic acid in the presence of 10 or 50 mM unlabeled [L-glutamic acid] in the perfusion media Upper bars represent 1 SE for n = 5 - 6 rats.

BNB Transport of Acidic Amino Acid L-[^{14}C]glutamic Acid

In contrast to the PA for L-leucine and L-arginine, the PA of the BNB to L-[^{14}C]glutamic acid was low [(1.6±0.3) x 10^{-5} ml/s.g; n=5] and did not change when perfusion media contained either 10 or 50 mM unlabeled [L-glutamic acid] (Figure 4).

DISCUSSION

Our results demonstrate that the unidirectional transport of L-leucine, a neutral amino acid and L-arginine, a basic amino acid, from blood into the mammalian peripheral nerve is facilitated, saturable, and mediated by separate transport systems.

BCH, a model substrate for L-system transporters, inhibits L-[^{14}C]leucine influx from blood to nerve as does unlabeled L-leucine. This suggests that L-[^{14}C]leucine crosses the BNB by an L-type carrier system, and is consistent with our previous finding that influx of another neutral amino acid, L-phenylalanine, from blood into rat sciatic nerve is mediated by an L-type carrier (16), analogous to that at the blood-brain barrier (7,13).

Unidirectional uptake of L-arginine into the rat nerve is not inhibited by BCH, MeAIB [a model substrate for the A-system (3,4)], or by Na-free Tris-Ringer. L-Homoarginine, a model substrate for the basic amino acid carrier system (the y+ system), however, inhibits L-[^{14}C]arginine influx into the rat's nerve at the same extent as does the unlabeled [L-arginine]. Thus, it is likely that L-arginine crossed the BNB by a mechanism different from that used by the neutral amino acids, such as L-phenylalanine and L-leucine, and that a basic amino acid carrier system, which can be inhibited by L-homoarginine, exists at the BNB of the rat sciatic nerve.

BNB PA for L-[^{14}C]glutamic acid is low and is not altered when unlabeled L-glutamic acid is added to the perfusion media. Thus, our results provide no evidence for saturable uptake of this amino acid at the endoneurial capillaries of the rat nerve. In the central nervous system, it is speculated that an acidic amino acid carrier system is located at the abluminal side of cerebral capillaries rather than on the side facing the blood (8) and that this carrier system pumps acidic amino acids out of the brain (8).

The estimated Michaelis-Menten kinetics and passive diffusion constants for neutral (e.g., L-leucine) and basic (e.g., L-arginine) amino acid transporters at the BNB are: for L-leucine, K_m=0.12 mM, V_{max}=8.0 nmol/min.g, and K_d=4.1 x 10^{-5} ml/s.g; for L-arginine, K_m=6.2 mM, V_{max}=1.50 nmol/min.g., and K_d=4.5 x 10^{-5} ml/s.g. Because the estimated K_ms are either within or exceed the normal concentrations of these amino acids in the rat plasma (~ 0.1 mM), (15), most neutral and basic amino acids probably enter the nerve from blood via these carrier-mediated systems in normal physiological conditions. In the brain, estimated V_{max}s for most neutral and basic amino acids at the BBB are in the ranges of 10 to 60 nmol/min.g (7,13). Thus, the carrier's capacity at the BNB is severalfold lower than that at the BBB.

In conclusion, this study demonstrates that the BNB of the rat peripheral nerve exhibits two distinct carrier transport systems for neutral and basic amino acids, similar to those found at the BBB. This finding and reports that the BNB also possesses saturable transport systems for glucose (10,11) and manganese (19) further support the notion that the BNB is not only a passive barrier but also exhibits a regulatory function similar to that found at the BBB. Regulation may be necessary for nerve function and metabolism. It would be of interest to examine the role of these transport mechanisms in the etiology of certain peripheral neuropathies, such as those found in inflammatory, traumatic, toxic, or metabolic lesions, and in aging.

REFERENCES

1. Buse, M.G., Jursinic, S., and Reid, S.S., Regulation of branched-chain amino acids oxidation in isolated muscles, nerves and aortas of rats, *Biochem. J.* 148:363-374, 1975.

2. Caston, J. D., and Singer, M., Amino acid uptake and incorporation into macromolecules of peripheral nerves, *J. Neurochem.* 16:1309-1318, 1969.

3. Christensen, H.N., Exploiting amino acid structure to learn about membrane transport, *Adv. Enzymol.* 49:41-101, 1979.

4. Christensen, H.N., On the strategy of kinetic discrimination of amino acid transport systems, *J. Membr. Biol.* 84:97-103, 1985.

5. Greene, D.A., Winegrad, A.I., Carpentier, J.L., Fukuma, M.J., and Orci, L., Rabbit sciatic nerve fascicle and "endoneurial" preparations for in vitro studies of peripheral nerve glucose metabolism, *J. Neurochem.* 33:1007-1018, 1979.

6. Olsson, Y., Microenvironment of the peripheral nervous system under normal and pathological conditions, *Crit. Rev. Neurobiol.* 5:265-311, 1990.

7. Pardridge, W.M., Brain metabolism: A perspective from the blood-brain barrier, *Physiol. Rev.* 63:1481-1535, 1983.

8. Pardridge, W.M., Regulation of amino acid availability to brain: Selective control mechanisms for glutamate, *in:* "Glutamic Acid: Advances in Biochemistry and Physiology," L.J. Filer, Jr. et al., eds., Raven Press, New York, 1979.

9. Rechthand, E., and Rapoport, S.I., Regulation of the microenvironment of peripheral nerve: role of the blood-nerve barrier, *Progr. Neurobiol.* 28:303-343, 1987.

10. Rechthand, E., Smith, Q.R., and Rapoport, S.I., Facilitated transport of glucose from blood into peripheral nerve, *J. Neurochem.* 45:957-964, 1985.

11. Rechthand, E., Smith, Q.R., and Rapoport, S.I., Structural specificity of sugar transport at the blood-nerve barrier, *J. Neurochem.* 53:119-123, 1989.

12. Segel, I.H., "Biochemical Calculations," 2nd edition, John Wiley & Sons, Inc., New York, 1976.

13. Smith, Q.R., and Takasato, Y., Kinetics of amino acid transport at the blood-nerve barrier studied using an in situ brain perfusion technique, *Ann. N.Y. Acad. Sci.* (The Neuronal Microenvironment) 481:186-201, 1986.

14. Stillway, L.W., Weigand, D.A., and Buse, M.G., Leucine as an in vitro precursor to lipids in rat sciatic nerve, *Lipids* 14:127-131, 1979.

15. Valdivieso, F., Ugarte, M., Maties, M., Gimenez, C., and Mayor, F., Free amino acids in the tissues of rats with experimentally induced phenylketonuria, *J. Ment. Defic. Res.* 21:95-102, 1977.

16. Wadhwani, K.C., Smith, Q.R., and Rapoport, S.I., Facilitated transport of L-phenylalanine across the blood-nerve barrier of the rat peripheral nerve, *Am. J. Physiol.* 258:R1436-R1444, 1990.

17. Wadhwani, K.C., Murphy, V.A., and Rapoport, S.I., Transfers of 45Ca and 36Cl at the blood-nerve barrier of the sciatic nerve in rats fed low and high calcium diets, *J. Neurosci. Res.* 28:563-566,1991.

18. Wadhwani, K.C., Caspers-Velu, L.E., Murphy, V.A., Smith, Q.R., Kador, P.K., and Rapoport, S.I., Prevention of nerve edema and increased blood-nerve barrier permeability-surface area products in galactosemic rats by aldose reductase or thromboxane synthetase inhibitors, *Diabetes* 38:1469-1477, 1989.

19. Wadhwani, K.C., Murphy, V.A., Smith, Q.R., and Rapoport, S.I., Saturable transport of manganese (II) across blood-nerve barrier of rat peripheral nerve, *Am. J. Physiol.* 262:R284-R288, 1992.

20. Weerasuriya, A., Permeability of endoneurial capillaries to K, Na and Cl and its relation to peripheral nerve excitability, *Brain Res.* 419:188-196, 1987.

21. Weerasuriya, A., Rapoport, S.I., and Taylor, R.E., Ionic permeabilities of the frog perineurium, *Brain Res.* 191:405-415, 1980.

22. Welch, K., and Davson, H., The permeability of capillaries of the sciatic nerve of the rabbit to several materials, *J. Neurosurg.* 36:21-26, 1972.

23. Zanakis, M.F., Chakraborty, G., Sturman, J.A., and Ingoglia, N.A., Posttranslational protein modification by amino acid addition in intact and regenerating axons of the rat sciatic nerve, *J. Neurochem.* 43:1286-1294, 1984.

EFFECT OF INTRA-UTERINE GROWTH RETARDATION ON CEREBRAL AMINO ACID TRANSPORT FROM BIRTH TO ADULTHOOD

Olivier Rabin, Jeanne-Marie Lafauconnier, Gabrielle Bernard, Claude Chanez and Jean-Marie Bourre

INSERM U 26, Hôpital F. Widal
200 rue du Faubourg Saint Denis 75475 Paris Cedex 10, France

INTRODUCTION

Undernutrition during perinatal life, caused by restriction of blood supply to the fetus by ligating the uterine vessels produces intra-uterine growth retardation (IUGR) [8] and has a critical influence on brain function and development. In this case, it has been reported that vascular concentrations of some amino acids were modified several days after undernutrition trauma occurred [6] and that modifications in cerebral monoamine or amino acid concentrations were observed in several brain regions [1,2]. To further investigate the origin of the brain alterations, we examined the possible role of the blood-brain barrier at the interface between the two modified compartments.

MATERIALS AND METHODS

IUGR was induced by blood supply restriction on the 17th day of gestation. The artery and vein of one uterine horn were ligated and the opposite horn was left untouched and served for control animals. At birth, we defined as IUGR animals those that had at least a 25% weight reduction when compared with controls of the same litter that had the same weight as animals of the same age.

Radioactive amino acids (^{14}C-L-alanine, ^{3}H-taurine), previously reported modified in IUGR brain, or two neurotransmitter precursors (^{14}C-tryptophan and ^{3}H-L-tyrosine), were injected as a bolus into the saphenous or femoral vein of 7-, 21-, and 60-day-old rats. Blood was withdrawn regularly in a heparinized catheter from the brachial (in 7-day-old rats) or the femoral artery to assess radioisotope distribution in the vascular space as a function of time. A vascular space correction had been previously determined with ^{14}C-sucrose. Experiments

Frontiers in Cerebral Vascular Biology: Transport and Its Regulation
Edited by L.R. Drewes and A.L. Betz, Plenum Press, New York, 1993

107

were performed in random order for sex, age and weight. After 90 sec, animals were decapitated and the brain immediately removed. Thirteen cerebral regions were dissected on chilled moistened filters over ice. Transfer coefficients (K_{in}) for each amino acid were calculated from the amount of radioactive tracer in the brain region minus the amount of tracer in plasma in brain over the integral of the plasma concentration of the tracer during the 90 sec.

Free plasma amino acid concentrations were assessed by reversed-phase high performance liquid chromatography after precolumn derivatization with O-pthaldialdehyde. Amino acid influx (J_{in}) across the blood-brain barrier was calculated by multiplying transfer coefficients by plasma amino acid concentrations.

RESULTS

Amino Acid Transfer Coefficients

In the 13 cerebral structures examined, the transfer coefficients did not significantly differ, except for cerebellum, which had a much higher coefficient than the other regions in 7-day-old animals. Parietal cortex was chosen as a representative structure for time course of the four amino acids tested.

Whatever the age, transfer coefficient values did not show alterations for IUGR animals compared to controls except for alanine, which was modified in the olfactory bulbs, striatum and superior colliculus in 60-day-old animals. As already reported (4), the transfer coefficients decreased with age.

Plasma Amino Acids

The only difference for total amino acids was a higher concentration of alanine in plasma of 7-day-old IUGR animals. Although total tryptophan was the same in IUGR animals and control animals, it has been reported (1), that free tryptophan is much higher in plasma of young IUGR animals.

We observed a marked trend to lower plasma amino acid concentrations in young adults compared to newborns except for taurine, which was constant throughout development.

Amino Acid Influx

As observed for transfer coefficients, influx of taurine, alanine and tyrosine did not present significant differences between IUGR animals and controls. It was not possible to assess tryptophan influx, as tryptophan is partly free and partly bound to albumin. We do not know if the radioactive tryptophan used to measure transport had the same distribution as cold tryptophan. It has been reported that increases of tryptophan in brain are associated with increases of free tryptophan in plasma (3). However, albumin-bound tryptophan seems also to participate in transport (5). We have thus not indicated on the figure the influx of tryptophan, but the product of K_{in} by free and total tryptophan. Free tryptophan was estimated from the results of Chanez et al. (1). The real influx is in between, but it is probably higher in young IUGR animals.

A marked decrease with age was found for influx, though the time course differs from one amino acid to another. Thus, the principal decrease for alanine is found after weaning, whereas it appears before this period for tryptophan and taurine. Tyrosine follows a regular decline, with no critical period for a major decrease in influx.

(Parietal cortex)

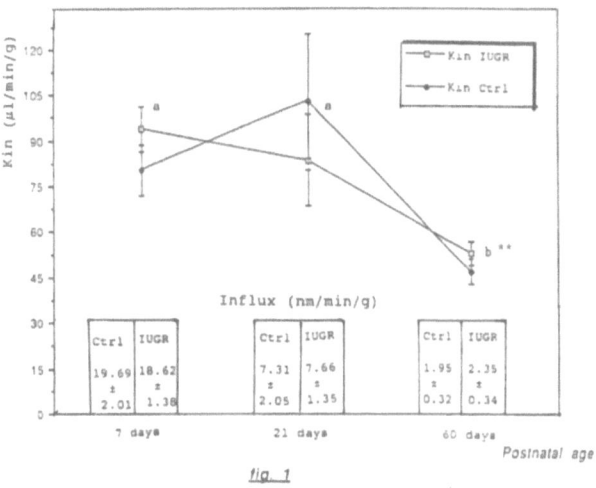

fig. 1

Alanine transport coefficient and influx trends with age

(parietal cortex)

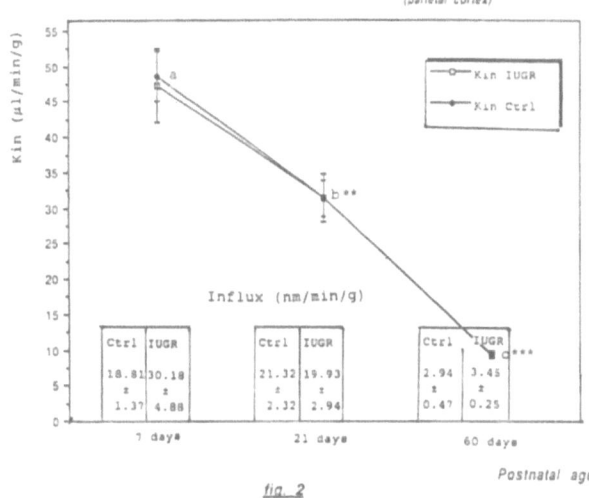

fig. 2

Figure 1 and 2. Changes in tyrosine and alanine blood-brain transfer coefficients: K_{in} (curves) and influx in IUGR animals and control (Ctrl) rats. Parietal cortex was chosen as a representative structure for each amino acid for 7-, 21- and 60-day-old animals. Each value is the mean ± SEM of 4 to 9 independent determinations in each group. Statistical analysis was performed with Student's t-test to compare values between IUGR animals and control groups (no statistical difference was observed between the two groups). Analysis of variance with Bonferroni's adjustment for multiple comparisons was used to analyze developmental changes with age: K_{in} values with a different letter (a, b or c) are significantly different ($P<0.05$). Significant threshold was set at **$P<0.05$ or ***$P<0.01$ for 21- and 60-day-old animals compared to the 7-day-old animals group.

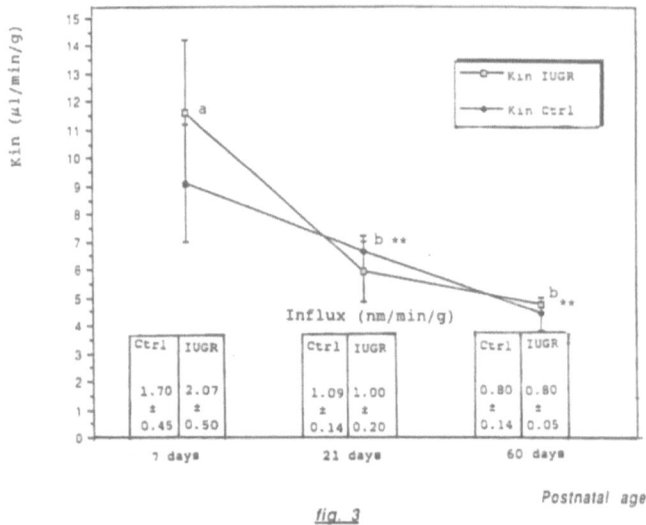

Taurine transfer coefficient and influx trends with age

(parietal cortex)

fig. 3

Tryptophan transfer coefficient and Influx trends with age

(parietal cortex)

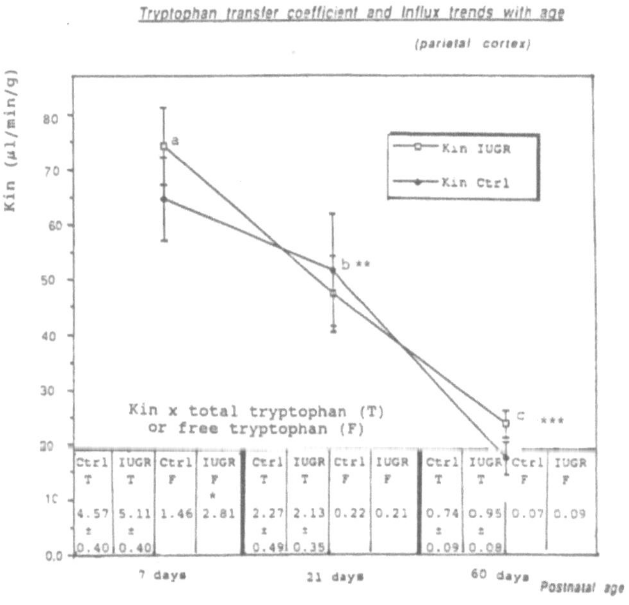

fig. 4

Figure 3 and 4. Changes in taurine and tryptophan blood-brain transfer coefficients: K_{in} (curves) and influx in IUGR animals and control (Ctrl) rats. Parietal cortex was chosen as a representative structure for each amino acid for 7-, 21- and 60-day-old animals. Each value is the mean ± SEM of 4 to 9 independent determinations in each group. Statistical analysis was performed with Student's t-test to compare values between IUGR animals and control groups (no statistical difference, except for free tryptophan in 7-day-old animals, was observed between the two groups). Analysis of variance with Bonferroni's adjustment for multiple comparisons was used to analyze developmental changes with age: K_{in} values with a different letter (a, b or c) are significantly different (P<0.05). Significant threshold was set at **P<0.05 or ***P<0.01 for 21- and 60-day-old animals compared to the 7-day-old animals group.

DISCUSSION

No major differences between IUGR animals and controls were observed for amino acid coefficients or influx, except probably for the influx of tryptophan. This higher influx of tryptophan in IUGR animals was only a result of the higher plasma concentration of the free amino acid. The higher cerebral tryptophan and serotonin concentration observed by Chanez et al. (1) in IUGR animals can probably be explained by the higher influx as a result of the higher plasma concentration of free tryptophan observed in young IUGR animals. Thus, it appears that the blood-brain barrier could not be involved, as initially thought, in the higher amines or amino acids concentrations. The blood-brain barrier, because of a good resistance to undernutrition, compared to other organs, could, to a certain extent, be a part of the mechanism leading to the "brain growth sparing" (7) reported in an undernutrition model.

We confirm, in our model, the general declining trend in transport parameters (transfer coefficients and influx) with age. We also report a differential decrease between alanine, taurine and neurotransmitter precursors tryptophan and tyrosine K_{in} and influx. This suggests a correlation between brain requirements and the trend of amino acid transfer coefficients and influx with age.

REFERENCES

1. Chanez, C., Priam, M., Flexor, M.A., Hamon, M., Bourgoin, S., Kordon, C., and Minkowshi, A., Long lasting effects of intrauterine growth retardation on 5-HT metabolism in the brain of developing rats, *Brain Res.* 207:397-408, 1981.
2. Chanez, C., Giguere, J.F., Rabin, O., Heroux, M., and Butterworth, R.F., Cerebral amino acid changes in an animal model of intra-uterine growth retardation, *Metab. Brain Dis.*, in press.
3. Knott, P.J., and Curzon, G., Free tryptophan in plasma and brain tryptophan metabolism, *Nature* 239:452-453, 1972.
4. Lefauconnier, J.M., Blood-brain amino acid transport during development, *in:* "Amino Acid Availability and Brain Function in Health and Disease," NATO ASI Series, Springer-Verlag, 1988.
5. Pardridge, W.M., and Fierer, G., Transport of tryptophan into the brain from the circulating albumin-bound pool in rats and rabbits, *J. Neurochem.* 54:971-976, 1990.
6. Roux, J.M., and Jahchan, Th., Plasma level of amino acids in the developing young rat after intra-uterine growth retardation, *Life Sci.* 14:1101-1107, 1974.
7. Seidler, F.J., Bell, J.M., and Slotkin, T.A., Undernutrition and overnutrition in the neonatal rat: long-term effect on noradrenergic pathways in brain regions, *Pediatr. Res.* 27:191-197, 1990.
8. Wigglesworth, J.S., Experimental growth retardation in the foetal rat, *J. Path. Bact.* 88:1-13, 1964.

DRUG TRANSFER ACROSS THE BLOOD-BRAIN BARRIER: COMPARISON OF IN VITRO AND IN VIVO MODELS

C. Chesné*, M.P. Dehouck*, P. Jolliet-Riant§, F. Brée§, J.P. Tillement§, B. Dehouck*, J.C. Fruchart*, and R. Cecchelli*

*BIOPREDIC Co, Technopole Rennes-Atalante Villejean, 35000 Rennes, France
*SERLIA et INSERM U 325, Institut Pasteur, 59019 LILLE Cédex, France
§Laboratoire Hospital-Universitaire de Pharmacologie, UFR de Médecine de Paris XII, 8 rue du Général Sarrail, 94010 Créteil, France

The passage of substances across the blood-brain barrier (BBB) is regulated in cerebral capillaries, which possess certain distinct different morphological and enzymatic properties compared with the capillaries of other organs. The endothelial cells of brain capillaries are sealed together by continuous tight junctions and have little transcellular vesicular transport (1). In addition, a number of proteins are specifically expressed by brain endothelial cells that may be required for metabolic protection or transport activities at the BBB interface. Because of these characteristics, the BBB is a regulatory interface that may limit or impair the delivery of drugs to the central nervous system as compared to other tissues.

During the past decade, several laboratories have developed methods for culturing endothelial cells from brain microvessels. Mechanical homogenization was the most widespread procedure used at the beginning of the development of brain microvessel endothelial cell cultures. With the work of Williams et al. (2), who demonstrated that microvessels originating from mechanical dispersion techniques showed a viability of only 5%, while cultures from microvessels isolated using the enzymatic procedure showed a viability of 84%, most of the laboratories have introduced the enzymatic procedure to isolate endothelial cells from brain microvessels for cultures.

Nevertheless, several lines of evidence suggest that cultured brain endothelial cells rapidly lose the characteristics of the differentiated BBB in vitro and it has been claimed that long-term cultures of brain endothelial cells may not provide good model systems for the BBB in vitro (3). Reasons for this could first of all be that the enzymatic method used for the preparation of microvessel endothelial cells makes it impossible to totally clear all arterioles and venules from the capillary preparation. So the possibility exists that migrating and proliferating cells might be derived from the endothelial wall of arterioles and venules that contaminated the capillaries. Several authors have shown that capillary endothelial cells differ from the endothelial cells of large arteries in their biochemical and functional characteristics, a point to be taken into consideration since it is now well known that the BBB

Frontiers in Cerebral Vascular Biology: Transport and Its Regulation
Edited by L.R. Drewes and A.L. Betz, Plenum Press, New York, 1993

113

is situated in capillaries and not in larger vessels (4,5). Secondly it has been suggested that barrier properties of the BBB may be the consequence of perivascular astrocytes acting on the endothelium they ensheath (6).

For the reasons mentioned above we initiated another strategy to prepare an in vitro model for the BBB (7-9). This technique allowed us to circumvent the culture limitations of primary cultures and quickly provide a large quantity of monolayers. Furthermore, disruption of brain tissue by mechanical dispersion and filtration techniques, without enzymatic treatment, enabled us to isolate a microvascular network. The fact that only capillaries adhere on the extracellular matrix, permits us to clone endothelial cells emerging from these capillaries. In contrast to the enzymatic digestion technique, the obvious advantage of the use of cloned endothelial cells emerging from capillaries is that the culture is not contaminated by endothelial cells of arteriolar or venular origin and pericytes. These cells retain all of the endothelial cell markers including i) factor VIII-related antigen, ii) the non-thrombogenic properties of the monolayer, iii) the production of PGI_2 cells in response to bradykinin, iv) the presence of angiotensin converting enzyme. Furthermore, at confluence, the cells express the characteristics of the BBB including tight junctions, a low rate of pinocytotic vesicles and monoamine oxidase activity, but they lose gamma-glutamyl transpeptidase activity.

In order to reinduce the gamma-glutamyl transpeptidase in bovine brain capillary endothelial (BBCE) cells and to reconstruct some of the complexity of the cellular environment that exits in vivo, we further developed our in vitro model of the BBB by growing endothelial cells on one side of a filter and astrocytes on the other (8). In these conditions, the activity of gamma-glutamyl transpeptidase is more than ten times higher in the two-sided filter coculture compared with BBCE cells cultured alone (35±3 nmol/min/mg of protein, n=9, versus 2.9±0.5 nmol/min/mg of protein, n=12).

Satisfied with the observations that monolayers of BBCE cells in coculture possess many of the properties that are known to have important barrier functions in vivo, we investigated the integrity of our monolayers by their ability to resist the transendothelial passage of electric current. After one week, the resistance of these monolayers in four experiments averaged 661±48 $\Omega.cm^2$ (n=20). When BBCE cells were cultured alone, the resistance was 416±58 $\Omega.cm^2$ (n=9), which is in the same range as that found by Rutten et al. (12).

In order to assess drug transport across the BBB, we have compared the maximal brain extraction E(o) using the in vivo Oldendorf method (13) with the permeability of our two-sided filter coculture (9). We have chosen different compounds transferred by a passive diffusion: some, such as sucrose and inulin with low brain extraction, propranolol for its high extraction. Glucose and leucine were chosen because they cross the BBB by a carrier-mediated system. Moreover, we tested diclofenac and four oxicam-related drugs, isoxicam, meloxicam, piroxicam and tenoxicam because they all have the same pharmacokinetic pattern, which involves high plasma binding percentage and low apparent distribution volumes indicating a poor, if any, brain transfer.

The in vivo E(o) values were plotted against the in vitro Pe values. The results demonstrated a strong correlation as shown by the Spearman's correlation coefficient (r=0.83; p<0.007). The occurrence of tight junctions between capillary endothelial cells could explain the lower values measured in vitro (and in better agreement with those calculated in vivo) than those published by Pardridge et al. (14). The in vitro permeability of glucose and leucine, i.e., compounds that traverse the BBB via carrier mediation performed by specific transport proteins, is in the same range as that of the BBB permeability in vivo, indicating that, in our model, the specific transporters are always present. Furthermore, this in vitro model allows better resolution with compounds that have a low brain extraction. For example, the in vivo E(o) for sucrose and inulin are equal (8%). When we analyzed the permeability of these molecules in vitro, we can see that the permeability of sucrose (0.63×10^{-3} cm/min) is twice that of the permeability of inulin (0.29×10^{-3} cm/min). This point is very important since several carefully performed in vivo studies have shown that peptides can cross the BBB in small amounts (15). Because of the potency of these peptides, such small amounts might indeed be biologically significant.

In conclusion, the strong correlation between the in vivo E(o) and in vitro Pe values demonstrated that this in vitro model will be an important tool for investigations of the role of the BBB in the delivery of drugs to the brain. Furthermore, the relative ease with which such cocultures can be produced in large quantities and the reproducibility of the system should facilitate not only the study of the biochemical mechanisms regulating peptide delivery to the central nervous system at the cellular level, but also provide a more efficient and selective system for the screening of new centrally active peptides.

ACKNOWLEDGMENTS

We thank Isabelle Desmaret for her skillful help and patience in the preparation of the manuscript and Paul Kelly for his editorial help.

REFERENCES

1. Reese, T.S., and Karnovsky, M.J., Fine structural localization of a blood-brain barrier to exogenous peroxidase, *J. Cell Biol.* 34:207-211, 1967.
2. Williams, S.K., Gillis, J.F., Matthew, M.A., Wagner, R.C., and Bitensky, M.W., Isolation and characterization of brain endothelial cells: morphology and enzyme activity, *J. Neurochem.* 35:374-381, 1980.
3. Risau, W., and Wolburg, H., Development of the blood-brain barrier, *TINS* 13:174-178,1990.
4. Zetter, B.R., Endothelial heterogeneity: influence of vessel size, organ localization, and species specificity on the properties of cultured endothelial cells, *in:* "Endothelial Cells," Vol II, U. Ryan, ed., CRC Press Inc., Boca Raton, pp. 63-79, 1988.
5. Joó, F., The cerebral microvessels in culture, an uptake, *J. Neurochem.* 58:1-17, 1992.
6. Davson, H., The blood-brain barrier, *J. Physiol.* (London), 255:1-28, 1976.
7. Méresse, S., Dehouck, M.P., Delorme, P., Bensaïd, M., Tauber, J.P., Delbart, C., Fruchart, J.C., and Cecchelli, R., Bovine brain capillary endothelial cells express tight junctions and monoamine oxidase activity in long-term culture, *J. Neurochem.* 53:1363-1371, 1989.
8. Dehouck, M.P., Méresse, S., Delorme, P., Fruchart, J.C., Cecchelli, R., An easier, reproducible, and mass-production method to study the blood-brain barrier in vitro, *J. Neurochem.* 54:1798-1801, 1990.
9. Dehouck, M.P., Jolliet-Riant, P., Brée, F., Fruchart, J.C., Cecchelli, R., and Tillement, J.P., Drug transfer across the blood-brain barrier: correlation between in vitro and in vivo models, *J. Neurochem.* 58:1790-1797, 1992.
10. Rutten, M.J., Hoover, R.L., and Karnovsky, M.J., Electrical resistance and microvascular permeability of brain endothelial monolayer culture, *Brain Res.* 425:301-310, 1987.
11. Oldendorf, W.H., Measurement of brain uptake of radiolabeled substance using a tritiated water internal, *Brain Res.* 24:372-376, 1971.
12. Pardridge, W.M., Triguero, D., Yang, J., and Cancilla, P.A., Comparison of in vitro and in vivo models of drug transcytosis through the blood-brain barrier, *J. Pharmacol. Exp. Ther.* 253:884-891, 1990.
13. Banks, W.A., and Kastin, A., Permeability of the blood-brain barrier to neuropeptides: the case for penetration, *Psychoneuroendocrin.*10:385-399, 1985.

DIFFERENTIAL BRAIN PENETRATION OF
CEREBROPROTECTIVE DRUGS

Berislav V. Zlokovic, J. Gordon McComb, Milo N. Lipovac,
Thomas C. Chen, Jasmina B. Mackic, John Schneider,
Steven L. Gianotta and Martin H. Weiss

Department of Neurological Surgery and Division of
Neurosurgery, Childrens Hospital Los Angeles,
USC School of Medicine, Los Angeles, CA 90033

INTRODUCTION

The use of cerebroprotective drugs has become a ubiquitous practice in neurosurgery, but our understanding of their pharmacokinetics and mode of action in the brain is limited. It has been speculated that beneficial therapeutic effects of cerebroprotective drugs may depend on their transport across the blood-brain barrier (BBB). To test this hypothesis, we studied two structurally different, but pharmacologically potent, cerebroprotective compounds, namely, a calcium-entry blocker, nimodipine, and a glucocorticoid, methyl-prednisolone. Clinical and experimental evidence suggests that nimodipine may act as a protective agent in ischemic brain damage secondary to cardiac arrest (1). Methyl-prednisolone is commonly used in the treatment of cerebral edema secondary to neoplasia (2,3). In this study, an in situ vascular brain perfusion technique in guinea-pigs (4), and a capillary-depletion method (5) were used to determine pharmacokinetics of nimodipine and methyl-prednisolone at the BBB. The results described below are derived from our ongoing studies.

MATERIALS AND METHODS

Perfused Brain Preparation

The vascular brain perfusion method and a capillary-depletion technique in guinea pigs were previously described (4,6). Briefly, Hartley guinea pigs of either sex and 250- to 300-gm body weight were anesthetized intramuscularly with 6 mg/kg xylazine (Rompun; Mobay KS) and 30 mg/kg ketamine (Ketacet; Aveco, IA) before surgical exposure of the neck vessels. The right common carotid artery was cannulated by a fine polyethylene or nylon catheter. The contralateral carotid artery was ligated and both jugular veins were cut at the time of perfusion to allow free drainage of the perfusate. The perfusion medium

Frontiers in Cerebral Vascular Biology: Transport and Its Regulation
Edited by L.R. Drewes and A.L. Betz, Plenum Press, New York, 1993

117

consisted of 20% washed sheep red blood cells suspended in artificial plasma. The medium was warmed to $37.6°C$ and gassed with 4% CO_2 and 96% O_2 to adjust the pH to 7.36. The perfusion medium was delivered from a reservoir by means of a peristaltic pump. The perfusion pressure was within the physiological range of the arterial blood pressure, and animals maintained spontaneous respiration, normal heart rate, arterial blood pressure, acid-base balance and cerebral flow rate in the perfused ipsilateral brain. The ultrastructural integrity of brain parenchyma and endothelial capillary wall remained unchanged (7). The water content, Na/K ratio, ATP, energy charge potential and lactate levels were normal.

Brain Uptake of Cerebroprotective Drugs

[3H]Nimodipine or [3H]methyl-prednisolone were introduced into cerebral circulation by vascular brain perfusion technique within the 10-min period. Either of two reference standards, a cerebrovascular space marker, [14C]sucrose, or cerebral blood flow marker, [14C]iodoantipyrine, were also included. The effect of unlabeled drug and plasma proteins were also examined.

Mathematical Treatment

Kinetic data were analyzed by previously published mathematical treatments, and rates of drug entry into different BBB regions (e.g., cortex, hippocampus, caudate) and/or brain compartments (capillary-depleted brain, vascular) were expressed as unidirectional transport constants, K_{in}. Cerebroprotective drug K_{in} values in the presence of unlabeled drug were used to calculate Michaelis-Menten parameters, V_{max} and K_m, for the saturable component of cerebral uptake, and K_d, for the non-saturable diffusion component (6).

RESULTS AND DISCUSSION

Figure 1 illustrates cerebral tissue uptake of [3H]nimodipine (A) and [3H]methyl-prednisolone (B) determined in different brain compartments after 1 and 10 min of brain perfusion. It can be seen that both drugs exhibit time-dependency, but their volume of distributions were markedly different. Nimodipine uptake in brain homogenates amounted to about 27% of its free drug plasma concentration after 10 min, as opposed to only about 1% found for methyl-prednisolone. This has been confirmed by regional analysis, and brain uptake values of nimodipine in the parietal cortex, caudate nucleus and hippocampus ranged between 30 and 40% of its plasma concentration; that was only 2 to 2.7 times less than the BBB uptake of the highly penetrating cerebral blood flow marker, iodoantipyrine (data not shown). Regional rates of nimodipine entry across the BBB determined from multiple-time brain uptake plots were 100 to 250 times higher than for sucrose (data not shown). K_{in} values estimated for [3H]nimodipine (2 nM) in the cortex, hippocampus and caudate nucleus were within a range between 22.4 ± 1.30 and 28.0 ± 1.90 µl/min/g. Regional brain uptake of free [3H]methyl-prednisolone (4 nM) was significantly lower in comparison to nimodipine (30 to 45 times), but progressive and time-dependent relative to the cerebrovascular space marker, sucrose (~3-6 times higher).

As indicated by brain compartmental analysis and shown in Figure 1, a cerebral capillary sequestration was a relatively significant portion in the cerebral uptake of both drugs, although the absolute values of nimodipine and methyl-prednisolone distribution into this compartment differ markedly. It seems that capillary sequestration precedes the entry of cerebroprotective drugs into brain parenchyma, as illustrated by their distribution in the capillary-depleted brain compartment.

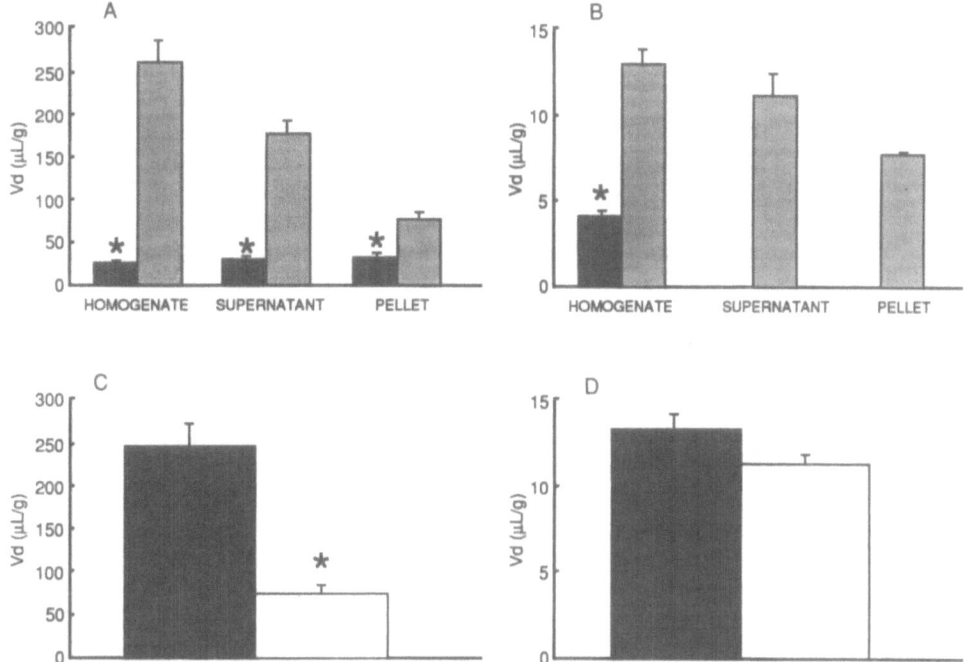

Figure 1. The volume of distribution, V_d, of radiolabeled **A.** nimodipine (2 nM) and **B.** methyl-prednisolone (4 nM) determined in brain homogenate, capillary-depleted brain tissue (supernatant) and isolated cerebral capillaries (pellet), after 1 min (black bars) and 10 min (patterned bars) of brain perfusion. Effect of 100% guinea-pig plasma on radiolabeled **C.** nimodipine and **D.** methyl-prednisolone brain uptake determined after 10 min of perfusion. Values are mean $\pm$ SEM, n = 3.

Figures 1C and D illustrate different effects of plasma proteins on cerebroprotective drug uptake. Namely, nimodipine uptake was highly sensitive to changes in plasma protein concentration, as indicated by the fivefold reduction of its volume of distribution in the presence of 100% plasma. In contrast, 100% plasma did not affect methyl-prednisolone uptake. Our ongoing experiments suggest that both nimodipine and methyl-prednisolone BBB in situ binding and transport exhibit saturation kinetics. A strong diffusion component was observed in the case of nimodipine, but non-saturable diffusion represented a negligible part of methyl-prednisolone brain uptake.

It is concluded that both cerebroprotective drugs penetrate into the brain, although their transport rates were found to be significantly different, i.e., slow for methyl-prednisolone and rapid and flow-limited for nimodipine, as determined from their free drug plasma pools. However, other factors, such as binding to plasma proteins, may significantly alter their blood-brain transport. The presence of nimodipine- and methyl-prednisolone-specific mechanisms at the luminal side of the BBB and significant binding to cerebral microvessels is suggested for both drugs.

ACKNOWLEDGMENTS

This work was supported by Childrens Hospital Los Angeles.

REFERENCES

1. Plum, F., Vulnerability of the brain and heart after cardiac arrest, *New Engl. J. Med.* 18:1278-1280, 1991.

2. French, L.A., The use of steroids in the treatment of cerebral edema, *NY Acad. Med.* 42:301-311, 1966.

3. Giannotta, S.L., Weiss, M.H., Apuzzo, M.L., and Martin E., High dose glucocorticoids in the management of severe head injury, *Neurosurg.* 15:497-501, 1984.

4. Zlokovic, B.V., Begley, D.J., Djuricic, B.M., and Mitrovic, D.M., Measurement of solute transport across the blood-brain barrier in the perfused guinea pig brain: method and application to N-methyl-α-aminoisobutyric acid, *J. Neurochem.* 46:1444-1451, 1986.

5. Triguero, D., Buciak, J.B., and Pardridge, W.M., Capillary depletion method for quantifying blood-brain barrier transcytosis of circulating peptides and plasma proteins, *J. Neurochem.* 54:1882-1888, 1990.

6. Zlokovic, B.V., In vivo approaches for studying peptide interactions at the blood-brain barrier, *J. Controlled Release* 13:185-202, 1990.

7. Zlokovic, B.V., McComb, J.G., Perlmutter, L., Weiss, M.H., and Davson, H., Neuroactive peptides and amino acids at the blood-brain barrier: possible implications to drug abuse, *in:* "Drug Availability and the Blood-Brain Barrier," J. Frankenheim, and R. Brown, eds., NIDA Research Monographs, Washington DC, Government Publication, pp. 26-42, 1992.

P-GLYCOPROTEIN IS STRONGLY EXPRESSED IN BRAIN CAPILLARIES

Lucie Jetté and Richard Béliveau

Laboratoire de Membranologie
Groupe de Recherche en Biothérapeutique Moléculaire
Département de Biochimie, Université du Québec à Montréal
C.P. 8888, Succ. A, Montréal (Québec), Canada, H3C 3P8

INTRODUCTION

The development of resistance to multiple chemotherapeutic drugs is a major obstacle in the clinical treatment of many human cancers. Cultured cells grown in medium containing selected drugs can acquire cross resistance to a remarkably wide range of compounds that have no obvious structural or functional similarities (1). A multidrug resistance (MDR) phenotype has been described, which consists of the amplification of members of a small gene family and the increased synthesis of a 4.5 - 6.0 kb mRNA (2,3). These alterations are associated with the overexpression of a glycoprotein (P-glycoprotein or P-gp) of relative molecular mass (Mr) 170,000-180,000 (4). P-gp is likely to function as an energy-dependent drug efflux pump (5) and to possess notable similarities with bacterial transport proteins (6).

The human and mouse P-glycoprotein consist of 1280 and 1276 homologous amino acid sequences respectively (6,7), with 12 predicted transmembrane segments and potential N-terminal glycosylation sites (8). P-gp is encoded by three mdr genes in rodents and two mdr genes in humans (9,10). In humans, only the mdr1 gene has been reported to be expressed in cancers (11). Single amino acid substitution in human mdr1 (12) can alter the activity and substrate specificity of this efflux pump. Recently, a possible role for the human mdr3 gene has been proposed to explain cytotoxic drug resistance in B-cell lymphocytic leukemias (13).

The mdr gene products are expressed not only in MDR tumor cells, but also in normal tissues such as hepatocytes, brush border membrane (BBM) of renal proximal tubules, adrenal glands, intestine, pancreas and gravid uterus (14,15). In this paper, we have used the monoclonal antibody C219 (mAb C219) in Western blot analysis to demonstrate clearly the presence of P-gp in isolated brain capillaries from different species.

Frontiers in Cerebral Vascular Biology: Transport and Its Regulation
Edited by L.R. Drewes and A.L. Betz, Plenum Press, New York, 1993

121

MATERIALS AND METHODS

Isolation of Brain Capillaries

Fresh bovine brains were obtained from a local abattoir and were transported to the laboratory in ice-cold physiological buffer (PB) containing of 147 mM NaCl, 4 mM KCl, 3 mM CaCl$_2$, 1.2 mM MgCl$_2$, 5mM glucose and 15 mM Hepes, pH 7.4. Normal human brains were obtained post mortem from patients less than one year old. The brain capillaries were isolated the same day of the death. Brain capillaries were also isolated from male Sprague-Dawley rats. For all these preparations, the brains were cleared of meninges, superficial large blood vessels and choroid plexus. The cerebral cortex was homogenized in five volumes of PB with a Polytron (Brinkman Instruments). The homogenates were mixed with an equal volume of 26%, 28% or 30% (w/v) Dextran T-70 in PB, for bovine, human and rat cortex, respectively. Brain capillary purification was then carried out according to the procedure of Dallaire et al. (16). The final pellets containing isolated microvessels were resuspended in PB and stored in liquid nitrogen until use.

Detection of P-Glycoprotein

P-Glycoprotein was detected by Western blot analysis. Sodium dodecyl sulfate (SDS) gel electrophoresis was performed according to the method of Laemmli (17). Capillaries were suspended in sample buffer to a final protein concentration of 1 mg/ml and loaded onto 7.5% acrylamide:bis-acrylamide (29.2:0.8) gels, with or without prior heating. The gels were electrophoretically transferred to Immobilon-P membranes (Millipore, 0.45mm). Blots were blocked overnight at 4°C, with 3% bovine serum albumin (BSA) in Tris-buffered saline (TBS) containing 50 mM Tris, 150 mM-NaCl, 2 mM-NaN$_3$, 0.05% (v/v) Tween-20, pH 7.0. The membranes, washed three times with 0.1% BSA in TBS, were incubated with the monoclonal antibody C219, 400 ng/ml, for 2 hr at 37°C. The membranes were then incubated with ^{125}I-labeled goat anti-mouse IgG, at a concentration of 10^6 cpm/ml, for 2 hr at room temperature. These two antibodies were diluted in TBS containing 1% BSA. The blots were exposed to Fuji film at -80°C for 1-3 days, and scanned with an LKB Ultroscan XL laser densitometer. Molecular weight determination was performed using the following standards: myosin (200 kDa), β-galactosidase (116 kDa), phosphorylase b (97 kDa), BSA (66 kDa), and ovalbumin (43 kDa).

RESULTS

The monoclonal antibody C219 has been shown to be directed toward peptide epitopes representing positions 568-574 and 1213-1219 in the amino acid sequence of P-gp (18). We used this antibody to determine the presence of P-glycoprotein in human, bovine and rat brain capillaries. A membrane preparation from a colchicine-resistant CHO cell line (CH^{R}C5, Centocor Diagnostics) was used as control (1). Figure 1 shows the immunodetection of P-gp in the human, bovine and rat isolated capillaries and in the positive membrane control. A protein with a molecular weight of 180,000 kDa was detected in all samples. Levels of the P-gp were the highest in the human capillaries (Figure 1, H) and the lowest in bovine capillaries (Figure 1, B).

A crossreacting protein was detected in BBM prepared from rat kidney cortex (Figure 2). The molecular weight, as determined by SDS gel electrophoresis, was 155 kDa. The level of expression of P-gp detected by mAb C219 was much higher in brain capillaries than in kidney cortex BBM isolated from these two species. A longer incubation time with the antibody C219 (16 hr) was necessary to detect P-gp in rat kidney cortex BBM. Under these conditions, no reactivity was observed with bovine kidney cortex BBM.

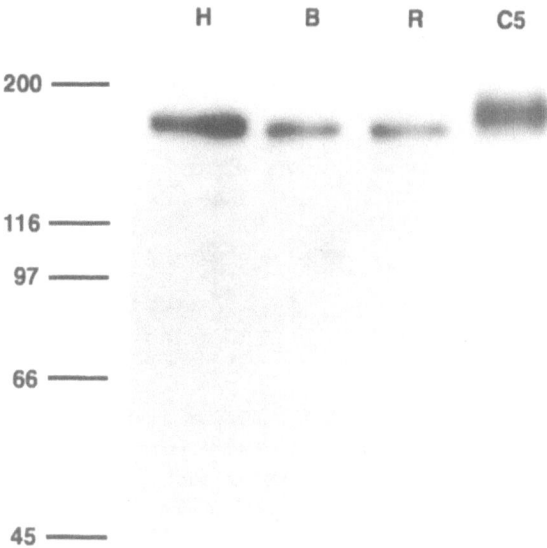

Figure 1. Detection of P-gp in isolated brain capillaries. Immunoblots were performed as described in Materials and Methods. Protein samples (10 mg) from human (H), bovine (B) and rat (R) isolated brain capillaries and CH^{R}C5 membranes (C5) were resolved by SDS gel electrophoresis using a 7.5% polyacrylamide gel. Monoclonal antibody C219 binding to the antigen was detected using ^{125}I-goat anti-mouse IgG. The molecular weight markers indicated to the left of the figure are myosin (200 kDa), β-galactosidase (116 kDa), phosphorylase b (97 kDa), BSA (66 kDa), and ovalbumin (45 kDa). (n=3).

DISCUSSION

We have detected a high level of P-gp expression in human, bovine and rat isolated brain capillaries by Western blot analysis (Figure 1), in agreement with the immunohistochemical studies reporting its presence in brain capillaries (20). A molecular weight of 180,000 kDa has been determined for these immunoreactive proteins. The protein detected in the C5 plasma membrane extract corresponds to the P-glycoprotein that is greatly enriched in the plasma membrane of this drug-resistant cell line (4). Our results show clearly that P-glycoprotein is expressed at high levels in isolated brain capillaries. The detected levels of P-gp vary from one species to another, even after correction for enrichment factors of each preparation (not shown). Brain capillaries contain much more of the immunoreactive protein than BBM prepared from renal cortex from the same species. A molecular weight of 155,000 kDa has been obtained for the P-gp detected in rat renal cortex BBM, which is in agreement with the molecular weight already reported for P-gp expressed in this tissue (21). It was necessary to over-develop the blot with respect to the P-glycoprotein expressed in the isolated brain capillaries in order to detect the cross-reacting protein in the rat renal cortex BBM.

Human mdr1 and mdr3 mRNAs have been detected in normal tissues such as liver, kidney, adrenal gland and spleen (22). Furthermore, Fojo et al. (14) detected higher levels of human mdr1 mRNA in kidney than in brain samples. Our Western blot results show that P-gp is expressed at higher levels in isolated rat brain capillaries than in kidney cortex BBM isolated from the same species. Therefore, it would be of great interest to establish whether P-gp class I isoform is the only mdr gene product expressed in the capillaries of the blood-brain barrier.

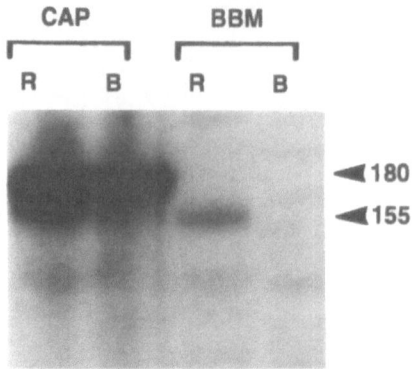

Figure 2. Comparison of P-gp expression in brain capillaries and in BBM isolated from kidney cortex. Protein samples (20 mg) from rat (R) and bovine (B) isolated brain capillaries (CAP) and renal brush border membranes (BBM) were resolved by SDS gel electrophoresis using 7.5% polyacrylamide gel. BBM were isolated according to the method of Booth and Kenny (19). Immunoblots were performed as described in Materials and Methods and antibody binding was detected using ^{125}I-goat anti-mouse IgG. The arrows indicate position of the P-glycoprotein from the isolated brain capillaries (180 kDa) and from the rat kidney cortex BBM (155 kDa). The molecular weight markers used were the same as in Figure 1.

In conclusion, our results clearly show that P-gp is endogenously expressed in brain capillaries and that the level of expression of this glycoprotein varies among the species. The strong P-glycoprotein expression in this non-cancerous tissue may indicate a physiological role for P-gp in regulating the entry of certain molecules into the central nervous system or in the secretory functions of the blood-brain barrier. P-gp would thus contribute to the barrier function of these endothelial cells. Understanding the role of P-gp in brain capillaries may help to design new drugs to circumvent chemotherapeutic resistance to treatment of infectious diseases and brain tumors.

ACKNOWLEDGMENTS

This work was supported by grants from the Natural Sciences and Engineering Research Council of Canada and by Sandoz. L. J. is a recipient of a fellowship of the Natural Sciences and Engineering Research Council of Canada. Many thanks to Dr. Jean-Marie Leclerc.

REFERENCES

1. Ling, V., and Thompson, L.H., Reduced permeability in CHO cells as a mechanism of resistance to colchicine, *J. Cell. Physiol.* 83:103, 1974.

2. Croop, J.M., Gros, P., and Housman, D.E., Genetics of multidrug resistance, *J. Clin. Invest.* 81:1303, 1988.

3. Fojo, A.T., Whang-Peng, J., Gottesman, M.M., and Pastan, I., Amplification of DNA sequences in human multidrug-resistant KB carcinomas cells, *Proc. Natl. Acad. Sci. U.S.A.* 82:7661, 1985.

4. Kartner, N., Riordan, J.R. and, Ling, V., Cell surface P-glycoprotein associated with multidrug resistance in mammalian cell lines, *Science* 221:1285, 1983.

5. Hamada, H., and Tsuruo, T., Characterization of the ATPase activity of the Mr 170,000 to 180,000 membrane glycoprotein (P-glycoprotein) associated with multidrug resistance in K562/ADM cells, *Cancer Res.* 48:4926, 1988.

6. Chen, C.-J., Chin, J.E., Ueda, K., Clark, D.P., Pastan, I., Gottesman, M.M., and Roninson, I.B., Internal duplication and homology with bacterial transport proteins in the mdr1 (P-glycoprotein) gene from multidrug-resistant human cells, *Cell* 47:381, 1986.

7. Gros, P., Croop, J., and Housman, D., Mammalian multidrug resistance gene: complete cDNA sequence indicates strong homology to bacterial transport proteins, *Cell* 47:371, 1986.

8. Yoshimura, A., Kuwazuru, Y., Sumizawa, T., Ichikawa, M., Ikeda, S.-I., Uda, T., and Akiyama, S.-I., Cytoplasmic orientation and two-domain structure of the multidrug transporter, P-glycoprotein, demonstrated with sequence-specific antibodies, *J. Biol. Chem.* 264:16282, 1989.

9. Ng, W.F., Sarangi, F., Zastawny, R.L., Veinot-Drebot, L. and Ling, V., Identification of members of the P-glycoprotein multigene family, *Mol. Cell. Biol.* 9:1224, 1989.

10. Croop, J.M., Raymond, M., Haber, D., Devault, A., Arceci, R.J., Gros, P., and Housman, D.E., The three mouse multidrug resistance (mdr) genes are expressed in a tissue-specific manner in normal mouse tissues, *Mol. Cell. Biol.* 9:1346, 1989.

11. Van der Bliek, A.M., Baas, F., Van der Velde-Koerts, T., Biedler, J.L., Meyers, M.B., Ozols, R.F., Hamilton, T.C., Joenje, H., and Borst, P., Genes amplified and overexpressed in human multidrug-resistant cell lines, *Cancer Res.* 48:5927, 1988.

12. Safa, A.R., Stern, R.K., Choi, K., Agresti, M., Tamai, I., Mehta, N.D., and Roninson, I.B., Molecular basis of preferential resistance to colchicine in multidrug-resistant human cells conferred by gly-185 Æ val-185 substitution, *Proc. Natl. Acad. Sci. U.S.A.* 87:7225, 1990.

13. Herweijer, H., Sonneveld, P., Baas, F., and Nooter, K., Expression of mdr1 and mdr3 multidrug-resistance genes in human acute and chronic leukemias and association with stimulation of drug accumulation by cyclosporine, *J. Natl. Cancer Inst.* 82:1133, 1990.

14. Fojo, A.T., Ueda, K., Slamon, D.J., Poplack, D.G., Gottesman, M.M., and Pastan, I., Expression of a multidrug-resistance gene in human tumors and tissues, *Proc. Natl. Acad. Sci. U.S.A.* 84:265, 1987.

15. Arceci, R.J., Croop, J.M., Horwitz, S.B., and Housman, D., The gene encoding multidrug resistance is induced and expressed at high levels during pregnancy in the secretory epithelium of the uterus, *Proc. Natl. Acad. Sci. U.S.A.* 85:4350, 1988.

16. Dallaire, L., Tremblay, L., and Béliveau, R., Purification and characterization of metabolically active capillaries of the blood-brain barrier, *Biochem. J.* 276:745, 1991.

17. Laemmli, U.K., Cleavage of structural proteins during the assembly of the head of bacteriophage T4, *Nature* (London) 227:680, 1970.

18. Georges, E., Bradley, G., Gariepy, J., and Ling, V., Detection of P-glycoprotein isoforms by gene-specific monoclonal antibodies, *Proc. Natl. Acad. Sci. USA.* 87:152, 1990.

19. Booth, A., and Kenny, A.J., A rapid method for the preparation of microvilli from rabbit kidney, *Biochem. J.* 142:575, 1974.

20. Cordon-Cardo, C., O'Brien, J.P., Casals, D., Rittman-Grauer, L., Biedler, J.L., Melamed, M.R., and Bertino, J.R., Multidrug-resistance gene (P-glycoprotein) is expressed by endothelial cells at blood-brain barrier sites, *Proc. Natl. Acad. Sci. U.S.A.* 86:695, 1989.

21. Lieberman, D.M., Reithmeier, R.A.F., Ling, V., Charuk, J.H.M., Goldberg, H., and Skorecki, K.L., Identification of P-glycoprotein in renal brush border membranes, *Biochem. Biophys. Res. Commun.* 162:244, 1989.

22. Chin, J.E., Soffir, R., Noonan, K.E., Choi, K., and Roninson, I.B., Structure and expression of the human mdr (P-glycoprotein) gene family, *Mol. Cell. Biol.* 9:3808, 1989.

Protein Transport

TRANSPORT OF PROTEINS ACROSS
THE BLOOD-BRAIN BARRIER VIA
THE TRANSFERRIN RECEPTOR

Phillip M. Friden and Lee R. Walus

Alkermes, Inc.
64 Sidney Street
Cambridge, MA 02139

INTRODUCTION

Unlike most other organs in the body, the brain is separated from the blood by a protective cellular barrier known as the blood-brain barrier (BBB). The BBB, although essential in maintaining a defined biochemical environment within the brain, represents a formidable obstacle to the effective delivery of neuropharmaceutical agents from the bloodstream. The capillaries that supply blood to the tissues of the brain constitute this barrier (1,2). Brain capillary endothelial cells are joined together by tight intercellular junctions that form a continuous wall against the passive movement of substances from the blood to the brain. Also characteristic of these cells is a paucity of pinocytic vesicles, which limits the amount of non-selective fluid-phase transport across the capillary wall. Together, these features limit the penetration of blood-borne hydrophilic molecules into brain tissue.

However, in order to function properly the brain does require, in a controlled manner, a variety of compounds such as glucose, amino acids and neuroactive peptides. To provide these and other compounds to the brain, the capillary endothelial cells that comprise the BBB possess a variety of receptors and transport systems that are responsible for providing controlled access to the brain for a variety of specific compounds (3-6). In many instances, these transport systems consist of membrane-associated receptors that, upon binding of their respective ligands, are internalized by the cell. The internalized ligands, either as receptor-ligand complexes or free molecules, are then directed to the abluminal membrane and released into the interstitial fluid of the brain. We have taken advantage of one of these transport systems, that for the iron transport protein transferrin, to develop a drug delivery system for the brain.

Serum transferrin is a monomeric glycoprotein with a molecular weight of 80,000 daltons that is involved in the transport of iron throughout the body (7,8). The uptake of iron by individual cells is mediated by the transferrin receptor, an integral membrane glycoprotein consisting of two identical 95,000 dalton subunits linked by a disulfide bond (9,10). In the cell lines studied to date, iron is released from transferrin in an acidified intracellular compartment following internalization of the transferrin-receptor complex (11). The

Frontiers in Cerebral Vascular Biology: Transport and Its Regulation
Edited by L.R. Drewes and A.L. Betz, Plenum Press, New York, 1993

apotransferrin remains bound to the receptor and the complex recycles to the cell surface where the apotransferrin is released. Because they are thought to be involved in the transport of iron-transferrin across the BBB, brain capillary endothelial cells must possess a pathway in which iron is transferred through the cell and released into the brain parenchyma, as opposed to the defined pathway in which iron is deposited within the cell (3,4,6). This would suggest that molecules that bind to the transferrin receptor could be transported across the BBB. The number of transferrin receptor molecules expressed on the surface of a cell is typically related to the metabolic activity of the cell, with actively dividing cells having higher receptor levels than terminally differentiated cells. However, it has been reported (12) that brain capillary endothelial cells have a high density of transferrin receptor on their surface even though they are not in a proliferative state. This observation is consistent with the role that these cells play in serving as a conduit for substances that are transported into the brain parenchyma. Sufficient numbers of receptors must be present at the BBB to transport from the blood the iron that is required by the cells of the brain. As a result, selective targeting to the brain via the transferrin receptor may be possible.

We have demonstrated that it is possible to utilize the transferrin receptor to deliver small molecules across the BBB (13). To avoid competition with the relatively high levels of transferrin in serum, antibodies that bind to the transferrin receptor at a site distinct from that for transferrin were utilized as carrier molecules. It was shown that following in vivo administration into rats via the tail vein, the OX-26 murine monoclonal antibody, which binds to the rat transferrin receptor, selectively accumulated in the vasculature of the brain in a dose-dependent manner. The localization of the antibody in the brain vasculature, which was uniform in appearance at 1 hr after injection, became quite punctate in appearance between 2 and 4 hr after injection. This time-dependent alteration in staining pattern suggested that the injected antibody had become sequestered in an intracellular compartment within brain capillary endothelial cells. The results of capillary depletion experiments, in which the distribution of radiolabeled antibody within the capillary and parenchyma fractions of the brain were examined over time, suggested that the antibody underwent transcytosis across the BBB. To demonstrate the ability of the antibody to act as a carrier, methotrexate (MTX) was conjugated to the antibody. Experiments with the OX-26 - MTX conjugate demonstrated that the antibody could still target to the brain vasculature following conjugation and that it could deliver MTX across the BBB. Based on these results, we have examined the ability of this carrier to deliver therapeutic proteins across the BBB.

DELIVERY OF NERVE GROWTH FACTOR ACROSS THE BBB

Nerve growth factor (NGF) is a 26,000 dalton protein that has been shown in vitro and in vivo to support the growth of basal forebrain cholinergic neurons (14-18). Degeneration of cholinergic neurons is one of the principal pathologies associated with Alzheimer's disease (19). These cells, which are localized in the nucleus basalis of Meynert and the medial septal nucleus, innervate the neocortex, hippocampus and amygdala in response to the NGF produced by these target areas. NGF binds to surface receptors on cholinergic neurons, is internalized and undergoes retrograde transport to the cell body where it, or a second messenger produced in response to the NGF-receptor interaction, mediates the trophic activity associated with this growth factor (14,20). It has been shown, both in rodents and primates, that if this process is interrupted by transection of the fimbria fornix, which results in a lesion in the septo-hippocampal pathway, the cholinergic neurons in the basal forebrain will degenerate (16,21-23). If exogenous NGF is administered by intracerebroventricular (ICV) infusion to the lesioned animals, the atrophy of cholinergic neurons can be prevented. These results clearly demonstrate the importance of NGF in the survival of these neurons.

Although changes in other neurotransmitter systems are associated with Alzheimer's disease, the loss of cholinergic function is most closely associated with impairment of cognitive and memory processes (24,25). This suggests that the loss of cholinergic neurons is responsible, at least in part, for the clinical symptoms characteristic of Alzheimer's disease and that NGF could play a therapeutic role in the treatment of the disease. In support of this theory, experiments in aged rats have shown that continuous intraventricular infusion of NGF leads to the amelioration of cholinergic neuron atrophy and spatial memory impairment (26).

Taken together, these data suggest a strong scientific rationale for the testing of NGF as a therapeutic in the treatment of Alzheimer's disease. Currently, a significant obstacle to clinical testing of NGF is the lack of an efficient, noninvasive means for delivering the protein across the BBB to the target cells. The only methodology currently in use for the delivery to the brain of proteins such as NGF is intraventricular infusion using an external pump. We have examined the feasibility of utilizing an anti-transferrin receptor antibody as a carrier to transport NGF across the BBB. If successful, this technology would provide a means for the delivery of NGF, as well as other protein therapeutics, to the brain via an intravenous route of administration and could enhance significantly their therapeutic potential.

Conjugate Synthesis

Our approach to the synthesis of OX-26 - NGF conjugates involved the introduction of a protected sulfhydryl group onto one of the constituents and a heterobifunctional cross-linker containing a thiol reactive group onto the other. The distinct advantage in using a heterobifunctional cross-linking agent is that the coupling is performed in a step-wise manner, thus avoiding the formation of homoprotein polymers.

The first step in the conjugation strategy is the modification of NGF through carboxylic acid groups (27). This was achieved using PDP (pyridyldithiopropionate)-hydrazide. This compound is a derivative of SPDP [N-succinimidyl 3-(2-pyridyldithio)propionate] (28) in which the NHS ester has been replaced by a hydrazide group. The carboxylates are first modified with EDC (1-ethyl-3-(3-dimethylaminopropyl)carbodiimide) to make them susceptible to attack by the hydrazide group. The second step in the conjugation process was the introduction of a protected thiol group onto the carrier antibody. This was achieved using SATA (N-succinimidyl S-acetylthioacetate), which reacts with the ε amines of lysine side chains of the antibody. Treatment with hydroxylamine removes the protecting group and reveals the free sulfhydryl. The extent of protein modification was carefully monitored and the reaction conditions were such that the number of groups attached to both the antibody and NGF were kept to a minimum (approximately 3 and 1, respectively). OX-26 - NGF conjugates were prepared by reacting derivatized NGF with the deprotected OX-26. In general, the reaction conditions used have generated a 1:1 antibody-NGF conjugate.

Purification of the OX-26 - NGF conjugate away from unreacted starting materials was achieved using a two-step affinity chromatography procedure. The crude conjugation mixture was first passed over a protein-A Sepharose column that retained antibody and antibody-NGF conjugate and allowing unreacted NGF to flow through. The material eluted from this column was then applied to an anti-NGF affinity column. Unreacted antibody flowed through this column while antibody-NGF conjugate was retained.

The conjugation and purification processes were monitored by analyzing samples taken at various steps using SDS-polyacrylamide gel electrophoresis. Gels run under reducing conditions, in which the disulfide linkage between the antibody and NGF was cleaved, were used to demonstrate the presence of both NGF and antibody in the conjugate eluted from the NGF affinity column. Gels run under nonreducing conditions, in which the disulfide linkage between the antibody and NGF was not cleaved, were used to show that NGF (MW 26,000 daltons) was now migrating on the gel at a molecular weight corresponding to ~180,000 daltons. NGF was specifically identified on these gels in two ways. In the first, the gel was blotted to a nylon membrane and then probed using an anti-NGF antibody. In addition to the immunoblots, autoradiography of gels was used to identify NGF in cases when the conjugate was synthesized using ^{3}H-NGF with a high specific activity. A 2-site ELISA was also used to demonstrate the composition of the purified conjugate. This ELISA was formatted such that only conjugate consisting of NGF and antibody would give rise to a positive signal. Together, these results demonstrated that the purified conjugate consisted of NGF that was covalently attached to the antibody via a reducible linkage.

In Vitro Biological Activity

A critical criterion for evaluating the OX-26 - NGF conjugates is the retention of the biological activity of the two components. To ascertain the effects of conjugation on the biological activity of NGF, the PC-12 cell neurite outgrowth assay was employed. The PC-12 cell line was originally derived from a rat pheochromocytoma (29). In response to stimulation with NGF, these cells undergo a reversible differentiation in which the cells take on a neuronal phenotype. The morphological and biochemical changes observed in these cells after exposure to NGF include the extension of axonal-like processes (neurites) and an increase in the levels of choline acetyltransferase (ChAT) and acetyl cholinesterase.

The extension of neurites in response to exogenous NGF can be used as the basis for a semi-quantitative assay for NGF biological activity (30). The percentage of cells that extend neurites can be determined and expressed as a function of NGF concentration, thus generating a dose-response curve. A comparison of the curves for the modified NGF and the OX-26 - NGF conjugates to that for an unaltered NGF standard was used to ascertain the effects of the conjugation process on the biological activity of the NGF.

In this assay, PC-12 cells are plated onto bovine type IV collagen-coated 96-well microtiter plates. Six hours after plating, the cells were exposed to native NGF, linker modified NGF or OX-26 - NGF conjugate. To generate a dose response curve, the samples were diluted in growth medium serially in 2-fold increments to cover a wide range of concentrations. After 5 days of exposure to the samples, the plates were scored by counting the total number of cells and the number of cells with neurites longer than 2 cell diameters in length.

When PDP-hydrazide-modified NGF and OX-26 - NGF conjugates prepared according to the scheme outlined above were compared to unmodified NGF in the PC-12 neurite outgrowth assay, there was no difference in activity (27). Thus, we were able to prepare a conjugate for the delivery of NGF across the blood-brain barrier in which NGF retained its biological activity.

In Vivo Analysis

Previous work has shown that the anti-rat transferrin receptor antibody OX-26 can be detected immunohistochemically in the brain vasculature following i.v. administration (13). Similar studies were performed using the OX-26 - NGF conjugate to demonstrate the ability of the carrier antibody to target proteins to the brain vasculature (27). For these experiments, 200 mg of conjugate per rat was injected i.v. via the tail vein. At 1 hr following injection, the animals were perfused with phosphate buffered saline (PBS) to flush out the blood and the animals sacrificed. The brains were removed immediately and frozen in liquid nitrogen. Brain sections were cut on a cryostat, melted onto gelatin-coated slides, air dried and fixed in acetone at room temperature. Immunohistochemistry was performed as described using an avidin/biotin/HRP system (ABC) with 3,3'-diaminobenzedine as the chromogenic substrate (13). Anti-mouse IgG was used to detect the carrier antibody and anti-NGF antibody was used to detect the passenger protein.

The immunohistochemistry results obtained with the conjugate were similar to those obtained with antibody alone, indicating that the attachment to the antibody of NGF did not appreciably alter its ability to localize to the brain vasculature following i.v. injection (13). When these sections were probed with the anti-NGF antibody, staining of the vasculature was also observed, demonstrating the co-localization of NGF and the antibody in the brain. When brain sections from animals injected with an equivalent amount of unconjugated NGF were stained for NGF, no labeling of the vasculature was observed. Thus, the localization of NGF to the brain capillaries was dependent on the covalent attachment of the protein to the carrier antibody, OX-26.

The results of the immunohistochemistry experiments described above demonstrate qualitatively that the OX-26 antibody can deliver NGF to the brain vasculature following i.v. administration. To ascertain the quantity of NGF that is delivered to the brain, as well as whether the NGF is taken across the BBB via the transferrin receptor, capillary depletion experiments were performed (31). This type of experiment was used previously to show that unconjugated anti-transferrin receptor antibody and antibody-methotrexate conjugates undergo transcytosis across the BBB (13). For the experiments described here, OX-26 - NGF conjugate was prepared in which the NGF was radiolabeled with ^{3}H to a high specific activity. The radiolabeled conjugate was injected into rats i.v. via the tail vein and animals were sacrificed at various times post injection. The brain was then removed and homogenized, and an aliquot of the homogenate was taken for analysis. By scintillation counting of the brain homogenate sample, the total amount of NGF associated with the brain as a function of time post injection could be determined. It was found that the peak amount of conjugate associated with the brain was between 0.6 and 0.7% of the injected dose (27).

The remaining brain homogenate was then subjected to density gradient centrifugation through 12% dextran. This centrifugation causes the brain capillary fragments to pellet, leaving the remaining brain tissue components in the supernatant. An examination of the radioactivity associated with the pellet (brain capillaries) and the supernatant (brain parenchyma) as a function of time can allow one to establish whether transcytosis is occurring. If the conjugate is simply binding to or accumulating in the capillary endothelial cells, radioactivity would accumulate in the pellet fraction of the brain with little or no radioactivity detected in the supernatant. However, if transcytosis is occurring, the radioactivity would initially be associated with the capillary pellet and would then decrease over time while the amount of radioactivity associated with the parenchyma would increase.

The results that we obtained with the OX-26 - NGF conjugate are consistent with the mechanism of transcytosis across the BBB (27). Initially after injection, the majority of the radiolabel is associated with the capillary fraction. This represents material bound to and internalized by rat brain capillary endothelial cells. Between 30 min and 4 hr following injection, the amount of radiolabel associated with the capillary fraction of the brain decreases substantially while the amount associated with the brain parenchyma fraction increases. These data indicate that between 0.4 and 0.5% of the injected dose of NGF (in the form of conjugate) reaches the brain parenchyma and that the NGF appears to remain in the brain for at least 24 hr post injection. This is in comparison to unconjugated NGF which, when injected i.v., gives rise to a very low level (~0.08 % of the injected dose) in the brain of protein, which appears to be cleared more quickly than the conjugate NGF. Thus, it appears that the anti-transferrin receptor antibody is an effective vehicle for the delivery of NGF to the brain.

An in vivo efficacy model has been utilized to evaluate the ability of this carrier system to deliver therapeutically relevant amounts of NGF to its target cells within the brain. This experimental paradigm involves examination of the effect of the OX-26 - NGF conjugate on fetal medial septal nucleus tissue that has been implanted into the anterior chamber of the eye (32-34). As mentioned above, the medial septum is rich in cholinergic neurons, which are dependent on NGF for their survival. This experimental system provides a means for studying the effects of NGF on the developing cholinergic neurons in the septum in isolation from other CNS influences. The advantages of this model are that the tissue is completely isolated from all sources of NGF other than that administered experimentally and that cholinergic neurons in fetal tissue are more sensitive to NGF than are those in adult tissue. This experimental system has been used to demonstrate that the survival during development of target-deprived basal forebrain cholinergic neurons in the anterior chamber of the eye is promoted by NGF administered into the intraocular fluid (35,36).

For the implant experiments, 2 mm^3 pieces of medial septal nucleus were dissected from rat fetuses at gestational day E18 and bilaterally grafted to the anterior chamber of the eye of adult rats (33,34). Treatment was not begun until two weeks after grafting to allow vascularization of the implants and establishment of the BBB. During the dosing period, the effects of the conjugate-derived NGF on the growth of the tissue transplants were monitored by observations through the cornea of the living host using a stereo-microscope equipped with an eyepiece micrometer. The growth of basal forebrain tissue grafts in the anterior chamber of the rat eye has been shown to be enhanced by the intraocular administration of NGF (35,36). At the completion of the dosing period, the animals were sacrificed and the tissue fixed, sectioned and processed for ChAT-immunostaining to identify cholinergic cells. Intraocular administration of NGF increased the number of ChAT-positive cells in basal forebrain tissue grafts by approximately 80% over that observed in untreated control grafts.

The septal tissue transplants in animals treated with the anti-transferrin receptor antibody-NGF conjugate displayed enhanced growth and survival of cholinergic neurons in comparison to those in animals treated with either NGF or OX-26 alone (27). These results indicate that NGF is delivered across the BBB by the OX-26 - NGF conjugate in a biologically active form and in amounts sufficient to enhance the growth of the transplanted tissue and prevent the atrophy of the septal cholinergic neurons.

CONCLUSIONS

The results presented here suggest that transport systems on brain capillary endothelial cells can be used as a means of delivering therapeutic compounds across the blood-brain barrier. An antibody that binds to the rat transferrin receptor has been shown to be transported into the brain parenchyma as a result of its interaction with the receptor protein and ability to undergo transcytosis. In addition, this antibody can function as a carrier protein and deliver both small drugs and larger proteins into the brain. Potentially, this technology will enable many neurotherapeutic compounds that do not readily cross the BBB to be administered systemically and reach their site of action within the brain.

REFERENCES

1. Brightman, M.W., Morphology of blood-brain interfaces, *Exp. Eye Res.* 25:1-25, 1977.
2. Reese, T.S., and Karnovsky, M.J., Fine structural localization of a blood-brain barrier to exogenous peroxidase, *J. Cell Biol.* 34:207-217, 1967.
3. Pardridge, W.M., Receptor-mediated peptide transport through the blood-brain barrier, *Endocrine Rev.* 7:314-330, 1986.
4. Fishman, J.B., Rubin, J.B., Handrahan, J.V., Connor, J.R., and Fine, R.E., Receptor-mediated transcytosis of transferrin across the blood-brain barrier, *J. Neurosci. Res.* 18:299-304, 1987.
5. Pardridge, W.M., Eisenberg, J., and Yang, J., Human blood-brain barrier insulin receptor, *J. Neurochem.* 44, 1771-1778 (1985).
6. Pardridge, W.M., Eisenberg, J., and Yang, J., Human blood-brain barrier transferrin receptor, *Metabolism* 36:892-895, 1987.
7. Aisen, P., and Listowsky, I., Iron transport and storage proteins, *Ann. Rev. Biochem.* 49:357-393, 1980.
8. MacGillivray, R.T.A., Mendez, E., Shewale, J.G., Sinha, S.K., Lineback-Zins, J., and Brew, K., The primary structure of human serum transferrin, *J. Biol. Chem.* 258:3543-3553, 1981.

9. McClelland, A., Kuhn, L.C., and Ruddle, F.H., The human transferrin receptor gene: Genomic organization, and the complete primary structure of the receptor deduced from a cDNA sequence, *Cell* 39:267-274, 1984.

10. Omary, M.B., and Trowbridge, I.S., Covalent binding of fatty acid to the transferrin receptor in human cells in vitro, *J. Biol. Chem.* 256:12888-12895, 1981.

11. Dautry-Varsat, A., Ciechanover, A., and Lodish, H.F., pH and recycling of transferrin during receptor-mediated endocytosis, *Proc. Natl. Acad. Sci. USA* 80:2258-2262, 1983.

12. Jefferies, W.A., Brandon, M.R., Hunt, S.V., Williams, A.F., Gatter, K.C., and Mason, D.Y., Transferrin receptor on endothelium of brain capillaries, *Nature* 312:162-163, 1984.

13. Friden, P.M., Walus, L.R., Musso, G.F., Taylor, M.A., Malfroy, B., and Starzyk, R.M., Anti-transferrin receptor antibody and antibody-drug conjugates cross the blood-brain barrier, *Proc. Natl. Acad. Sci. USA* 88:4771-4775, 1991.

14. Greene, L.A., and Shooter, E.M., The nerve growth factor receptor: biochemistry, synthesis and mechanism of action, *Annu. Rev. Neurosci.* 3:353-402, 1980.

15. Hartikka, J., and Hefti, F., Development of septal cholinergic neurons in culture: Plating density and glial cells modulate effects of NGF on survival, fiber growth and expression of transmitter-specific enzymes, *J. Neurosci.* 8:2967-2985, 1988.

16. Hagg, T., Manthorpe, M., Vahlsing, H.L., and Varon, S., Delayed treatment with nerve growth factor reverses the apparent loss of cholinergic neurons after acute brain damage, *Exp. Neurol.* 101:303-312, 1988.

17. Kromer, L.F., Nerve growth factor treatment after brain injury prevents neuronal death, *Science* 235:214-216, 1987.

18. Whittemore, S.R., and Seiger, A., The expression, localization and functional significance of beta-nerve growth factor in the central nervous system, *Brain Res. Rev.* 12:439-464, 1987.

19. Coyle, J.T., Price, D.L., and Delong, M.R., Alzheimer's disease: a disorder of cortical cholinergic innervation, *Science* 219:1184-1190, 1983.

20. Hefti, F., Hartikka, J., and Knusel, B., Function of neurotrophic factors in the adult and aging brain and their possible use in treatment of neurodegenerative diseases, *Neurobiol. Aging* 10:515-533, 1989.

21. Junard, E.O., Montero, C.N., and Hefti, F., Long-term administration of mouse nerve growth factor to adult rats with partial lesions of the cholinergic septohippocampal pathway, *Exp. Neurol.* 110:25-38, 1990.

22. Hagg, T., Vahlsing, H.L., Manthorpe, M., and Varon, S., Nerve growth factor infusion into the denervated adult rat hippocampal formation promotes its cholinergic reinnervation, *J. Neurosci.* 10:3087-3092, 1990.

23. Hoffman, D., Wahlberg, L., and Aebischer, P., NGF released from a polymer matrix prevents loss of ChAT expression in basal forebrain neurons following a fimbria fornix lesion, *Exp. Neurol.* 110:39-44, 1990.

24. Price, D. L., New perspectives on Alzheimer's disease, *Annu. Rev. Neurosci.* 9:489-512, 1986.

25. Whitehouse, P.J., Price, D.L., Stuble, R.G., Clar, A.W., Coyle, J.T., and Delong, M.R., Alzheimer's disease and senile dementia: Loss of neurons in the basal forebrain, *Science* 215:1237-1239, 1982.

26. Fischer, W., Wictorin, K., Bjorklund, A., Williams, L.R., Varon, S., and Gage, F.H., Amelioration of cholinergic neuron atrophy and spatial memory impairment in aged rats by nerve growth factor, *Nature* 329:65-68, 1987.

27. Friden, P.M., Walus, L.R., Watson, P., Doctrow, S.R., Kozarich, J.W., Backman, C., Bergman, H., Hoffer, B., Bloom, F., and Granholm, A.-C., NGF-anti-transferrin receptor antibody conjugate crosses the blood-brain barrier and enhances survival of medial septal nucleus neurons, submitted, 1992.

28. Carlsson, J., Drevin, H., and Axen, R., Protein thiolation and reversible protein-protein conjugation, *Biochem. J.* 173:723-737, 1978.

29. Greene, L.A., and Tischler, A.S., Establishment of a noradrenergic clonal line of rat adrenal pheochromocytoma cells which respond to nerve growth factor, *Proc. Natl. Acad. Sci. USA* 73:2424-2428, 1976.

30. Buxser, S., et al., Single-step purification and biological activity of human nerve growth factor produced from insect cells, *J. Neurochem.* 56:1012-1018, 1991.

31. Triguero, D., Buciak, J., and Pardridge, W.M., Capillary depletion method for quantification of blood-brain barrier transport of circulating peptides and plasma proteins, *J. Neurochem.* 54:1882-1888, 1990.

32. Giacobini, M.M.J., Olson, L., Hoffer, B., and Sara, V.R., Truncated IGF-I exerts trophic effects on fetal brain tissue grafts, *Exp. Neurol.* 108:33-37, 1990.

33. Olson, L., and Seiger, A., Brain tissue transplanted to the anterior chamber of the eye. I. Fluorescence histochemistry of immature catecholamine and 5-hydroxytryptamine neurons reinnervating the rat iris, *Z. Zellforsch. Mikrosk. Anat.* 135:175-194, 1972.

34. Olson, L., Seiger, A., and Stomberg, I., Intraocular transplantation in rodents: a detailed account of the procedure and examples of its use in neurobiology with special reference to brain tissue grafting, *in:* "Advances in Cellular Neurobiology," S. Federoff, L. Hertz, eds., Academic Press, New York, vol. 4, pp. 401-442, 1983.

35. Eriksdotter-Nilsson, M., Skirbol, S., Ebendal, T., Hersh, L., Grassi, J., Massoulie, J., and Olson, L., NGF treatment promotes development of basal forebrain tissue grafts in the anterior chamber of the eye, *Exp. Brain Res.* 74:89-98, 1989.

36. Eriksdotter-Nilsson, M., Skirboll, S., Ebendal, T., and Olson, L., Nerve growth factor can influence growth of cortex cerebri and hippocampus: evidence from intraocular grafts, *Neurosci.* 30:755-766, 1989.

ENDOTHELIAL CELL BIOLOGY AND THE
ENIGMA OF TRANSCYTOSIS THROUGH
THE BLOOD-BRAIN BARRIER

Richard D. Broadwell

Division of Neurological Surgery
University of Maryland School of Medicine
Baltimore, Maryland 21201

INTRODUCTION

The mammalian blood-brain barrier (BBB) to non-lipid soluble macromolecules is associated morphologically with two basic cell types: the cerebral, non-fenestrated endothelium and perivascular phagocytes; the latter is represented by macrophages, microglia and pericytes surrounded by a basal lamina and located on the abluminal surfaces of cerebral arterioles, venules, and less so capillaries (1). Characteristics of the cerebral endothelium serving to define the BBB include circumferential belts of interendothelial tight junctional complexes that preclude the bidirectional, extracellular movement of macromolecules between blood and brain, and acid hydrolase-containing secondary lysosomes (4). The perivascular phagocytes also contain populations of secondary lysosomes and represent the first line of defense once the BBB is breached normally, experimentally, or pathologically (1,10). In vitro and in vivo data implicate a third cell type, the astrocyte, in influencing the formation of interendothelial tight junctions; however, astrocytes and their perivascular endfeet are not a physical or morphologically defined-barrier to the extracellular movement of macromolecules between the endothelium and brain parenchyma.

The endothelium of the BBB is no different from other mammalian cell types; it contains a prominent endomembrane system of organelles (e.g., endoplasmic reticulum, Golgi complex, vesicles, mitochondria, primary and secondary lysosomes, endosomes, etc.) and engages in normal cell functions associated with recycling or internalization of the plasmalemma, endocytosis, and vesicular transport. The latter under normal conditions, contrary to much of the BBB literature, does occur, if not transendothelially then intraendothelially among constituents of the endomembrane system of organelles. Indeed, the concept of transendothelial vesicular transport or transcytosis of macromolecules (e.g., proteins, peptides) bidirectionally through the BBB has been and continues to be a controversial topic within BBB literature. This concept, initially reserved for the experimentally manipulated BBB and the BBB in pathological states, now is in receipt of widespread recognition for potential delivery of blood-borne peptides, proteins, and chemotherapeutic agents into the CNS under normal conditions. Data advocating

Frontiers in Cerebral Vascular Biology: Transport and Its Regulation
Edited by L.R. Drewes and A.L. Betz, Plenum Press, New York, 1993

137

transcytosis through the "normal" or intact BBB is, for the most part, pharmacologically/biochemically based and associated with specific processes of adsorptive and receptor-mediated endocytoses (13). Morphological affirmation of adsorptive and receptor-mediated transcytoses of proteins/peptides through BBB endothelia is available (4,7,9,14, and Friden in this volume) but, as will be considered below, is equivocal from a cell biological perspective.

TRANSCYTOSIS THROUGH THE BBB APPEARS VECTORIAL

Endothelia of the BBB readily recycle the luminal plasmalemma and, hence, participate in fluid phase, adsorptive and receptor-mediated endocytic processes. Examples of blood-borne macromolecules entering the endothelium by endocytosis include the tracer horseradish peroxidase and albumin (fluid phase), wheatgerm agglutinin and cationized probes (adsorptive phase), and ferro-transferrin, insulin, and vasopressin (receptor-mediated phase). The common denominators among these three endocytic processes in BBB endothelia are 40-70 nm wide endocytic vesicles, endosomes, and dense body secondary lysosomes. Blood-borne macromolecules enter the endothelium within plasmalemma-derived endocytic vesicles directed to the endosome (prelysosome) compartment and/or to secondary lysosomes for eventual degradation (5). Blood-borne substances entering BBB endothelia by adsorptive and receptor-mediated endocytosis also are channeled by transporting vesicles to the Golgi complex either directly from the luminal plasmalemma and/or indirectly from the endosome compartment. These endothelial organelles are suspected of yielding additional transporting vesicles for ferrying blood-borne adsorptive/receptor-mediated phase macromolecules to the abluminal front for exocytosis (Figure 1A). Available morphological data do not support a direct transendothelial vesicular transport of any blood-borne macromolecule or the transendothelial transfer of fluid phase macromolecules. The latter are believed to undergo lysosomal degradation within BBB endothelia (4).

Exposure of the abluminal plasmalemma of BBB endothelia to fluid, adsorptive, or receptor-mediated phase macromolecules delivered into the brain parenchyma by ventriculocisternal perfusion or injection into the subarachnoid space demonstrates that the endocytic activity of the abluminal membrane is virtually non-existent or extremely minor compared to that of the luminal plasmalemma. Populations of endothelial organelles quickly and readily sequestering blood-borne macromolecules are not observed when the same macromolecules are introduced to and bathe the abluminal plasmalemma for 5 min through 18 hr (4,8,14). Suspected endocytic vesicles derived from the abluminal plasma membrane and that appear filled with probe molecules assessed in randomly selected ultrathin sections are revealed in ultrathin serial sections to be predominantly static invaginations or pits in the abluminal plasmalemma (4,8). These abluminal pits have been misinterpreted in the BBB literature as vesicles participating in transendothelial transport bidirectionally through the BBB (4). The data suggest that BBB endothelia are polarized regarding the retrieval/recycling of cell surface membrane and endocytosis vis-a-vis the luminal versus abluminal plasmalemma. This endothelial polarity further suggests that if transcytosis of macromolecules through the BBB is demonstrable the process is vectorial, from blood to brain but not from brain to blood. Potential transcytosis of blood-borne macromolecules through BBB endothelia indicates that the BBB can be circumvented and, therefore, the barrier is not absolute. Combined absence of demonstrable endocytic activity at the abluminal front and transcytosis through these endothelia from brain to blood argues for a brain-blood barrier that appears a priori to be more absolute than is the BBB.

Similar investigation of the potential for transcytosis through epithelia of the choroid plexus of the blood-cerebro-spinal fluid (CSF) barrier indicates that the choroid epithelium engages in endocytosis and recycling of its plasmalemma circumferentially and that adsorptive transcytosis through this cell is bidirectional between the blood and CSF (2,14).

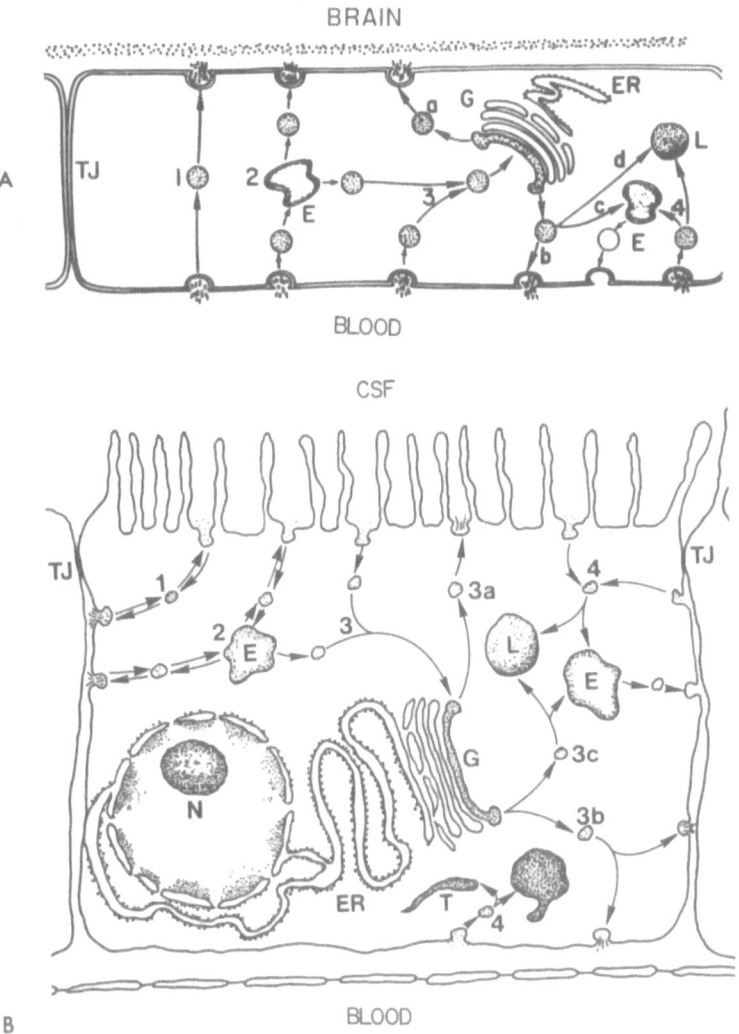

Figure 1. Potential transcellular transfer of adsorptive/receptor-mediated phase macromolecules through endothelia (A) of the BBB and choroid plexus epithelia (B) of the blood-CSF barrier may involve direct vesicular transport (1) or indirect vesicular transport through the endosome compartment (E; 2) and the Golgi complex (G; 3). These macromolecules and those associated with fluid phase endocytosis (4) also are channeled to secondary lysosomes (L). Macromolecules entering BBB endothelia do so predominantly from the blood and less so, if at all, from the brain side; hence, transcytosis through the BBB is vectorial from blood to brain. Choroid epithelia engage in endocytosis circumferentially, suggesting potential transcytosis of macromolecules through this cell type can be bidirectional. ER, endoplasmic reticulum; N, nucleus; T; tubules;TJ, tight junctional complex.

Organelles within choroid epithelia sequestering blood-borne or CSF-borne protein/peptide by the three different endocytic processes are identical to those organelles in BBB endothelia that sequester blood-borne protein/peptide (Figure 1B). Although transcytosis of fluid phase probes (e.g., HRP) fails to occur in choroid epithelia, the potential for adsorptive transcytosis is documented to involve transepithelial vesicular transport directly or indirectly, with the latter by way of the endosome compartment independent of the Golgi complex (2,14).

THE ENIGMA OF TRANSCYTOSIS THROUGH THE BBB ENDOTHELIUM

Evidence, whether biochemical, pharmacological or morphological, advocating transcytosis of proteins and peptides through endothelial cells in general implies that the endothelium participates in the cellular secretory process (12) and, as such, is anticipated to manifest membrane behavior compatible with that process. Noteworthy among membrane dynamics of the cellular secretory process are the complementary events of exocytosis and endocytosis, with the latter serving to retrieve cell surface membrane as a compensatory response to the addition of exocytic vesicle membrane to the plasmalemma. Retrieval or internalization of cell surface membrane associated with endocytosis ensures the overall surface area of the cell remains static with each exocytic event. The complementary events of exocytosis and endocytosis are no better exemplified than in the axon terminal with secretion of neurotransmitter(s) or in exocrine and endocrine cells secreting hormones, peptides or proteins.

Data from our laboratory support but also question the significance of transcytosis for selected proteins and peptides entering BBB endothelia by adsorptive and receptor-mediated endocytic processes (e.g., wheatgerm agglutinin, ferro-transferrin, antibody to the transferrin receptor). The absence of demonstrable endocytic activity, whether fluid phase, adsorptive phase or receptor-mediated, and absence of membrane retrieval at the abluminal front of BBB endothelia are viewed by us as the enigma of protein and peptide transcytosis through the BBB from blood to brain. The absence of a demonstrable endocytic activity associated with the abluminal plasmalemma seriously questions transcytosis as a significant event through the BBB. The potential for transcellular transfer of blood-borne macromolecules into the CNS is greatest for cells in which membrane retrieval and endocytosis are demonstrated to occur readily at each of the opposite poles of the cell (e.g., choroid plexus epithelia, neurosecretory and motor neurons).

Delivery of blood-borne proteins/peptides into the CNS by transendothelial transport within vesicles, endosomes and the Golgi complex may be augmented by two additional routes, neither of which encompasses endocytosis. The first utilizes only the plasmalemma and considers that the outer surface of the luminal membrane is continuous with the outer surface of the abluminal membrane. Peptide binding to the luminal plasmalemma could diffuse freely into the membrane within which the peptide may move to the abluminal side where dissociation from the plasmalemma and free entry to the brain would take place (3). The second possible route is associated with patent extracellular pathways circumventing the BBB. These extracellular pathways involve leaky blood vessels supplying the circumventricular organs (e.g., median eminence, area postrema, subfornical organ, etc.) and the subarachnoid space/pial surface. Such leaky vessels in vivo permit blood-borne protein the size of endogenous IgG and IgM and exogenous HRP and IgG to enter adjacent brain parenchyma and perivascular clefts for eventual exposure to the abluminal surface of BBB endothelia and perivascular phagocytes throughout the CNS (6,11). For this reason, blood-borne macromolecules inhabiting the perivascular clefts or labeling perivascular phagocytes are not accurate signals heralding the transcytosis of proteins and peptides through BBB endothelia.

CONCLUSIONS

Absence of demonstrable membrane recycling and endocytic activity at the abluminal plasmalemma of BBB endothelia is an enigma if transcytosis, whether adsorptive or receptor-mediated, is supported as a significant event. Involvement of the endosome compartment and/or the Golgi complex in the transendothelial, secretory pathway suggests transcytosis through the BBB is possible. The question begging an answer is, how significant is the process? A definitive answer is complicated by patent extracellular avenues into the brain allowing blood-borne macromolecules to circumvent the endothelial BBB for entry to the CNS.

ACKNOWLEDGEMENTS

The author's studies are supported by grant NS18030 from the Stroke and Trauma Program, NINDS, NIH, Bethesda, Maryland.

REFERENCES

1. Baker, B., Broadwell, R., and Wolf, A., Cellular line of defense upon breachment of the blood-brain barrier, *Soc. Neurosci. Abstr.*, in press.
2. Balin, B.J., and Broadwell, R.D., Transcytosis of protein through the mammalian cerebral epithelium and endothelium. I. Choroid plexus and the blood-cerebrospinal fluid barrier, *J. Neurocytol.* 17:809, 1988.
3. Banks, W.A., and Kastin, A.J., Review: Interactions between the blood-brain barrier and endogenous peptides: Emerging clinical implications, *Amer. J. Med. Sci.* 31:459, 1988.
4. Broadwell, R.D., Transcytosis of macromolecules through the blood-brain barrier. A critical appraisal and cell biological perspective, *Acta Neuropath.* 79:117, 1989.
5. Broadwell, R.D., Transcytosis of macromolecules through the fluid-brain barriers in vivo. "Biological Barriers to Protein Delivery," Vol. 5, in press, 1992.
6. Broadwell, R.D., Baker, B.J., Ebert, P.S., Hickey, W.F., and Villegas, J., Intracerebral grafting of solid tissues and cell suspensions: The blood-brain barrier and host immune response, *Prog. Brain Res.*, in press, 1992a.
7. Broadwell, R., Baker, B., Friden, P., Tangoren, M., and Wolf, A., Receptor-mediated transcytosis through the blood-brain barrier: Ferro-transferrin and its receptor, *Soc. Neurosci. Abstr.*, in press, 1992b.
8. Broadwell, R.D., Balin, B.J., Salcman, M., and Kaplan, R.S., A brain-blood barrier? Yes and no, *Proc. Nat. Acad. Sci. USA* 80:7352, 1983.
9. Broadwell, R.D., Balin, B.J., and Salcman, M., Transcytosis of blood-borne protein through the blood-brain barrier, *Proc. Nat. Acad. Sci. USA* 85:632, 1988.
10. Broadwell, R.D., and Salcman, M., Expanding the definition of the blood-brain barrier, *Proc. Natl. Acad. Sci. USA* 78:7820, 1981.
11. Broadwell, R.D., and Sofroniew, M.V., Serum proteins by-pass the blood-brain barrier for extracellular entry to the central nervous system, submitted to *J. Neurosci.* 1992.
12. Palade, G., Intracellular aspects of the process of protein secretion, *Science* 189:347, 1975.
13. Pardridge, W.M., "Peptide Drug Delivery to the Brain," Raven Press, New York, 1991.
14. Villegas, J., and Broadwell, R.D., Transcytosis of protein through the mammalian cerebral epithelium and endothelium. II: Absorptive transcytosis and the blood-brain and brain-blood barriers, submitted to *J. Neurocytol.*

BLOOD-BRAIN TRANSPORT OF VASOPRESSIN

Berislav V. Zlokovic, Jasmina B. Mackic, Milo N. Lipovac,
J. Gordon McComb and Martin H. Weiss

Department of Neurological Surgery and Division of Neurosurgery
Childrens Hospital Los Angeles, USC School of Medicine
Los Angeles, CA 90033

INTRODUCTION

Arginine-vasopressin (AVP) is a hormone/neuropeptide that regulates a number of peripheral (e.g., antidiuresis, glycogenolysis) and central (e.g., memory, learning) functions (1). AVP is also involved in control of brain water metabolism (2), brain edema (3), cerebrospinal fluid (CSF) formation (4), and secretion of pituitary peptides (1). A potential therapeutic application of AVP is that of memory disorders caused by brain trauma, Alzheimer's disease and senile dementia (5). Also V_1-receptor antagonists may be helpful in prevention and therapy of brain edema (6). However, the lack of adequate experimental evidence has discouraged a wider clinical application of AVP and its neuroactive analogs.

A number of studies have examined the transport of AVP in the CNS. Earlier experimental work claimed total impermeability of the blood-brain barrier (BBB) to AVP (7), but more recent investigations indicated that there is a significant bidirectional transport of AVP across the BBB that is possibly mediated by a specific vasopressinergic transporter and/or receptor (8,9).

We have recently tested a hypothesis that brain AVP levels can be influenced by plasma AVP and/or AVP receptor antagonists by using a vascular brain perfusion (VBP) model in guinea pigs (10). This model has been previously used to characterize various peptide transport systems at the BBB (11). The molecular forms of the brain uptake of peptides are determined by HPLC analysis, and a capillary-depletion step (12) is added to the VBP technique to distinguish between peptide BBB binding and transport. The methods and results described below are limited to our reported (13-16) and on-going studies with AVP.

MATERIALS AND METHODS

Perfused Brain Preparation

The VBP model was previously described (10,11) and here will only be briefly summarized. Guinea pigs (Hartley strain) of either sex and 250 to 300 gm body weight were

Frontiers in Cerebral Vascular Biology: Transport and Its Regulation
Edited by L.R. Drewes and A.L. Betz, Plenum Press, New York, 1993

143

anesthetized intramuscularly with 6 mg/kg xylazine (Rompun; Mobay KS) and 30 mg/kg ketamine (Ketacet; Aveco, IA) before surgical exposure of the neck vessels. A fine polyethylene or nylon catheter connected to the perfusion system was inserted into the right common carotid artery. Immediately after the start of perfusion, the contralateral carotid artery was ligated and both jugular veins were cut to allow free drainage of the perfusate. The perfusion medium consisted of 20% washed sheep red blood cells suspended in mock plasma. The medium was gassed with 96% O_2 and 4% CO_2, warmed to about $37.6^{\circ}C$ and pumped from a reservoir by means of a peristaltic pump. During perfusion the animal maintained spontaneous respirations, normal heart rate, arterial blood pressure, acid-base status, EKG, EEG and cerebral flow rate in the ipsilateral perfused hemisphere. Biochemical status in the perfused hemisphere is normal as indicated by water content, Na^+/K^+ ratio, ATP and lactate levels, and energy charge potential. Also, ultrastructural integrity of brain parenchyma and endothelial capillary wall remained unchanged (17).

Brain AVP Uptake

[Phe-3,4,5^{3}H(N)]-AVP-(1-9) and [^{14}C]sucrose (cerebrovascular space marker) were exposed to the blood-brain interface (1 to 10 min), and the effects of unlabeled AVP-(1-9), peptide fragments: desGly-NH_2-AVP, pressinoic acid, [pGlu^4, Cyt^6]-AVP-(4-9), V_1-antagonist (TMeAVP), V_2-agonist (dDAVP), aminopeptidase inhibitor (bestatin) and L-amino acids, were examined as previously reported (13). Following brain perfusion the ipsilateral forebrain was quickly removed from the skull and the arachnoid membranes were peeled away. The choroid plexus and anterior pituitary were separated from the brain. The forebrain was then subjected to regional dissection and/or to the capillary/depletion procedure (12), and the radioactivity was determined in homogenate, vascular pellet, and capillary-depleted brain tissue. The percent contamination with blood vessels of capillary-depleted brain tissue was quantified by measuring gamma-glutamyl transpeptidase, GGTP activity (18).

Mathematical Treatment

Rates of [^{3}H]AVP blood-brain transport and capillary sequestration were estimated by computing regional and/or compartmental unidirectional transfer constants, K_{in}, from multiple-time and/or single-time tissue uptake series as described (10,11). Corrections for the peptide isotopic exchange rates within the brain vascular space was made by subtracting sucrose uptake values from AVP tissue uptake. [^{3}H]AVP K_{in} values in the presence of unlabeled peptide and/or inhibitor were used to calculate Michaelis-Menten parameters, V_{max} and K_m, for the saturable component of AVP, and the inhibitory constant, K_i, by means of a 'velocity ratio' (14,15).

High Pressure Liquid Chromatography

Radiolabeled AVP and its products of proteolysis were determined in forebrain homogenates and postcapillary supernatants by reversed-phase, high-pressure liquid chromatography (HPLC) as previously reported (16).

RESULTS AND DISCUSSION

Figure 1 illustrates tissue uptake of simultaneously infused [^{3}H]AVP and [^{14}C]sucrose determined in different brain compartments at various perfusion times. A time-dependent and progressive uptake of [^{3}H]AVP relative to [^{14}C]sucrose was measured both in brain homogenate and postcapillary supernatant within the 10-min period of brain perfusion. During the same period of time, both tracers exhibited insignificant uptake into the vascular pellet. The GGTP supernatant/pellet ratio of about 4.5% indicated minimal contamination of capillary-depleted brain tissue by the vasculature. Unidirectional compartmental rate constants (K_{in}) for both tracers were computed as slopes of tissue uptake regression lines. This analysis suggested that cerebrovascular permeability to circulating AVP is about ten times higher than for sucrose. Although the capillary sequestration rate of AVP was somewhat higher than for sucrose, its contribution to total brain uptake of the peptide was less than 8%, and all of the rest of the AVP was found in the capillary-depleted brain parenchyma.

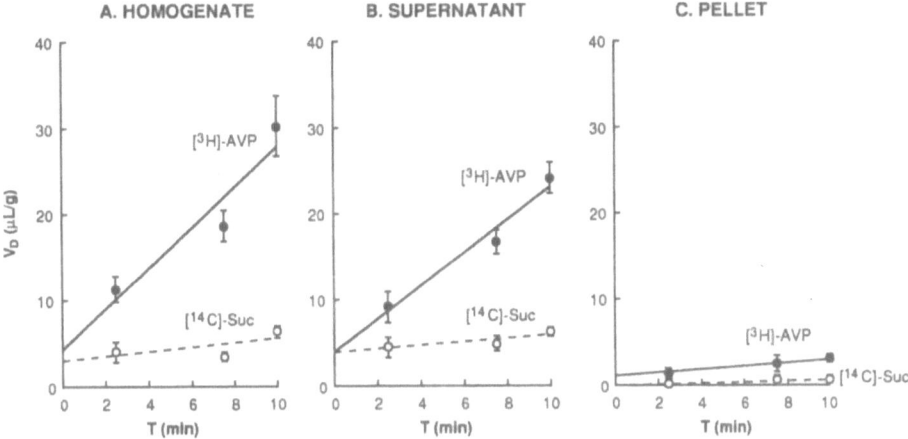

Figure 1. The volume of distribution, V_d, of radiolabeled AVP (3 nM) and sucrose determined in brain homogenate, capillary-depleted brain tissue, and isolated capillaries is plotted against brain perfusion time, T (16). Values are mean ± SE, n=3.

Bestatin-resistant saturation kinetics of AVP at the blood-brain interface has been demonstrated (13). A dose-dependent inhibition of [3H]AVP with unlabeled peptide was obtained at both BBB and non-BBB regions (13-15). The choroid plexus exhibited the highest affinity for circulating AVP and was about two orders of magnitude higher than at the BBB regions (Table 1).

With respect to low nanomolar concentrations of AVP in peripheral blood, the AVP BBB uptake system exhibits relatively high K_m, and therefore it will function in an almost linear fashion to direct AVP into brain parenchyma. Thus, its physiological relevance might be to continuously transport blood-borne AVP across the BBB. This in turn may represent a significant source of brain AVP.

A significant competitive inhibition of BBB AVP transfer was obtained only with TMeAVP. Regional K_i constants indicated that affinity of the V_1-antagonist was 1.7 to 2.5 times less than of AVP itself. Kinetic experiments revealed that influx of AVP into brain was not altered by the presence of the peptide fragments: AVP-(1-8), pressinoic acid, and [pGlu4,Cyt6]-AVP-(4-9), nor by aminopeptidase inhibitors and the L-amino acid transport system substrates.

HPLC of capillary-depleted brain tissue indicated that the intact [3H]AVP-(1-9) progressively declined with time, from 49% at 1 min to 11.9% at 10 min. Concomitantly, the major detectable metabolite, [3H]phenylalanine, accumulated in the brain reaching about 50% at 10 min. The most consistent of small radioactivity peaks were fractions 15-17, that rose from about 3% at 1 min to almost 11% after 10 min. All other small peaks were

Table 1. Kinetic parameters of AVP

BBB Regions	K_m (μM)	V_{max} (pmol/min/g)
Cortex	2.1 ± 0.3	5.5 ± 0.7
Caudate	2.6 ± 0.5	5.6 ± 0.9
Hippocampus	2.7 ± 0.3	4.9 ± 0.5
Hypothalamo-Pituitary		
Anterior Pituitary	0.79 ± 0.03	22.1 ± 0.2
Hypothalamus	0.76 ± 0.02	2.1 ± 0.1
Non-BBB Regions		
Pineal Gland	0.19 ± 0.02	1.6 ± 0.02
Choroid Plexus	0.03 ± 0.005	0.5 ± 0.06

Values are means ± SE for 5-6 K_{in} observations based on 24-27 individual experiments.

significantly lower than the 15-17 fraction, ranging only between 2% and 5% at 10 min. These peaks represent radioactivity associated with different AVP peptide fragments containing labeled phenylalanine, and according to previously reported AVP brain chromatograms (19), they possibly represent [Cyt[6]]-AVP-(3-9) and AVP-(3-9), while fraction 30 most likely represents AVP-(2-9). These results suggest that the predominant step of proteolysis of blood-borne AVP in the brain involves cleavage of the peptide bond from the N-terminus, which is characteristic for aminopeptidase activity. Since we were not able to demonstrate any changes in AVP BBB transfer in the presence of the potent aminopeptidase inhibitor bestatin (16), it is suggested that enzymatic hydrolysis of AVP does not take place at the luminal side of the BBB. Thus, peptide degradation may be at the abluminal BBB side, or at the level of pericytes, and/or in the brain parenchyma itself, as shown by local AVP brain microinjection studies (19). According to previous findings (19) the in vivo hydrolysis of blood-borne AVP may also result in formation of highly neuroactive fragment AVP-(4-9), and, in that sense, hydrolysis of AVP in the brain may not necessarily lead to its physiological and/or neuropharmaceutical inactivation.

Recent work from our laboratory has suggested that chronic nicotine treatment increases both blood-brain transport and capillary sequestration of AVP (20). It appears that the nicotine effects are not the result of an increase in non-specific BBB permeability, but are most likely mediated by an increase in AVP specific transport binding sites at the luminal side of the BBB.

In conclusion, our studies suggest that transport of AVP into the brain is mediated by a specific peptide carrier and/or V_1-receptor mediated mechanism at the luminal side of the BBB that can be regulated by nicotine. A rapid in vivo metabolism of AVP occurs after BBB transport, most likely in the brain parenchyma, with no evidence of significant capillary sequestration or degradation of AVP by the BBB.

ACKNOWLEDGMENTS

This work was supported by funds provided by the Cigarette and Tobacco Surtax Fund

146

of the State of California through the Tobacco-Related Disease Research Program of the University of California, grant 2RT0071.

REFERENCES

1. Segal, M.B., and Zlokovic, B.V., "The Blood-Brain Barrier, Amino Acids and Peptides," Kluwer Academic, Lancaster, pp. 103-105, 1990.
2. Raichle, M.E., and Grubb R.L., Jr., Regulation of brain water permeability by centrally-released vasopressin, *Brain Res.* 143:191-194, 1978.
3. Doczi, T., Joó, F., Szerdahelhy P., and Bodosi, M., Regulation of brain water and electrolyte contents: the opposite actions of central vasopressin and atrial natriuretic factor (ANF), *Acta Neurochir.* 43:186-188, 1988.
4. Davson, H. and Segal, M.B., The effects of some inhibitors and accelerators of sodium transport on turnover of ^{22}Na in the cerebrospinal fluid, *J. Physiol., London* 209:131-153, 1970.
5. Kastin, A.J., Ehrensing, R.H., Banks, W.A., and Zadina J.E., Possible therapeutic implications of the effects of some peptides on the brain, *Prog. Brain Res.* 72:223-235, 1987.
6. Nagao, S., Kagawa, M., Kuniyoshi, T., and Ogawa, T., Treatment of vasogenic brain edema by V_1 receptor antagonist of arginine vasopressin, The 60th Annual Meeting, *AANS Abstr.* p. 359, 1992.
7. Ermisch, A., Ruhle, H.J., Landgraf, R., and Hess, J., Blood-brain barrier and peptides, *J. Cereb. Blood Flow Metab.* 5:350-357, 1985.
8. Banks, W.A., Kastin, A.J., Horvath, A. and Michals, E.A., Carrier-mediated transport of vasopressin across the blood-brain barrier of the mouse, *J. Neurosci. Res.* 18:326-332, 1987.
9. Zlokovic, B.V., Hyman, S., McComb, J.G., Lipovac, M.N., Tang, G., and Davson, H., Kinetics of arginine-vasopressin uptake at the blood-brain barrier, *Biochim. Biophys. Acta* 1025:191-198, 1990.
10. Zlokovic, B.V., Begley, D.J., Djuricic, B.M. and Mitrovic, D. M., Measurement of solute transport across the blood-brain barrier in the perfused guinea-pig brain: method and application to N-methyl-a-aminoisobutyric acid, *J. Neurochem.* 46:1444-1451, 1986.
11. Zlokovic, B.V., In vivo approaches for studying peptide interactions at the blood-brain barrier, *J. Controlled Release* 13:185-202, 1990.
12. Triguero, D., Buciak, J.B., and Pardridge, W.M., Capillary depletion method for quantifying blood-brain barrier transcytosis of circulating peptides and plasma proteins, *J. Neurochem.* 54:1882-1888, 1990.
13. Zlokovic, B.V., Hyman, S., McComb, J.G., Tang G., Rezai, A.R., and Weiss, M.H., Vasopressin uptake by hypothalamopituitary axis and pineal gland in guinea pigs, *Am. J. Physiol.* 260:E633-E640, 1991.
14. Zlokovic, B.V., Segal, M.B., McComb, J.G., Hyman, S., Weiss, M.H., and Davson, H., Kinetics of circulating vasopressin uptake by choroid plexus, *Am. J. Physiol.* 260:F216-F224, 1991.
15. Zlokovic, B.V., Banks, W.A., Kadi, H.E., Erchegyi, J., Mackic, J.B., McComb, J.G., and Kastin, A.J., Blood-to-brain transport and metabolism of circulating vasopressin, *Soc. Neurosci. Abstr.* 17:240, 1991.
16. Zlokovic, B.V., Banks, W.A., ElKadi, H., Erchegyi, J., Mackic, J.B., J. McComb G., and Kastin, A.J., Transport, uptake, and metabolism of blood-borne vasopressin by the blood-brain barrier, *Brain Res.*, in press, 1992.
17. Zlokovic, B.V., McComb J.G., Perlmutter, L., Weiss, M.H., and Davson, H., Neuroactive peptides and amino acids at the blood-brain barrier: possible implications to drug abuse, *in:* "Drug Availability and the Blood-Brain Barrier," J. Frankenheim, and R. Brown, eds., NIDA Research Monographs, Washington DC, Government Publication, pp. 26-42, 1992.
18. Naftalin, L., Sexton, M., Whitaker, J.F., and Tracey, D., A routine procedure for estimating gamma-glutamyl transpeptidase activity, *Clin. Chim. Acta* 26:293-296, 1969.
19. Stark, H., Burbach, P.H., van Der Kleij, A.M., and De Wied, D., In vivo conversion of vasopressin after microinjection into limbic brain areas of rats, *Peptides* 10:717-720, 1989.
20. Lipovac, M.N., Barron, E., Perlmutter, L.S., McComb, J.G., Weiss, M.H., and Zlokovic, B.V. Chronic nicotine treatment increases blood-brain transport and capillary sequestration of vasopressin, *Soc. Neurosci. Abstr.* in press, (1992).

CEREBRAL PERICYTES — A SECOND LINE OF DEFENSE IN CONTROLLING BLOOD-BRAIN BARRIER PEPTIDE METABOLISM

Dorothee Krause, Jörg Kunz and Rolf Dermietzel

Institute of Anatomy, University of Regensburg
Universitätsstr. 31, 8400 Regensburg, Germany

INTRODUCTION

The production and use of antibodies specific for brain blood vessels represents a powerful approach for the evaluation of the molecular and functional properties of the blood-brain barrier (BBB). Such antibodies can be used as immunocytochemical probes to detect antigenic determinants of the cerebral blood vessels and to define the biochemical nature of these antigens. We therefore produced blood-brain barrier specific monoclonal antibodies by using isolated brain microvessels as a collective immunogen.

METHODS AND RESULTS

We first isolated rat brain microvessels by the method of Mrsulja et al. (4). The isolated microvessels consisted of endothelial cells, basement membranes, pericytes and a few smooth muscle cells. They were used as the collective antigen by injecting a suspension into mice for the production of monoclonal antibodies.

Supernatants of IgG-producing hybridomas were screened by indirect immunofluorescence for immunoreactivity with cerebral blood vessels on cryostat sections. We found 15 supernatants that stained specifically for cerebral microvessels.

One of the monoclonal antibodies showed a selective immunoreactivity toward brain microvessels, in particular arterioles, capillaries and postcapillary venules. Larger brain blood vessels and blood vessels of non-nervous tissues did not reveal any immunofluorescence staining (2).

To ensure vascular specificity of this antibody we performed double-immunolabelling, using an antibody to the factor-VIII associated antigen as an endothelial marker. This procedure showed a consistent co-localization of both immunoreactive sites.

Immunoreactivity to our antibody is absent in brain regions known to be provided with fenestrated capillaries and accordingly reveals a modified or leaky blood-brain barrier. The results clearly demonstrate that the antigen is restricted to brain microvessels of the tight type and thus constitutes a component of the specific molecular setup of the BBB.

Frontiers in Cerebral Vascular Biology: Transport and Its Regulation
Edited by L.R. Drewes and A.L. Betz, Plenum Press, New York, 1993

To define the cellular expression of the antigenic target further we performed pre- and post-embedding electronmicroscopical immunocytochemistry. We found the gold label distributed in irregular patches and chains at the extracytoplasmic side of the pericyte plasma membrane. The luminal and abluminal surface of the endothelium, the astrocytes, and the basement membranes were devoid of any immunoreactivity. High power electron micrographs suggested that the antigenic target consists of filamentous extracellular membrane proteins specifically expressed by cerebral pericytes.

In non-nervous tissue we were able to detect the antigen in various transporting epithelia. In no case, however, was immunolabelling of vessel walls outside the nervous tissue detectable. Liver bile canaliculi were selectively stained at the luminal surfaces of the microvilli. Epithelial cells of the proximal tubules of the kidney showed an intense immunogold labelling on the outer surface of the microvilli up to the regions of the zonulae occludentes, where the labelling terminated abruptly. Similar to the labelling pattern in cerebral pericytes, the gold particles occurred in a kind of chain-like arrangement.

To elucidate the nature of the antigen, we performed a biochemical characterization of the protein by isolating membranes from rat brain microvessels and rat kidney brush borders. Western blots with the pericyte-specific monoclonal antibody revealed selective immunostaining of a 140 kDa protein. From Western blots it is clear that kidney brush border membranes are enriched with the 140 kDa protein, whereas in brain microvessels a less prominent but distinct band at the same position is discernible. For this reason we used kidney brush border preparations as a source for enrichment of the 140 kDa protein to perform microsequencing by Edman degradation.

A peptide sequence consisting of 18 amino acids from the N-terminus of the 140 kDa antigen was obtained. The sequence was screened for homologies on a protein data bank, and we found optimal scores with a 100% identity of our N-terminus fragment to the microsomal aminopeptidase of rat, known as "aminopeptidase N or M." Compared to aminopeptidase N (AP-N) of pig, there is only one conservative exchange at sequence position 9. This is also the case for human AP-N, which is identical with the cell surface glycoprotein CD 13.

Until now we have not been able to get information on the AP-N sequence of the cerebral pericytes because of N-terminus blockage. This may be a result of differences in glycosylation and/or acetylation or of blockage of the N-terminus during the isolation process of cerebral microvessels. We assume, however, that the cerebral pericytic 140 kDa protein is identical with the kidney AP-N because of the identical immunoreactive pattern on Western blots.

To extend our investigations further into the functional properties of pericytic AP-N, we acquired in vitro methods for culturing rat pericytes alone and in co-culture with endothelial and astrocytic cells. Different stages of the cultured rat cerebral pericytes were immunolabelled with the monoclonal antibody. The same cells were counterstained with a polyclonal antibody against γ-glutamyl-transpeptidase (γGT), an endothelial and pericyte-specific marker (1,3). Anti-γGT labelled some freshly isolated microvascular cells including the AP-N positive cells. Cell specificity was confirmed by electron microscopic studies using pre-embedding labelling with anti-AP-N and subsequent immunogold labelling. The immunogold labelling was distributed on the external surface of cell plasma membranes, which, because of their topological relationship to endothelial cells, can be regarded as pericytes. The efficiency of our antibody in labelling isolated pericytes makes it a selective marker for these cells within the first two days of culture. When pericytes became adherent, after 24 to 48 hr, AP-N was dramatically down-regulated. This is comparable with the situation described previously for the expression of different endothelial blood-brain barrier proteins in culture, e.g., glucose transporter and γGT (5). As yet we have not succeeded in getting a re-expression of AP-N in adherent pericytes. Neither co-culturing with astrocytes nor supplementation of specific substrates like met-encephalin has proven successful.

To investigate the functional properties of AP-N in vitro, it is therefore essential that the microvessel cells are freshly prepared. AP-N activity was, in fact, detected by hydrolysis of L-alanine-substituted methoxynaphthylamide substrate and subsequent fast blue B reaction. We also obtained inhibition of AP-N activity by pre-incubation of the microvascular preparation with our anti-AP-N antibody. The relative reduction of AP-N activity was about 22% of the AP-N activity in the microvascular cell suspension. The specific peptidase inhibitor Bestatin reduced catalytic activity of AP-N by nearly 100%.

DISCUSSION

The existence of a pericyte-specific AP-N implies an active functional role of the pericytes in regulative mechanisms between the vascular milieu and the fluid environment of the brain. The presence of AP-N has been described in various organs and in different regions of the brain, e.g., in synaptic complexes, in the choroid plexus epithelium and in microvessel walls (6). It has, however, never been attributed to the pericytic investment of cerebral microvessels.

AP-N is an ectopeptidase that apparently removes N-terminal residues from oligopeptides derived mainly from neurons or endocrine cells. The enzyme is not organ- or tissue-specific, nor does it carry peptide specificity. It has been suggested that cerebral AP-N participates in neurotransmitter- and neuropeptide inactivation.

Because our data show that AP-N is highly expressed in cerebral pericytes, the idea that these perivascular constituents play a role in homeostatic regulation of neuropeptide metabolism seems appealing. The lack of AP-N in leaky vascular segments implies a difference in the regulative capacities of different vascular segments. The observed lack of immunoreactivity in the neuropil, where AP-N has also been described, strongly suggests that different isoforms of this enzyme do exist in the brain. Apparently, cerebral AP-N plays a dual role: first, in preventing the access of circulating blood-born peptides to the brain and, second, in inactivating neuropeptides released from synapses.

SUMMARY

The above data clearly indicate that cerebral pericytes are equipped with a specific aminopeptidase (AP-N). Functionally, this finding suggests that cerebral pericytes are involved in amino acid and peptide catabolism of the brain and that they constitute an essential carrier of the enzymatic blood-brain barrier.

ACKNOWLEDGMENTS

We would like to express our gratitude for excellent technical assistance to D. Puchner, V. Könecke and D. Schünke.

REFERENCES

1. Frey, A., Meckelein, B., Weiler-Güttler, H., Möckel, B., Flach, R., and Gassen, H. G., Pericytes of the brain microvasculature express γ-glutamyl transpeptidase, *Eur. J. Biochem.* 202:421-429, 1991.
2. Krause, D., Vatter, B., and Dermietzel, R., Immunochemical and immunocytochemical characterization of a novel monoclonal antibody recognizing a 140 kDa protein in cerebral pericytes of the rat, *Cell Tissue Res.* 252:543-555, 1988.

3. Meyer, J., Mischeck, U., Veyhl, M., Henzel, K., and Galla, H.-J., Blood-brain barrier characteristic enzymatic-properties in cultured brain capillary endothelial cells, *Brain Res.* 514:305-309, 1990.

4. Mrsulja, B.B., Mrsulja, B.J., Fujimoto, T., Klatzo, I., and Spatz, M., Isolation of brain capillaries: a simplified technique, *Brain Res.* 110:361-365, 1976.

5. Risau, W., Dingler, A., Albrecht, U., Dehouck, M.P., and Cecchelli, R., Blood-brain barrier pericytes are the main source of γ-glutamyltranspeptidase activity in brain capillaries, *J. Neurochem.* 58:667-672, 1992.

6. Solhonne, B., Gros, C., Pollard, H., and Schwartz, J.-C., Major localization of aminopeptidase M in rat brain microvessels, *Neuroscience* 22:225-232, 1987.

Receptors and
Intracellular Messengers

THE ROLE OF SECOND MESSENGER MOLECULES
IN THE REGULATION OF PERMEABILITY IN THE
CEREBRAL ENDOTHELIAL CELLS

Ferenc Joó

Laboratory of Molecular Neurobiology, Institute of Biophysics
Biological Research Center, 6701-Szeged, Hungary

INTRODUCTION

The endothelium is a single-cell layer lining the blood vessels and represents an active interface between blood and tissue. It acts as a selective permeability barrier, regulates coagulation, and contributes to the behavior of cells both in the circulation and in the vessel wall (20).

It is now generally believed that, apart from a few exceptions where the structural basis of restricted movement of certain substances is the result of an impermeable glial sheath (1,2), the endothelial cells comprise the "blood-brain barrier" (BBB). This term was first used by Stern and Peyrot (43) to summarize a number of different regulatory mechanisms that control the exchange of materials between blood and brain and cerebrospinal fluid. It is worthy of mention that, from a phylogenetic point of view, the endothelial blood-brain barrier can be regarded as an old structure: a living vertebrate Lamperta fluviatilis, which branched off from the general evolutionary "tree" more than 500 million years ago, has cerebral microvessels with structural characteristics similar to those of vertebrates with a much shorter evolutionary history (7,8). It was demonstrated in electron microscopic studies (6) that the barrier, which prevented the passing of proteins from blood to brain, consisted of tight junctions between cerebral endothelial cells. Another equally important feature of the cerebral endothelial cells, in contrast to those of most other tissues, is the relative lack of pinocytotic vesicles. From a physiological point of view, the cerebral endothelial cells represent a very tight cellular barrier with high (1900 ohm/cm^2) membrane resistance (9,10,14). This potent restraint delays the penetration of solutes to the brain at the capillary endothelium. On the other hand, a number of polar metabolic substrates cross the plasma membranes of cerebral endothelium via highly specific, self-saturable and inhibitable transport systems (5,35,36). These particular structural and functional characteristics of cerebral capillaries are determined by the surrounding central nervous tissue, especially by the influence deriving from astrocytes (3,44,45,48). It became evident, mainly from studies carried out on isolated brain microvessels, that the cerebral endothelial cells are endowed with many different receptors (for review see 22-24). It remained to be seen, however, which

Frontiers in Cerebral Vascular Biology: Transport and Its Regulation
Edited by L.R. Drewes and A.L. Betz, Plenum Press, New York, 1993

155

cellular transduction system(s), if any, operate in the cerebral endothelial cells possibly mediating the effect of vasoactive substances.

The aim of our studies was to check if the second messenger molecules were produced in the cerebral endothelial cells and had a role in the regulation of transendothelial permeability.

THE EFFECTS OF CYCLIC NUCLEOTIDES ON THE PERMEABILITY OF CEREBRAL ENDOTHELIAL CELLS

The first results indicating that the lipid-soluble derivative of cyclic AMP, the dibutyryl (dibu-) cyclic AMP, can increase the number of pinocytotic vesicles and the albumin penetration in mature brain microvessels were published in 1972 (21) (Figure 1). Likewise, the non-hydrolyable cGMP analogue, dibu-cyclic GMP, was found (27) to increase pinocytosis (Figure 2) and to open the BBB for albumin after infusion into the common carotid artery. In these experiments, different concentrations (25, 50, 100 and 200 μg) of dibu-cGMP, the lipid-soluble derivative of cGMP, were infused in a volume of 0.5 ml by an Infumat driver (Kutesz, Hungary) in 5 min at a rate of 0.1 ml/min into the common carotid artery of adult rats (27). Control animals were infused with butyric acid or guanosine-3',5'-cyclic monophosphoric acid (Sigma, St. Louis) - corresponding to the proportion of these substances in 200 μg concentration of dibu-cGMP- or with Krebs-Ringer buffer only. Brains were washed out by perfusion with a buffered Krebs-Ringer solution prior to fixation and processed for the immunohistochemical detection of albumin. Serum albumin was not detected in the control sections. Infusion of cGMP seemed to induce the uptake of serum albumin by some microvessels. However, after infusion with 25, 50, 100 and 250 μg of dibu-cGMP, strong immunoreactivity was seen in the wall of capillaries and venules indicating the uptake and accumulation of serum albumin by the endothelial cells. In many cases, strong perivascular staining was also observed, being indicative of the transendothelial transport of the macromolecule in question. The results of our investigations have clearly shown that the given concentrations of dibu-cGMP could induce the macromolecular transport for albumin in a dose-dependent manner with an accompanying activation of pinocytosis in the endothelium of brain microvessels (Figure 2).

Later, the synthesizing and degrading enzymes of both cAMP and cGMP were detected in the cerebral endothelial cells (for review see 22,24), indicating that these second messenger molecules are produced and metabolized locally. Recently, a correlation was found (4) between the activation of adenylate cyclase in cerebral endothelial cells and the induction of transcapillary albumin transport, suggesting that elevated cAMP may trigger macromolecule permeation through the microvessels (Figure 3). Specific histamine H_2-receptor blockers were seen (18) to prevent the activation of this transport possibly by binding to receptors linked to the adenylate cyclase in the brain microvessels (29). In agreement with our findings, the data of Sen and Campochiaro (41) also suggest that stimulation of intracellular cAMP accumulation may be a common feature of mediators that cause breakdown of the blood-retinal barrier.

In contrast to these results, Kempski et al. (30) found no increase of permeability to trypan blue-albumin complex in a new model of cultured cerebromicrovascular endothelial cells when the endothelial synthesis of cAMP was stimulated directly by forskolin. Recently, Rubin et al. (40) reported that increasing cAMP levels in the cerebral endothelial cells already treated with astrocyte-conditioned medium caused a striking change in cell morphology. In addition, agents that increased cAMP levels produced a rapid increase in electrical resistance as well, which was considerably higher in the presence of astrocyte-conditioned medium. The results of Stelzner et al. (42) showed that an increase in cAMP content was in fact associated with a 3-10-fold reduction in albumin transfer across monolayers from bovine

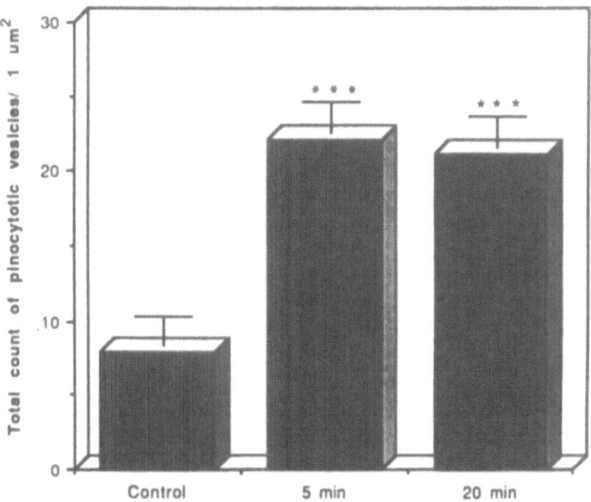

Figure 1. Treatments with dibu-cAMP increased significantly the number of pinocytotic vesicles in the cytoplasm of cerebral endothelial cells, P < 0.001.

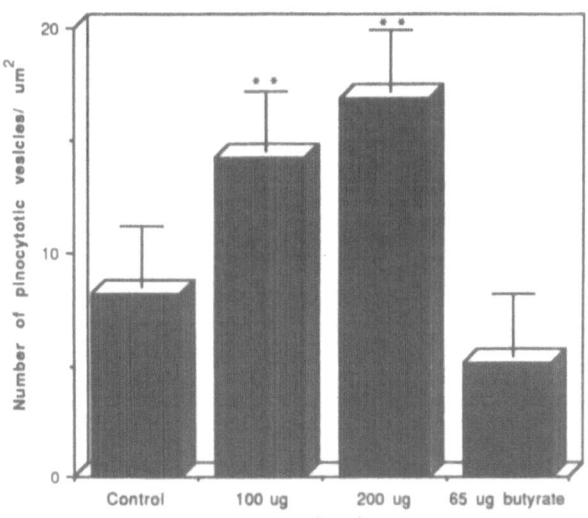

Figure 2. Different concentrations (100 and 200 μg) of dibu-cGMP, infused into the carotid artery, increased significantly the number of pinocytotic vesicles in the cytoplasm of cerebral endothelial cells, P < 0.05.

pulmonary arterial endothelial cells. By working also with endothelial cells of peripheral origin, Carson et al. (11) reported on similar results showing that agents which increase cAMP in endothelial cells prevent the increase in microvascular permeability to albumin that

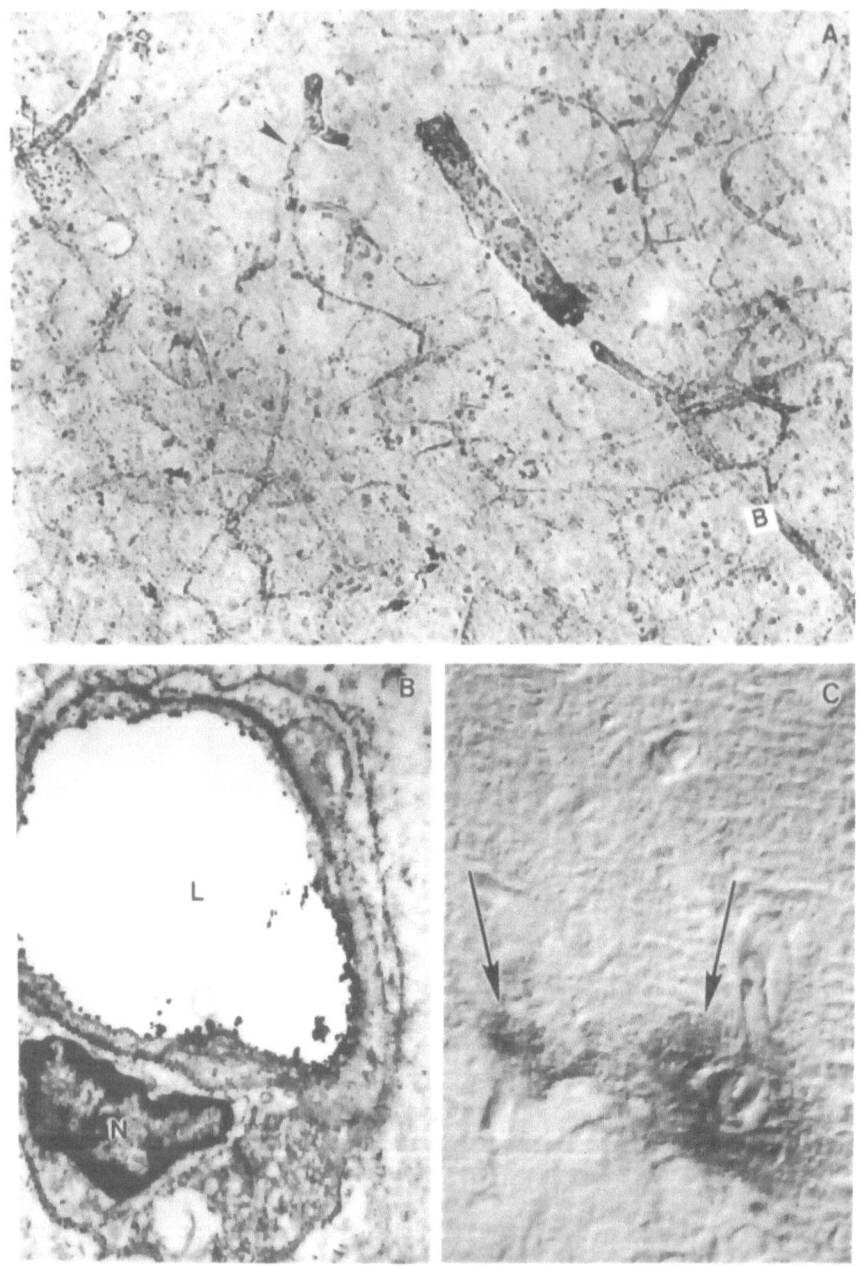

Plate 1. A: Light microscopic visualization (arrowhead) of guanylate cyclase activity in the brain capillaries, x 120. B: Electron dense deposits indicate the fine structural localization of guanylate cyclase activity in the luminal and abluminal membranes of cerebral endothelial cells, L = lumen, N = nucleus of a pericyte, x 19,000. C: Light microscopic demonstration of albumin extravasation (arrows) induced by dibu-cGMP (200 µg) infusion, x 400.

follows histamine exposure. The results of Rotrosen and Galline (39) have, however, elucidated that the permeability enhancing effect of histamine appeared to be dependent on histamine H_1-receptor occupancy in endothelial cells of peripheral origin, causing increases in intracellular calcium. Recently, evidence has been provided (38) for a rise in intracellular calcium on the effect of vasoactive agents in the cerebral endothelial cells as well. The important differences between endothelial cells from brain tissue and peripheral vessels should, however, be kept in mind during the interpretation of these seemingly contradictory

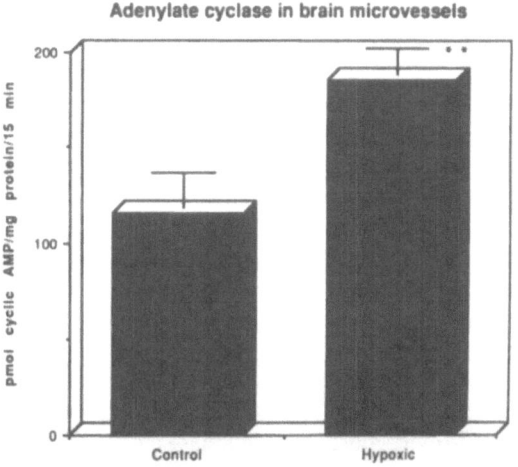

Figure 3. Adenylate cyclase activity was significantly higher in microvessels isolated from brains with edema, $P < 0.05$.

results. As we have learned by now, cultured cells are often unable to express fully the properties of mature cerebral endothelium with BBB characteristics because of the absence of astrocytic influences, in vitro. For example, various enzymes (e.g., γ-glutamyl transpeptidase), (15,16) and receptors (e.g., acetylcholine-, serotonin- and histamine H_2-receptors) (28) are missing as a rule from the cultured cerebrovascular endothelial cells, making these cells similar, at least in this respect, to peripheral vessels.

Fluid-phase endocytosis within primary cultures of brain microvessel endothelial cell monolayers, an in vitro BBB model, has been shown recently (19) to be significantly stimulated by nanomolar concentrations of phorbol myristate acetate. This effect of phorbol esters was not mediated by prostaglandins. Since the protein kinase C (PK C) is a prime target for actions of the phorbol esters (33), the involvement of PK C in the activation of blood-brain barrier opening seems to be established.

The presence and translocation of PK C, a key regulatory enzyme involved in both signal transduction and cellular proliferation, were observed (32) to phorbol esters as well as to transforming growth factor beta, a widely distributed regulatory peptide that promotes endothelial cell differentiation in microvessels isolated from rat brain. Similarly, stimulation of PK C translocation was found by Catalán et al. (13) in isolated brain microvessels on the effect of Substance P. The finding suggested that Substance P may be involved in the regulation of processes underlying protein phosphorylation in the BBB.

The presence in the cerebral endothelial cells of calmodulin-dependent kinases was revealed by studying protein phosphorylation (34). Calmodulin stimulated the phosphorylation of 58(57)-, 55-, and 50 kDa proteins more than that in the control, and the phosphorylation peaked at approximately 4 min. Sometimes a short lag period could be observed during the rising phase. Dephosphorylation, carried out by phosphoprotein phosphatases, followed relatively rapid kinetics, in contrast to the phosphorylation induced by cyclic nucleotides. The most prominent substrates of calmodulin-dependent phosphorylation were the 50- and 55 kDa polypeptides, which are most likely identical to the β- and α-subunits of the calmodulin-dependent protein kinase II (31). These subunits are known to be phosphorylated usually by asymmetric kinetics. The similar substrates, evaluated together with proteins of relative molecular mass of 58 (57) kDa, probably correspond to the α- and β-subunits of tubulin (31).

MODULATION OF PROTEIN PHOSPHORYLATION BY SECOND MESSENGERS AT THE CEREBRAL ENDOTHELIUM

The fact that protein phosphorylation occurs in brain microvessels was first shown by Pardridge et al. (37). Later, the isolation and partial characterization of a 56,000-dalton phosphoprotein from the BBB was published by Weber et al. (49). A detailed study on the kinetics of protein phosphorylation and its modulation by second messengers revealed (34) the presence in the cerebral endothelium of Ca^{2+}-calmodulin, Ca^{2+}/phospholipid (PK C)-, cyclic GMP-, and cyclic AMP-dependent protein kinases. Since protein phosphorylation is thought to be a common effector process in the action of many second messengers, it was proposed that these protein substrates may participate in the coupling of signal transduction to endothelial metabolism.

Recently, evidence has been provided (12) on the regulatory action of vanadate on protein phosphorylation in brain microvessels. The authors proposed that vanadate activated the phosphorylation of cAMP-dependent protein kinase and, thereby, may modulate transport processes across the BBB.

In order to elucidate the molecular interactions between different second messenger systems, further studies are warranted.

RELATIONS TO BRAIN EDEMA FORMATION

In another experiment, decompression brain edema was produced (26) by removing a 3 x 8 mm piece of the parietal bone from the skull with a high speed dental drill, between the coronal and lambdoid sutures near the sagittal suture. The dura mater was excised under the operating microscope and the pial surface of the brain was covered with an artificial fibrin sponge (Spongostan) soaked in physiological saline of 0.26 Osm. The animals were sacrificed 15 min later for microvessel isolation. Brain capillaries isolated from animals with decompression brain edema showed significantly higher adenylate cyclase activity than those isolated from controls. A similar increase in adenylate cyclase activity was observed (17) in the cerebral microvessels of rats subjected to a prolonged hypobaric-hypoxic treatment.

Prevention of Brain Edema Formation by Histamine H_2-Receptor Antagonists

To check if histamine receptor blockers had any effect on brain edema formation, [90]Yttrium cubes (approx. 4 mm^3) of varying strength (from 2.5 mCi to 0.1 mCi) were implanted into the surface of the parietal cortex on the right side in 4 adult dogs and 2 cats (25). For the quantitative expression of edema extent, the animals were given 2.5 ml/kg^{-1} of 1% Evans blue 24 hr before investigation, and the area of blue staining was measured planimetrically on symmetrical coronal slices (approx. 5 mm thick) obtained from the hemispheres of different animals. Corresponding areas of the right and left hemispheres from 3 control and 3 metiamide-treated animals were averaged, and the means and S.D. were calculated. Half of the animals were treated intraperitoneally with metiamide in a maintenance dose of 50 µg/kg^{-1}. The animals subjected to [90]Yttrium irradiation were as a rule somnolent and apathetic throughout the entire period of observation, while those treated with metiamide did not show any clinical sign indicating the development of severe brain edema. At 24 or 72 hr after implantation, severe extravasation of Evans blue was observed in the untreated animals, whereas the extent of blue staining indicating the leakage of albumin was considerably reduced in metiamide-treated animals. The extent of edema in the control animals was found to be significantly greater (5.2 ± 2.5 cm^2) than that in metiamide-treated animals (1.8 ± 1.0 cm^2).

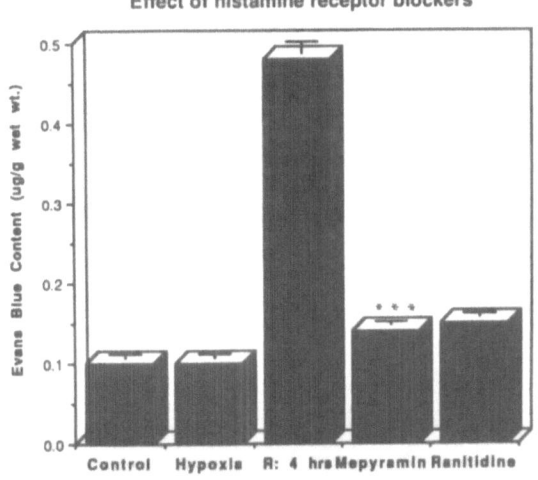

Figure 4. Mepyramine and ranitidine treatments attenuated hypoxia-induced albumin leakage.

Figures 4 and 5. Mepyramine and ranitidine treatments prevented water accumulation and albumin extravasation in an ischemic model of brain edema in newborn piglets, [*]P< 0.05; [**]< 0.01.

The possible brain edema preventing effect of different histamine receptor blockers was studied in detail using newborn piglets in another experimental model (18). General hypoxaemia was evoked experimentally by bilateral pneumothorax, and the development of severe brain edema of the vasogenic type was observed in the recirculation phase, 4 hr after the hypoxic challenge. Histamine receptor antagonists, mepyramine (H$_1$-receptor blocker), metiamide, cimetidine and ranitidine (H_2-receptor antagonists) were administered either intraperitoneally or intrathecally to check to what extent the formation of brain edema could be reduced. Mepyramine and ranitidine decreased the accumulation of water, sodium and albumin in the parietal cortex (Figures 4 and 5).

Subcutaneous pretreatment with a histamine H_2-receptor blocking agent, ranitidine, in a dose of 5 mg/kg given 2 hr before and at the time of kainic acid injection, partially decreased the edema formation in the thalamus (Figure 6) (46). It was assumed that, because of repetitive discharges evoked by the kainic acid (47), an excessive release of histamine from internal (mast cells and neuronal) sources may activate the histamine H_2-receptor-coupled adenylate cyclase in the brain microvessels and result in the induction of brain edema.

SUMMARY

The view that the cerebral endothelial cells represent the cellular analogue of the blood-brain barrier has been generally accepted. The regulation of transport processes operating in the cerebral endothelial cells is of great current interest. Different elements of the intracellular signaling messenger systems have been detected in the course of our studies in the cerebral endothelial cells. Our knowledge of these regulatory mechanisms is briefly reviewed here with special emphasis on the importance of second messenger molecules and phosphorylation of certain proteins of microvascular origin.

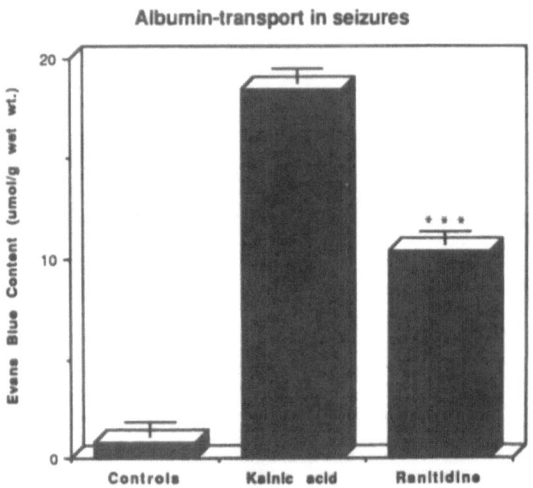

Figure 6. Ranitidine prevented the development of albumin extravasation in an epileptic model evoked by kainic acid treatment, $P < 0.001$.

ACKNOWLEDGMENTS

The valuable work of all co-workers and collaborators who have helped the accomplishment of this research is greatly acknowledged. The research was supported by Grants from the Hungarian Research Fund (OTKA) and the Ministry of Public Welfare.

REFERENCES

1. Abbott, N.J., Access of ferritin to the interstitial space of *carcinus* brain from intracerebral blood vessels, *Tissue and Cell*, 4:99, 1972.

2. Abbott, N.J., Bundgaard M., and Cserr, H., Fine-structural evidence for a glial blood-brain barrier to protein in the cuttlefish, *Sepia officinalis, J. Physiol.* 316:52P, 1981.

3. Arthur, F.E., Shivers, R.R., and Bowman, P.D., Astrocyte mediated induction of tight junctions in brain capillary endothelium: an efficient in vitro model, *Dev. Brain Res.* 36:155, 1987.

4. Ádám, G., Joó, F., Temesvári, P., Dux, E., and Szerdahelyi, P., Effects of acute hypoxia on the adenylate cyclase and Evans blue transport of brain microvessels, *Neurochem. Int.* 10: 529, 1987.

5. Bradbury, M.W.B., The blood-brain barrier in vitro, *Neurochem.Int.* 7:27, 1985.

6. Brightman, M.W., and Reese, T.J., Junctions between intimately apposed cell membranes in the vertebrate brain, *J. Cell Biol.* 40:648, 1969.

7. Bundgaard, M., Ultrastructure of frog cerebral and pial microvessels and their impermeability to lanthanum ions, *Brain Res.* 241:57, 1982.

8. Bundgaard, M., and Cserr, H., A glial blood-brain barrier in Elasmobranchs, *Brain Res.* 226:61, 1981.

9. Butt, A.M., and Jones, H.C., Effect of histamine and antagonists on electrical resistance across the blood-brain barrier in rat brain-surface microvessels, *Brain Res.* 569: 100, 1992.

10. Butt, A.M., Jones, H.C., and Abbott, N.J., Electrical resistance across the blood-brain barrier in anaesthetized rats: A developmental study, *J. Physiol.* 429:47, 1990.

11. Carson, M. R. Shasby, S.S., and Shasby, D.M., Histamine and inositolphosphate accumulation in endothelium: cAMP and a G protein, *Am. J. Physiol.* 257:L259, 1989.

12. Catalán, R.E., Martínez, A.M., Aragonés, M.D., and Díaz, G., Evidence for a regulatory action of vanadate on protein phosphorylation in brain microvessels, *Biochem. Biophys. Res. Comm.* 163:771, 1989.

13. Catalán, R.E., Martínez, A.M., Aragonés, M.D., and Fernández, I., Substance P stimulates translocation of protein kinase C in brain microvessels, *Biochem. Biophys. Res. Comm.* 164:595, 1989.

14. Crone, C., and Olesen, S.P., Electrical resistance of brain microvascular endothelium, *Brain Res.* 241:49, 1982.

15. DeBault, L.E., γ-Glutamyl transpeptidase induction mediated by glial foot processes-to endothelium contact in co-culture, *Brain Res.* 220:432, 1981.

16. DeBault, L.E., and Cancilla, P.A., γ-glutamyl transpeptidase in isolated brain endothelial cells: Induction by glial cells in vitro, *Science*, 207:653, 1979.

17. Dux, E., Temesvári, P., Joó, F., Ádám, G., Clementi, F., Dux, L., Hideg, J., and Hossmann, K.-A., The blood-brain barrier in hypoxia: ultrastructural aspects and adenylate cyclase activity of brain capillaries, *Neuroscience*, 12: 951, 1984.

18. Dux, E., Temesvári, P., Szerdahelyi, P., Nagy, Á., Kovács, J., and Joó, F., Protective effect of antihistamines on cerebral edema induced by experimental pneumothorax in newborn piglets, *Neuroscience*, 22:317, 1987.

19. Guillot, F.L., and Audus, K.L., Angiotensin peptide regulation of fluid-phase endocytosis in brain microvessel endothelial cell monolayer, *J. Cereb. Blood Flow Metab.* 10:827, 1990.

20. Jaffe, E.A., Cell biology of endothelial cells. *Hum.Pathol.* 18:234, 1987.

21. Joó, F., Effect of N^6,O^6-dibutyrl cyclic 3',5'-adenosine monophosphate on the pinocytosis of brain capillaries in mice, *Experientia*, 28:1470, 1972.

22. Joó, F., The blood-brain barrier in vitro: ten years of research on microvessels isolated from the brain, *Neurochem. Int.* 7:1, 1985.

23. Joó, F., New aspects to the function of cerebral endothelium, *Nature*, 321:197, 1986.

24. Joó, F., The cerebral microvessels in tissue culture, an update, *J. Neurochem.* 58:1, 1992.

25. Joó, F., Szúcs, A., and Csanda, E., Metiamide-treatment of brain edema in animals exposed to ^{90}Yttrium irradiation, *J. Pharm. Pharmacol.* 28:162, 1976.

26. Joó, F., Mihály, A., Temesvári, P., and Dux, E., Basic molecular events underlying transendothelial transport in brain capillaries, *in*: "Recent Progress in the Study and Therapy of Brain Edema," K.G. Go and A. Baethmann, eds., Plenum Publishing Corporation, New York, pp. 107-116, 1984.

27. Joó, F., Temesvári, P., and Dux, E., Regulation of the macromolecular transport in the brain microvessels: the role of cyclic GMP, *Brain Res.* 278:165, 1983.

28. Karnushina, I., Spatz, M., and Bembry, J., Cerebral endothelial cell culture. I. The presence of β_2 and α_2-adrenergic receptors linked to adenylate cyclase activity, *Life Sci.* 30:849, 1982.

29. Karnushina, I.L., Palacios, J.M., Barbin, G., Dux, E., Joó, F., and Schwartz, J.C., Studies on a capillary-rich fraction isolated from brain: histaminic components and characterization of the histamine receptors linked to adenylate cyclase, *J. Neurochem.* 34:1201, 1980.

30. Kempski, O., Villacara, A., Spatz, M., Dodson, R.F., Corn, C.,Merkel, N., and Bembry, J., Cerebromicrovascular endothelial permeability. In vitro studies, *Acta Neuropathol.* 74:329, 1987.

31. Mackie, K., Lai, Y., Nairn, A.C., Greengard, P., Pitt, B.R., and Lazo, J.S., Protein phosphorylation in cultured endothelial cells, *J. Cell Physiol.* 128:367, 1986.

32. Markovac, J., and Goldstein, G.W., Transforming growth factor beta activates protein kinase C in microvessels isolated from immature rat brain, *Biochem. Biophys. Res. Comm.* 150:575, 1988.

33. Nishizuka, Y., Studies and perspectives of protein kinase C. *Science*, 233:305, 1984.

34. Oláh, Z., Novák, R., Lengyel, I., Dux, E., and Joó, F., Kinetics of protein phosphorylation in microvessels isolated from rat brain: modulation by second messengers, *J. Neurochem.* 51:49, 1988.

35. Oldendorf, W.H., The blood-brain barrier. *Exptl. Eye Res.* 25 Suppl.:177, 1977.

36. Pardridge W.M., Recent advances in blood-brain barrier transport, *Annu. Rev. Pharmacol. Toxicol.* 28:25, 1988.

37. Pardridge, W.M., Yang, J., and Eisenberg, J., Blood-brain barrier protein phosphorylation and dephosphorylation, *J. Neurochem.* 45:1141, 1985.

38. Revest, P.A., Abbott, N.J., and Gillespie, J.I., Receptor mediated changes in intracellular Ca^{2+} in cultured rat brain capillary endothelial cells, *Brain Res.* 549:159, 1991.

39. Rotrosen, D., and Galline, J.I., Histamine type I receptor occupancy increases endothelial cytosolic calcium, reduces F-actin, and promotes albumin diffusion across cultured endothelial monolayers, *J. Cell Biol.* 103:2379, 1986.

40. Rubin, L.L., Hall, D.E., Porter, S., Barbu, K., Cannon, C., Horner, H.C., Janatpour, M., Liaw, C.W., Manning, K., Morales, J., Tanner, L.I., Tomaselli, J., and Bard, F., A cell culture model of the blood-brain barrier. *J. Cell Biol.* 115:1725, 1991.

41. Sen, H.A., and Campochiaro, P.A., Stimulation of cyclic adenosine monophosphate accumulation causes breakdown of the blood-retinal barrier, *Invest. Ophthalmol. and Visual Science* 32:2006, 1991.

42. Stelzner, T.J., Weil, J.V., and O'Brien, R.F., Role of cyclic adenosine monophosphate in the induction of endothelial barrier properties, *J. Cell Physiol.* 139:157, 1989.

43. Stern, L., and Peyrot, R., Le fonctionnement de la barrière hématoencéphalique aux divers stades de développement chez diverses espèces animales, *C. r. séance. Soc. Biol.* 96: 1124, 1927.

44. Stewart, P.A., and Hayakawa, E.M., Interendothelial junctional changes underlie the developmental "tightening" of the blood-brain barrier, *Dev. Brain Res.* 32:271, 1987.

45. Stewart, P.A., and Wiley, M., Developing nervous tissue induces formation of blood-brain barrier characteristics in invading endothelial cells: A study using quail-chick transplantation chimeras, *Devel. Biol.* 84:183, 1981.

46. Sztriha, L., Joó, F., and Szerdahelyi, P., Histamine H_2 receptors participate in the formation of brain edema induced by kainic acid in rat thalamus, *Neurosci. Lett.* 75: 334, 1987.

47. Sztriha, L., Joó, F., Szerdahelyi, P., Lelkes, Z., and Ádám, G., Kainic acid neurotoxicity: characterization of the blood-brain barrier damage, *Neurosci. Lett.* 55:233, 1985.

48. Tao-Cheng, J.H., Nagy, Z., and Brightman, M.W., Tight junctions of brain endothelium in vitro are enhanced by astroglia, *J. Neurosci.* 7:3293, 1987.

49. Weber, M., Mehler, M., and Wollny, E., Isolation and partial characterization of a 56,000-dalton phosphoprotein phosphatase from the blood-brain barrier, *J. Neurochem.* 49:1050, 1987.

PEPTIDERGIC INDUCTION OF ENDOTHELIN 1
AND PROSTANOID SECRETION IN HUMAN
CEREBROMICROVASCULAR ENDOTHELIUM

M. Spatz, D.B. Stanimirovic, F. Bacic, S. Uematsu,
J. Bembry, and R.M. McCarron

National Institute of Neurological Disorders and Stroke,
National Institutes of Health, Bethesda, MD 20892 USA

INTRODUCTION

Angiotensin (Ang II) and arginine vasopressin (AVP), known vasoconstrictive agents, have been implicated as playing a role in the pathogenesis of hypertension. These peptides are among a number of neurotransmitters, hormones or cytokines that have been shown to induce endothelin 1 (ET-1) (a potent vasoconstrictor) or prostaglandins, in a variety of cellular systems including endothelium (1-4). None of the previous investigations, using either in vitro or in vivo model systems, have demonstrated a concomitant Ang II- or AVP-stimulated production of ET-1 and prostanoids. Since vascular diseases such as hypertension and vasospasm may be the result of an altered vascular balance between vasodepressors [vasodilatory peptides, prostanoids, endothelial-derived relaxing factor (EDRF)] and vasopressors [vasoconstrictive peptides, prostanoids, endothelial-derived constrictive factor (EDCF)] produced by the endothelium (5-8), we studied the type and possible mechanism of some mediators simultaneously elicited by AVP or Ang II in endothelium derived from capillary and microvessels of human brain (HBEC). This report describes an interrelationship between AVP- and Ang II-induced formation of ET-1 and prostanoids which can be modulated by either inhibitors of phospholipase A_2 (PLA_2), cyclooxygenase, or lipoxygenase.

MATERIALS AND METHODS

Endothelial Cell Cultures

Samples of normal brain surgically removed for the treatment of idiopathic epilepsy served for separation, dissociation and cultivation of capillary and microvascular endothelium by a modified technique of Gerhart et al. (9). The purity of the endothelial cultures was

Frontiers in Cerebral Vascular Biology: Transport and Its Regulation
Edited by L.R. Drewes and A.L. Betz, Plenum Press, New York, 1993

165

>98% as determined by immunostaining for von Willebrand (Factor VIII)-related antigen and to 9 individual experiments that were performed in triplicate.

Experimental Procedures

Confluent propagated endothelial cell cultures (5th - 10th generation) grown in small (34 mm) Petri dishes were exposed to the tested substances in 1 ml of serum-free medium (Gibco M199) after removal of serum containing medium and washing with phosphate-buffered solution. Stable cellular conditions were achieved by preincubation of the HBEC in Gibco M199 in the presence or absence of blockers for cyclooxygenase [indomethacin (20 μM) or acetylsalicylic acid (ASA, 500 μM)] or lipoxygenase [nordihydroguaiateric acid (NDGA, 50 μM)] or phospholipase A_2 [(PLA$_2$), dexamethasone 50 μM] or Ang II or AVP receptor antagonists [Sar^1, Ala^8-Ang II (10^{-5} M) and l-(β-mercapto-β,β-cyclopententa-methylene propionic acid) POMT 10^{-6} M], respectively, for 30 min at 37°C. HBEC exposed to Ang II or AVP (10^{-8} - 10^{-10} M) were incubated for either 4 or 4-24 hr. Aliquots of unextracted medium from each culture were collected with or without addition of 25 μl/ml 1 M citric acid [for the respective determination of prostaglandins (PGF$_2$, TxB$_2$, PGD$_2$, PGE$_2$, and 6-keto PGI$_2$) and ET-1], quickly frozen and stored at -70°C until assayed. The prostaglandins and ET-1 were determined by radioimmunoassays with the use of respective antibodies as described previously. The individual prostaglandin antibodies with cross-reactivity < 1% except for PGE$_2$ (8%) were purchased from Dr. Levine, Waltham, MA; radiolabeled ET-1 kit and unlabeled Sar^1, Ala^8 - Ang II and POMT were obtained from Peninsula Laboratories, Belmont, CA; all other unlabeled substances were purchased from Sigma Chem. Co., St. Louis, MO.

For the IP$_3$ assays, the HBEC were prelabeled with myo-[^{3}H]inositol (5 μCi) for 24 hr, washed with Gibco M199, exposed to either Ang II (10 μM) or AVP (10 μM) and lithium chloride (20 mM) in the presence or absence of the respective peptide antagonists (100 μM). Trichloroacetic acid (15%) was used to terminate the reaction and ion exchange chromatography was utilized to separate the radioactivity incorporated in the IP$_3$ fraction as previously reported (10,11). Protein was determined by the method of Lowry et al. (12).

RESULTS

Production of ET-1

AVP or Ang II dose-dependently stimulated the HBEC secretion of ET-1. AVP was a more potent inducer of ET-1 than Ang II (Table 1). The respective antagonists of AVP (POMT) and Ang II (Sar^1, Ala^8-Ang II) blocked the ET-1 stimulated secretion induced by these agents (Figure 1). The HBEC pretreated with either dexamethasone or NDGA alone or in the presence of either AVP or Ang II, potentiated the stimulated formation of ET-1 (Table 2). Indomethacin elicited similar effects as those observed with dexamethasone or NDGA (data not shown).

Production of Prostanoids

PGD$_2$ the main prostaglandin formed by the HBEC under control conditions, was also dose-dependently enhanced by either AVP or Ang II (data not shown). The stimulatory effect of Ang II or AVP on PGD$_2$ was almost completely inhibited (80%) by their respective antagonists (Figure 1). Moreover, the AVP-inducible secretion of PGD$_2$ was abolished by pretreatment with ASA. Each of these peptides also stimulated the production of other prostanoids with the exception of 6-keto PGF$_1$ by HBEC (Figure 2). AVP, in contrast to Ang II, did not significantly affect the TxB$_2$ secretion in most of the experiments (Figure 2).

Table 1. Dose-dependent effects of AVP and Ang II on ET production by HBEC.

	AVP	Ang II
Control	2.39 ± 0.17(9)	2.39 ± 0.17(9)
0.01 μM	6.92 ± 0.71 (3)	2.78 ± 0.24 (3)
0.10 μM	22.14 ± 3.12(3)*	3.69 ± 0.14 (3)*
1.00 μM	128.13 ± 27.10 (2)*	17.12 ± 3.21 (2)*

Values are given in pg/mg protein, as means ± SE for the number of independent experiments performed in triplicate indicated in the parentheses.
* indicates significant difference (p<0.01; Student's t-test) from the previous (lower) concentration.

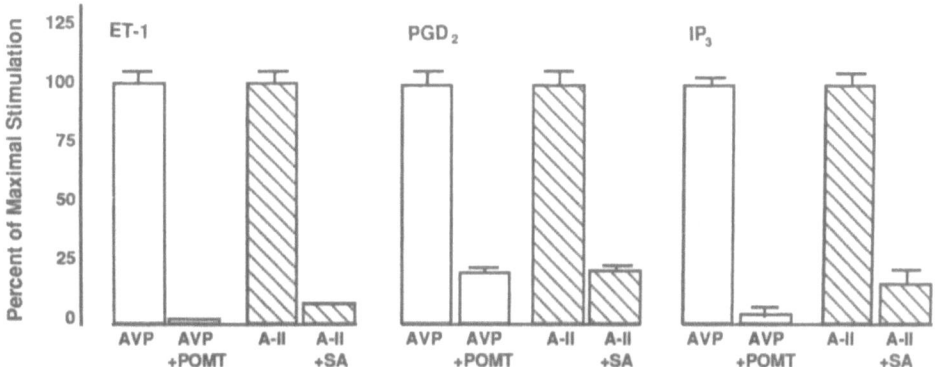

Figure 1. Effect of arginine vasopressin (AVP) or angiotensin II (Ang II) on endothelin I (ET-1), prostaglandin D_2 (PGD$_2$) or inositol triphosphate (IP$_3$) formation in the presence or absence of their respective antagonists (as described in the text). Results are expressed as mean ± SE in percent of maximal stimulation (100%) of 3-5 individual experiments, each performed in triplicate.

Dexamethasone completely abolished the AVP- or Ang II-enhanced formation of all the prostaglandins except for Ang II-induced secretion of PGF$_{2a}$ (Figure 2). The temporal dynamics of the individual prostanoid secretion stimulated by AVP was different from that of Ang II even though the degree of AVP (100 nM) enhanced production was similar to that of Ang II (1 μM). The maximal HBEC secretion of PGD$_2$ induced by AVP (2-fold over basal levels) was seen after 8 hr whereas that of Ang II (2.5-fold over basal levels) already at 4 hr after exposure to each substance. The highest TxB$_2$ Ang II-induced stimulation was observed after 8 hr. At the same time, a significant stimulated 6-keto PGF$_{1a}$ production was elicited with Ang II but not with AVP.

Table 2. Effects of dexamethasone (100 μM), NDGA (1 μM), AVP (100 nM), Ang II (1 μM), and their combination on ET-1 production by HBEC.

	Control	+AVP	+Ang II
M199	2.39 ± 0.17 (9)	22.14 ± 3.12 (3)*	17.12 ± 3.21 (2)*
Dxm	34.18 ± 0.51 (3)*	72.50 ± 2.65 (2)**	70.75 ± 4.40 (3)**
NDGA	12.50 ± 0.38 (2)*	79.25 ± 6.78 (2)**	50.00 ± 1.51 (2)**

Values are given in pg/mg protein (means ±SE) for the number of independent experiments performed in triplicate indicated in the parentheses.
* indicates significant difference (p<0.01; ANOVA) from (M199) values;
** indicates significant difference (p<0.01; ANOVA) from each substance alone.

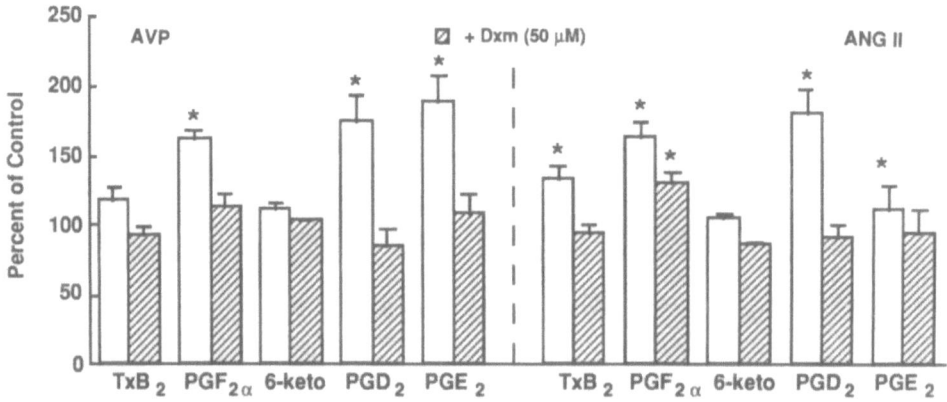

Figure 2. Effect of arginine vasopressin (AVP) or angiotensin II (Ang II) in the presence or absence of dexamethasone (DxM) on the production of prostanoids. The results are expressed as mean ± SE in percent of respective controls (incubated in M199 for 4 hr) of 3-8 individual experiments, each performed in triplicate. * p < 0.01 (Student's t-test as compared to control values).

Analysis of IP_3

AVP or Ang II induced an increase in IP_3 formation which was inhibitable by POMT and Sar[1], Ala[8]-Ang II, respectively (Figure 1). The half-maximal effective dose of AVP for IP_3 formation was lower (EC_{50} = 82 nM) than that of Ang II (EC50 = 970 nM).

DISCUSSION

These findings indicate that the AVP- or Ang II-induced augmented production of ET-1 and PGD_2 as well as stimulated IP_3 formation in HBEC is receptor-mediated since each of the peptidergic responses was blocked by the respective selective antagonists. The results also clearly show that the affinity of AVP receptor-mediated HBEC events is greater than that of Ang II since the same dose of each agent induced a greater accumulation of ET-1 (Table 1) and PGD_2 (4), and the EC_{50} of AVP was 10 times lower than the EC_{50} of Ang II required for IP_3 formation. Moreover, the observed inhibition of the AVP- or Ang II-inducible prostaglandin formation and the potentiation of the AVP or Ang II-stimulated secretion by dexamethasone indicate an involvement of PLA_2 in these events.

AVP or Ang II receptor-mediated stimulation of polyphosphoinositide metabolism of the cellular membrane has been shown to be important in generating intracellular signals responsible for initiating effects on vascular endothelium (1,11) and other cells such as vascular smooth muscle (13) and renal mesangial cells (14). A receptor-mediated stimulation of prostaglandin production inducible by either AVP or Ang II was also described in these cells. Similar effects mediated by endothelin receptors have been elicited by exogenous ET-1 in HBEC and other cells (11).

It is generally accepted that the membrane release of arachidonic acid (AA) required for prostanoid production can be activated by stimulation of PLA_2 and/or PLC (through activation of the phosphoinositol cascade, formation of diacylglycerol, and mobilization of Ca^{2+}). Although both events may occur in parallel, it is still debatable if the increased production of prostanoids induced by either of the tested peptides results from independent primary activation of PLA_2 or PLC and/or a consequence of PLC stimulation and phosphoinositol cascade. However, it has been shown that the process by which ET-1 receptors are coupled to PLA_2 and PLC is independently regulated in vascular smooth muscle (14).

The simultaneous inhibition of AVP- or Ang II-inducible prostaglandin production and potentiation of AVP or Ang II stimulated ET-1 secretion by dexamethasone (PLA_2 inhibitor) suggest an interchange between these events. We would like to propose that the receptor-mediated AVP or Ang II stimulated ET-1 secretion could activate the HBEC ET-1 receptor (autocrine) and in turn stimulate prostaglandin formation through activation of PLA_2 as one of the participating processes in the intricate interaction between peptides and prostaglandins in the HBEC. However, the mechanism responsible for the peptidergic interaction with prostanoids remains to be clarified. The incomplete inhibition of PGF_2 by dexamethasone strongly implies that PGF_2 represents a PGD_2 metabolite which can be formed from PGD_2 by HBEC (15).

Most importantly, this study indicates for the first time that dexamethasone (a glucocorticoid that was shown to induce ET-1 secretion in peripheral vascular smooth muscle cells but not in the endothelium (16)) and inhibitors of cyclooxygenase or lipoxygenase can stimulate ET-1 secretion in HBEC. These observations in addition to the demonstrated AVP or Ang II augmentation of ET-1 secretion that was associated with temporal selective stimulation of prostaglandin formation suggest that the interaction of these substances may play a role in autoregulation of microcirculation including the blood-brain barrier properties. These findings are also consistent with the concept that an endothelial imbalance between the production of the vasoconstrictive and vasodilatory substances could contribute to the pathogenesis of cerebrovascular diseases.

REFERENCES

1. Yanagisawa, M., and Masaki, T., Molecular biology and biochemistry of the endothelins, *TIPS* 10:374-369, 1989.
2. Scharschmidt, L.A., and Dunn, M.J., Prostaglandin synthesis by rat glomerular mesangial cells in culture, *J. Clin. Invest.* 71:1756-1764, 1983.
3. Vallotton, M.B., Gerber-Wicht, C., Dolci, W., and Wuthrich, R.P., Interaction of vasopressin and angiotensin II in stimulation of prostacyclin synthesis in vascular smooth muscle cells, *Am. J. Physiol.* 257:E617-E624, 1989.

4. Bakris, G.L., Fairbanks, R., and Traish, A.M., Arginine vasopressin stimulates human mesangial cell production of endothelin, *J. Clin. Invest.* 87:1158-1164, 1991.

5. Bacic, F., Uematsu, S., McCarron, R.M., and Spatz, M., Prostaglandin D_2 in cultured capillary and microvascular endothelium of human brain, *Prostaglandins, Leukotrienes Med.*, in press.

6 Luscher, T.F., Boulanger, C.M., Dohi, Y., and Yang, Z., Endothelium-derived contracting factors, *Hypertension* 19:117-130, 1992.

7. Ralevic, V., Lincoln, J., and Burnstock, G., Release of vasoactive substances from endothelial cells, *in*: "Endothelial Regulation of Vascular Tone," U.S. Ryan and G.M. Rubanyi, eds., Marcel Dekker, Inc., New York, 1992.

8. Carretero, O.A., and Scicli, A.G., Local hormonal factors (intracrine, autocrine, and paracrine) in hypertension, *Hypertension*, 18 (suppl I):I-58- I-69, 1991.

9. Gerhart, D.Z., Broderius, M.A., and Drewes, L.R., Cultured human and canine endothelial cells from brain microvessels, *Brain Res. Bull.* 21:785-793, 1988.

10. Berridge, J.M., Downes, P., and Hanley, R.M., Lithium amplifies agonist-dependent phosphatidyl-inositol responses in brain and salivary glands, *Biochem. J.* 206:587-595, 1982.

11. Stanimirovic, D.B., Yamamoto, T., Uematsu, S., and Spatz, M., Endothelin-1 receptor binding and cellular signal transduction in cultured human brain cells, submitted to *J. Cereb. Blood Flow Metab.*

12. Lowry, O.H., Rosenbough, N.J., Farr, A.L., and Randall, R.J., Protein measurement with Folin phenol reagent, *J. Biol. Chem.* 193:265-275, 1951.

13. Nabika, T., Velletri, P.A., Lovenbcrg, W., and Beaven, M.A., Increase in cytosolic calcium and phosphoinositide metabolism induced by angiotensin II and [Arg] vasopressin in vascular smooth muscle cells, *J. Biol. Chem.* 260:4661-4670, 1985.

14. Reynolds, E.E., Mok, L.L.S., and Kurokawa, S., Phorbol ester dissociates endothelin-stimulated phosphoinositide hydrolysis and arachidonic acid release in vascular smooth muscle cells, *Biochem. Biophys. Res. Comm.* 160:868-873, 1989.

15. Spatz, M., Stanimirovic, D., Uematsu, S., Roberts L.J., II, Bembry, J., and McCarron, R.M., Production of PGF_{2a}, 9a-11-epi PGF_2, PGE_2, and TxB_2 induced by PGD_2 in capillary endothelium of human brain, submitted to *Brain Res.*

16. Kanse, S.M., Takashashi, K., Warren, J.B., Ghatei, M., and Bloom, S.R., Glucocorticoids induce endothelin release from vascular smooth muscle cells but not endothelial cells, *Eur. J. Pharmacol.* 199:99-101, 1991.

ENDOTHELIN-1 BINDING TO HUMAN BRAIN MICROVASCULAR AND CAPILLARY ENDOTHELIUM: MEMBRANES vs. INTACT CELLS

Danica B. Stanimirovic[1], Toshifumi Yamamoto[1], Hideko Yamamoto[1], Sumio Uematsu[2], and Maria Spatz[1]

[1]Stroke Branch, NINDS, NIH, Bethesda, MD
[2]John Hopkins University Hospital, Baltimore, MD 21205

INTRODUCTION

Endothelins are a group of 21-amino acid peptides secreted by endothelial cells, and expressed as three isoforms in the human genome (endothelin-1, -2 and -3) (6). Endothelin-1 (ET-1) has potent vasoconstrictor activity both in vitro and in vivo (8,22). In addition to endothelium, presence and/or secretion of immunoreactive ET-1 has been shown in vascular smooth muscle cells, various glial cells, and neurons (5,12,15). Vasoactive actions of endothelins are mediated by two types of endothelin receptors characterized as ET_A, with high affinity for ET-1 and low affinity for ET-3, predominantly localized on vascular smooth muscle cells, and nonselective ET_B, mainly expressed on endothelial cells (7). Intracellular signaling events mediating vasoactive actions of endothelins depend on the type of receptors, tissue and species investigated (7,14,20), and comprise activation of phospholipase C (PLC), subsequent inositol triphosphate (IP_3)-mediated mobilization of intracellular calcium, diacyl glycerol (DAG)-mediated stimulation of protein kinase C (PKC), activation of phospholipase A_2 (PLA_2) and arachidonic acid release, and influence on ionic transport across the membranes (7,16,20).

Endothelins have been detected in cerebrospinal fluid of patients with various cerebrovascular disorders (subarachnoidal haemorrhage, stroke) (19) and are implicated in predisposing vascular diseases (hypertension) (19). Recently, we have detected secretion of immunoreactive endothelin-1 by human brain endothelial cells (1). This study demonstrates the expression of ET_A receptors on these cells coupled to PLC activation and changes in both cyclic adenosine monophosphate (cAMP) and cyclic guanosine monophosphate (cGMP) formation.

Frontiers in Cerebral Vascular Biology: Transport and Its Regulation
Edited by L.R. Drewes and A.L. Betz, Plenum Press, New York, 1993

171

MATERIALS AND METHODS

Cell Culture

Normal samples of temporal lobe removed surgically for the treatment of idiopathic epilepsy were used for isolation of microvessels and capillaries and subsequent dissociation and cultivation of endothelial cells (HBEC) by modified technique of Gerhart et al. (4). The purity of the cultures was assessed by immunostaining for von Willebrand (Factor VIII)-related antigen and glial fibrillary acidic protein (GFAP).

Endothelin Binding Assay

Binding assays were performed on both disrupted cellular membranes, homogenized in Tris-HCl buffer, pH 7.4, and on intact cells ("living" confluent cell culture) grown in 96-well microtiter plates. Binding of [^{125}I-ET-1] to the membranes was conducted in a final volume of 0.1 ml of 50 mM Tris-HCl, pH 7.4, containing 100 pM of radiolabeled ET-1, and various concentrations of displacing unlabeled ET-1, ET-2, or ET-3 (10^{-14}-10^{-2} M). Samples were incubated for 120 min at $25°$C and binding was terminated by rapid filtration through Whatman F/B glass fiber filters (presoaked in 0.2% bovine serum albumin (BSA) in 50 mM Tris-HCl buffer pH 7.4). Filters were washed three times with 0.2 ml of 0.2% BSA and counted in a gamma counter. The binding procedure for the intact cells was essentially the same as for the membranes, except that cell harvesting was performed by overnight incubation in Triton X-100. Nonspecific binding in both cases was determined in the presence of 500 nM of ET-1. Experiments were performed in triplicate and repeated three to four times for each condition. Binding parameters were calculated using the LIGAND program (13) for weighted nonlinear regression curve fitting. Protein content was determined according to Lowry et al. (11).

Determination of [^{3}H]IP$_3$ Levels

Cells were prelabeled with 5 μCi of myo-[^{3}H]inositol for 24 hr and, after washing with Gibco-M199 in the presence of 20 mM lithium chloride (LiCl), exposed to endothelins and other tested substances. The reaction was terminated with 0.5 ml of 15% trichloroacetic acid and the radioactivity incorporated in the IP$_3$ fraction was separated by DOWEX AG 1x8 anion exchange chromatography as described previously (2,17).

Determination of cAMP and cGMP

Both cAMP and cGMP were determined with the use of their respective radioimmunoassays in 0.3 M perchloric acid cell extracts. Experiments were performed in the presence of 1 mM iso-butyl-methyl xanthine (IBMX), using protein kinase C (regulatory unit) obtained from SIGMA and a radioimmunoassay kit obtained from New England Nuclear.

RESULTS

Binding of [^{125}ET-1] to either HBEC membranes or intact cells was time-dependent, reaching equilibrium after 60 min at $25°$C. Saturation studies performed in the picomolar range of concentrations demonstrated high affinity binding of ET-1 to a single class of receptor with apparent dissociation constants of 111 pM and 68 pM for membranes and intact cells, respectively. However, detailed analysis of displacement data performed in the wider concentration range (nanomolar) revealed the presence of both high and low affinity binding sites for ET-1 regardless of the preparation (membranes or cells) used (Table 1). In general, similar apparent dissociation constants were obtained for the membrane preparation and intact cells (Table 1). However, the density (B_{max}) of both high and low affinity binding sites was substantially higher on intact HBEC as compared to membranes. The order of potency for displacing radiolabeled ET-1, for both the membranes and intact cells, was ET-1>ET-2>ET-3, and ET-2>ET-1=ET-3 for high and low affinity ET-1 binding sites, respectively.

ET-1 induced marked (20 times), dose-dependent increases in IP_3 formation, reaching a maximum 15 min after stimulation at a concentration of 100 nM. In contrast to ET-1, ET-3 (100 nM) did not stimulate IP_3 formation in HBEC. ET-1-induced IP_3 stimulation was diminished by pretreatment with phorbol myristate ester (PMA, 10 μM), and pertussis toxin (Ptx, 500 ng/ml) was ineffective (Figure 1).

Table 1. Apparent dissociation constants (K_D) and density of binding sites (B_{max}) for ET-1, ET-2, and ET-3 binding to either intact HBEC or cellular membranes.

| | CELLS | | MEMBRANES | |
K_D	High	Low	High	Low
ET-1	0.122 ± 0.014	32 ± 6	0.197 ± 0.016*	39 ± 9
ET-2	0.187 ± 0.025	10 ± 2	0.230 ± 0.018	25 ± 9
ET-3	6.7 ± 1.3	34 ± 5	4.1 ± 1.6	46 ± 12
B_{max}	112 ± 20	909 ± 54	47 ± 9*	237 ± 35*

Values are given as Mean ± SE for three independent displacement experiments, each performed in triplicate. Values for K_D are presented in nM, whereas values for B_{max} were given as fmol/mg protein.
*-indicates significant difference (p<0.01; Student's t-test) between intact cells and their disrupted membranes.

cAMP production by HBEC was increased (80% above the control values) by ET-1 (10 nM), whereas it was inhibited (-35%) by ET-3 (10 nM). Phorbol myristate ester (PMA, 10 μM) potentiated the ET-1 (10 nM) effect on cAMP production by HBEC (Figure 1). cGMP production, contrary to cAMP, was decreased (-40%) by ET-1 (10 nM) and unchanged by ET-3 (Figure 1).

DISCUSSION

This study describes the presence of both high and low affinity binding sites for ET-1 on HBEC. Both types of binding were expressed on intact cells as well as on a membrane preparation. The apparent dissociation constants for the high affinity ET-1 binding site were in the picomolar range, whereas they were in the nanomolar range for the low affinity site. Although the dissociation constants were similar (somewhat lower affinity was found on the membrane preparation), the density of respective ET-1 binding sites was significantly lower on the membranes than on the living cells. The affinity constants obtained for both membranes and cells, as well as the order of potency for displacing ET-1 from high affinity binding site for various endothelins (ET-1>ET-2>ET-3) were similar to those described in the literature for rat brain capillary endothelial cells (20), glia (12) and neurohybrid NG 108-15 cells (24). The low affinity and high density ET-1 binding site may represent non-selective ET_B receptor, as suggested by characteristics of ET-3 binding to HBEC (not shown).

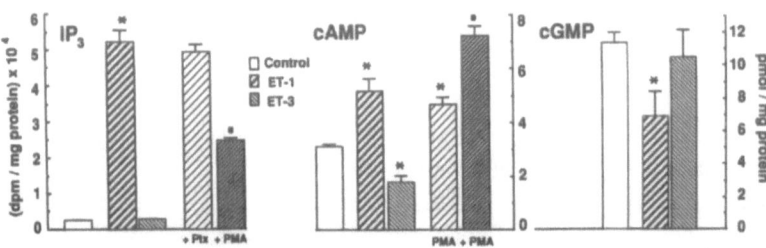

Figure 1. Effects of ET-1 (alone or in the presence of pertussis toxin (+Ptx) or phorbol myristate ester (+PMA)) and ET-3 on IP_3, cAMP, and cGMP production in HBEC. Experimental protocols were described in Materials and Methods. Results are expressed as Mean ± SE for three independent experiments, each performed in triplicate. * - p<0.01 (Student's t-test) as compared with control values; •- p<0.01 (Student's t-test) as compared with ET-1-stimulated group.

The differences in the affinity constants and density of binding sites found between intact cells and disrupted cellular membranes may be explained by existing different dynamic states of receptors during their assay conditions. Disrupted membrane preparations are, in general, considered more reliable for binding studies because a constant equilibrium can be attained between bound and dissociated ligand (i.e., presence of a constant number of receptors). However, in the living cells, receptors for some polypeptide hormones are known to be in a dynamic equilibrium between the surface membranes and the intracellular pools arising from the receptor recycling or de novo synthesis (10), as it has been demonstrated for ET-1 receptors mediating the release of luteinizing hormone from adenohypophysis, which undergo fast and rapid internalization after forming a complex with ET-1 (18). Furthermore, affinity states for receptor-agonist interactions on cell surface receptors, could be modulated by GTP and guanine nucleotide analogues present inside the "living" cells, causing a prevalence of lower affinity receptor-agonist interaction and underestimation of unoccupied receptors available in vivo (10). Nevertheless, it appears that the properties of the ET-1 receptors were not influenced by a "living" metabolic state of the HBEC since the kinetic parameters of ET-1 binding sites, as well as the order of potency for displacing ET-1, were similar for both membranes and living cells. It is possible, therefore, that some of the physically present receptors were lost during the preparation of the membranes, accounting for the lower number of binding sites (B_{max}) detected on the membranes as compared to the living cells.

Most importantly, the main advantage of binding studies performed on living cells is that receptor occupancy by ligands can be compared with in vivo receptor-mediated physiological effects under identical experimental conditions. We have demonstrated that occupancy of the high affinity ET-1 binding site on HBEC by ET-1 was associated with an increase in intracellular IP_3 and cAMP and a decrease in cGMP levels. The high affinity ET-1 receptor coupling to PLC activation, with a consequent breakdown of inositolphosphatides and accumulation of IP_3 have also been demonstrated on rat capillary endothelial cells (21) and neurohybrid cells (24). Activation of PLC in HBEC was shown to be down-regulated by PMA (activator of protein kinase C), and independent of PTx-sensitive G-proteins. ET-3, in contrast to ET-1, was ineffective on PLC.

Mechanisms involved in ET-1-mediated stimulation of cAMP may comprise PKC-mediated activation of adenylate cyclase (PMA alone increased cAMP levels by 50%), as well as inhibition of phosphodiesterases by nitric oxide-mediated increase in cGMP as suggested by Namiki et al. (14). However, the demonstrated decrease of cGMP in response to ET-1, as well as potentiation of the PMA effects on cAMP production by ET-1 in HBEC, suggest an involvement of some other regulatory mechanisms. IP3-mediated mobilization of intracellular calcium (18) as well as extracellular calcium entry (3) represent one of the possibilities. Judging from the presented data (cAMP, cGMP), it seems that the regulation of cyclic nucleotide messenger systems in HBEC upon ET-1 binding to the high affinity receptor is reciprocal in nature.

Stimulation of different messenger systems in HBEC by ET-1 can lead to activation of various secretory functions of the endothelium. Recently, it has been demonstrated that ET-1 induces secretion of vasoactive prostanoids (16), angiotensin II (9), and self-secretion (23) by endothelial cells. Thus, ET-1 evoked endothelial secretion of vasoactive substances may modulate the ET-1-induced responses of the cerebromicrovascular bed including properties of the blood-brain barrier.

REFERENCES

1. Bacic, F., Uematsu, S., McCarron, R.M., and Spatz, M., Secretion of immunoreactive endothelin-1 by capillary and microvascular endothelium of human brain, *Neurochem. Res.* 17:699, 1992.

2. Berridge, J.M., Downes, P., and Hanley, R.M., Lithium amplifies agonist-dependent phosphatidylinositol responses in brain and salivary glands, *Biochem J.* 206:587, 1982.

3. Chan, J., and Greenberg, A.D., SK&F 96365, a receptor-mediated calcium entry inhibitor, inhibits calcium responses to endothelin-1 in NG108-15 cells, *Biochem. Biophys. Res. Comm.* 177:1141, 1991.

4. Gerhart, D.Z., Broderius, M.A., and Drewes, L.R., Cultured human and canine endothelial cells from brain microvessels, *Brain Res. Bul.* 21:785, 1988.

5. Giald, A., Gibson, S.J., Herrero, M.T., Gentleman, S., Legon, S., Yanagisawa, M., Masaki, T., Ibrahim, N.B.N., Roberts, G.W., Rossi, M.L., and Polak, J.M., Topographical localization of endothelin mRNA and peptide immunoreactivity in neurones of the human brain, *Histochem.* 95:303, 1991.

6. Inoue, A., Yanagisawa, M., Kimura, S., Kasuya, V., Miyachi, T., Goto, K., and Masaki, T., The human endothelin family - three structurally and pharmacologically distinct isopeptides predicted by three separate genes, *Proc. Nat'l. Acad. Sci. USA* 86:2853, 1989.

7. Jones, C.R., Hiley, C.R., Pelton, J.T., and Miller, R.C., Endothelin receptor heterogeneity: structure activity, autoradiographic and functional studies, *J. Receptor Res.* 11:299, 1991.

8. Kobayashi, H., Hayashi, M., Kobayashi, S., Kabuto, M., Handa, Y., Kawano, H., and Ide, H., Cerebral vasospasm and vasoconstriction caused by endothelin, *Neurosurg.* 28:673, 1991.

9. Kawaguchi, H., Sawa, H., and Yasuda, H., Endothelin stimulates angiotensin I to angiotensin II conversion in cultured pulmonary artery endothelial cells, *J. Mol. Cell Cardiol.* 22:839, 1990.

10. Limbird, E.L., "Cell Surface Receptors: A Short Course on Theory and Methods," Martinus Nijhoff, Boston, 1986.

11. Lowry, H.O., Rosebrough, U.N., Farr, L.A., and Randall, J.R., Protein measurement with the Folin phenol reagent, *J. Biol. Chem.* 193:265, 1951.

12. MacCumber, M.W., Ross, C.A., and Snyder, H.S., Endothelin in the brain: Receptors, mitogenesis, and biosynthesis in glial cells *Proc. Natl. Acad. Sci. USA* 87:2359, 1990.

13. Munson, J.P., and Rodbard, D., LIGAND: A versatile computerized approach for characterization of ligand-binding systems, *Analytical Biochem.* 107:220, 1980.

14. Namiki, A., Hirata, Y., Ishikawa, M., Moroi, M., Aikawa, J., and Machii, K., Endothelin-1-and endothelin-3-induced vasorelaxation via common generation of endothelium-derived nitric oxide, *Life Sci.* 50:677, 1992.

15. Resink, T.J., Hahn, A.W.A., Scott-Burden, T., Powell, J., Weber, E., and Buhler, F., Inducible endothelin mRNA expression and peptide secretion in cultured human vascular smooth muscle cells, *Biochem. Biophys. Res. Comm.* 168:1303, 1990.

16. Spatz, M., Bacic, F., Stanimirovic, D., Uematsu, S., McCarron, R., and Bembry, J., Endothelin-1 and PGD$_2$ stimulates the production of PGF$_2$ and Thromboxane B$_2$ in human cerebrovascular endothelium, *Stroke* 23:156, 192.

17. Stanimirovic, D.B., Yamamoto, T., Uematsu, S., and Spatz, M., Endothelin-1 receptor binding and cellular signal transduction in cultured human brain endothelial cells, submitted to *J. Cereb. Blood Flow Metab.*

18. Stojiljkovic, S.S., Balla, T., Fukuda, S., Cesnjaj, M., Merelli, F., Krsmanovic, L.Z., and Catt, K.J., Endothelin ET$_A$ receptors mediate the signaling and secretory actions of endothelins in pituitary gonadotrophs, *Endocrin.* 130:465, 1992.

19. Suzuki, H., Sato, S., Takekoshi, K., Ishihara, N., and Shimoda, S., Increased endothelin concentration in CSF from patients with subarachnoidal hemorrhage, *Acta. Neurol. Scand.* 81:553, 1990.

20. Vigne, P., Ladoux, A., and Frelin, C., Endothelins activate Na$^+$/H$^+$ exchange in brain capillary endothelial cells via a high affinity endothelin-3 receptor that is not coupled to phospholipase C, *J. Biol. Chem.* 266:5925, 1991.

21. Vigne, P., Marsault, R., Breittmayer, P.J., and Frelin, C., Endothelin stimulates phosphatidylinositol hydrolysis and DNA synthesis in brain capillary endothelial cells, *Biochem. J.* 266:415, 1990.

22. Yanagisawa, M., Kurihara, H., Kimura, S., Tomobe, Y., Kobayashi, M., Mitsui, Y., Yazaki, Y., Goto, K., and Masaki, T., Endothelin: a novel potent vasoconstrictor peptide produced by vascular endothelial cells, *Nature* 332:411, 1988.

23. Yokokawa, K., Kohno, M., Yasunari, K., Murakawa, K., and Takeda, T., Endothelin-3 regulates endothelin-1 production in cultured human endothelium, *Hypertension* 18:304, 1991.

24. Yue, L.T., Nambi, P., Wu, L.H., and Feuerstein, G., Endothelin receptor binding and cellular signal transduction in neurohybrid NG108-15 cells, *Neurosci.* 44:215, 1991.

POSSIBLE INVOLVEMENT OF C-KINASE

IN OCCURRENCE OF CHRONIC CEREBRAL

VASOSPASM AFTER SUBARACHNOID HEMORRHAGE

Toru Matsui, Yoh Takuwa*, Hiroyuki Kaizu, and
Takao Asano

Department of Neurosurgery, Saitama Medical Center/School
1981 Kamoda, Kawagoe, Saitama, Japan 350
*Department of Vascular Biology, Faculty of Medicine
Tokyo University, 7-3-1 Hongo, Bunkyo-ku, Tokyo, Japan

INTRODUCTION

Rasmussen et al. (1) introduced the concept that the sustained phase of smooth muscle contraction is ascribable to activation of protein kinase C (PKC). Based upon their concept, we have demonstrated that phorbol 12,13-diacetate (PDA) induced sustained contraction of the canine basilar artery via activation of the PKC-dependent contractile system (2,3). This finding led us to the hypothesis that PKC activation might play a pivotal role in the development of chronic vasospasm after subarachnoid hemorrhage. Subsequent studies have revealed that the content of 1,2-diacylglycerol (DG), an intrinsic activator of PKC, in the canine basilar artery (BA) undergoing chronic vasospasm (VS; day 7) following experimental subarachnoid hemorrhage (SAH) increased significantly (Figure 1). H-7 or staurosporine, a PKC inhibitor, dilated the spastic basilar artery, but both calcium channel blocking agents and calmodulin antagonists failed (Figure 2) (4). The present study was designed to examine phospholipid metabolism in BA during chronic VS by determining the increases in DG content and by measuring phosphorylation of myosin light chain (20 kDa MLC) and activation of PKC in the occurrence of VS.

MATERIALS AND METHODS

SAH was produced according to the method of Varsos et al. (5). Briefly, autologous nonheparinized arterial blood (0.4 ml/kg weight) was injected into the cisterna magna following removal of isovolemic cerebrospinal fluid on day 1. On day 3, the second induction of SAH was repeated in the same fashion as on day 1.

Frontiers in Cerebral Vascular Biology: Transport and Its Regulation
Edited by L.R. Drewes and A.L. Betz, Plenum Press, New York, 1993

177

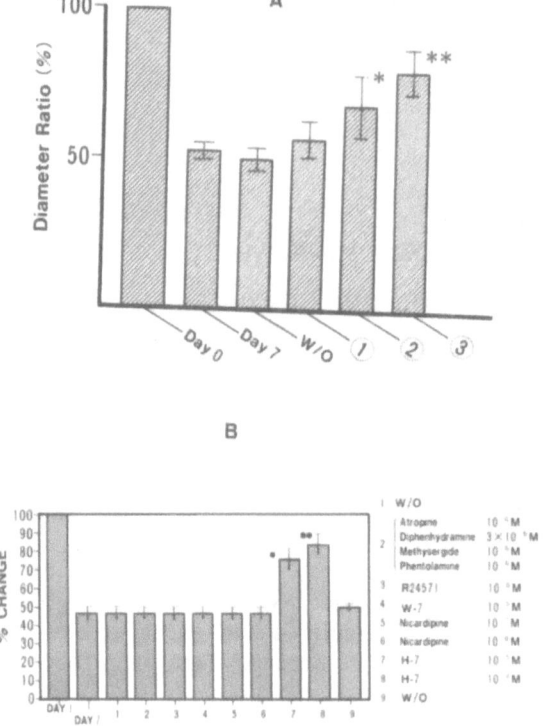

Figure 1. Effect of vasospasm on 1,2-diacylglycerol content of canine basilar artery. The content of 1,2-diacylglycerol (DG), an intrinsic acitvator of PKC, began to increase in the beagle basilar artery on day 2 and stayed elevated until day 7. On day 14, the DAG level returned to normal. * p<0.05, ** p<0.01 vs. control.

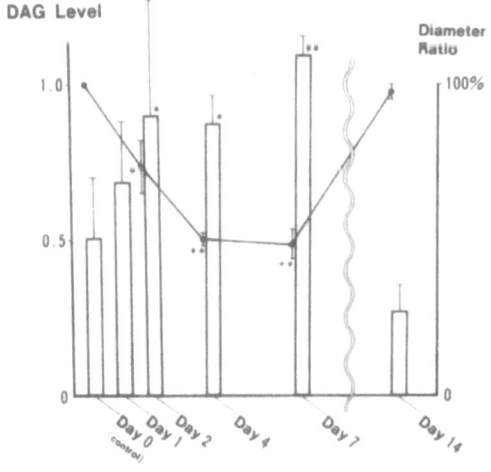

Figure 2. The basilar artery after 7 days of vasospasm could be dilated with the topical application of staurosporine (A) or H-7 (B). Staurosporine concentrations in A are ①: 10^{-9}, ②: 10^{-8}, ③: 10^{-7}M. *p<0.05; **p<0.01 vs. control.

Experiment One

Using the BA of day 7, incorporation of ^{3}H-choline, ^{14}C-ethanolamine and ^{3}H-myoinositol into phosphatidylcholine (PC), phosphatidylethanolamine (PE) and phosphoinositides (PI), respectively, was measured. Excised BA were incubated in modified Krebs-Henseleit solution containing ^{3}H-choline, ^{14}C-ethanolamine and ^{3}H-myoinositol for 20 min, and the reaction was terminated with trichloroacetic acid (TCA). After homogenization of the samples, lipids were extracted using the method of Bligh and Dyer (6) for PC and PE and the method of Schacht (7) for PIs. PC and PE were separated on heat-activated silica gel 60 plates using chloroform/methanol/acetic acid/water (65:43:1:3, v/v) as a solvent; PIs were separated on heat-activated, 1% (w/v) potassium oxalate-impregnated silica gel 60 plates using chloroform/methanol/acetone/acetic acid/water (40:15:13:12:8, v/v). Each spot was scraped and counted with a liquid scintillation spectrometer.

Experiment Two

The ratios of phosphorylated 20 kDa MLC to the total 20 kDa MLC of BA on day 6 were compared with those of the normal control BA. Excised BA were frozen in dry ice/acetone with 10% TCA and 20 mM dithiothreitol (DTT) and thereafter thawed to room temperature. After homogenization, followed by 5 min of boiling, the samples were centrifuged at 1500 x g for 3 min and the supernatant was decanted for analysis by isometric focusing followed by SDS-PAGE. Gels were stained with Coomassie brilliant blue and scanned with a densitometer.

Experiment Three

The PKC activity in the cytosol and membrane fractions of BA on days 0 (control), 4 and 7 was measured. Excised BA were weighed and homogenized in an ice-cold homogenization buffer, followed by centrifugation (100,000 x g for 60 min at 4°C). The supernatant was used as the cytosolic fraction. The pellet was rehomogenized in buffer containing 1% Triton X-100 and centrifuged at 100,000 x g for another 60 min. The supernatant was taken as the membrane fraction. Both fractions were applied to 1-ml DE-52 columns equilibrated in buffer A (20 mM Tris/HCl, pH 7.5; 0.5 mM EGTA; 0.5 mM EDTA; 1 mM DTT and 10% glycerol). After washing the columns with buffer A, the PKC activity was eluted with buffer A containing 400 mM NaCl. The PKC activity was assessed using a PKC enzyme assay kit from Amersham.

Experiment Four

The control BA was homogenized in the homogenization buffer containing 1% Triton X-100. The homogenate was centrifuged at 100,000 x g and the supernatant was subjected to SDS-PAGE, followed by electrophoretical transfer of proteins onto a polyvinylidene difluoride membrane. The immunoreactive proteins were detected by employing monoclonal antibodies specific to PKC α, β and γ (8), and a blotting detection kit using alkaline phosphatase.

RESULTS

Experiment One

The absolute masses of all phospholipids were similar between spastic (day 7) and control BA. The spastic BA showed significantly higher rates of incorporation of ^{3}H-choline, and ^{14}C-ethanolamine into PC and PE, in comparison with that of control BA (Figure 3).

Experiment Two

The extent of 20 kDa MLC phosphorylation was not significantly different between spastic and normal control BA (Figure 4).

Incorporation of ^{3}H-choline, ^{15}C-ethanolamine,
^{3}H-myoinositol

	PC	PE	PI	PIP	PIP$_2$
control	(**) 9300±3400	(**) 190±40	880±160	180±40	50±20
Day 7	18700±2200	330±60	740±360	160±40	60±20

(**) : p<0.01 vs. control

Figure 3. The incorporation of ^{3}H-choline into phosphatidylcholine (PC), ^{14}C-ethanolamine into phosphatidylethanolamine (PE), and ^{3}H-myoinositol into phosphoinositides (PI, PIP, PIP$_2$) by basilar arteries from control dogs and after 7 days of hemorrhage.

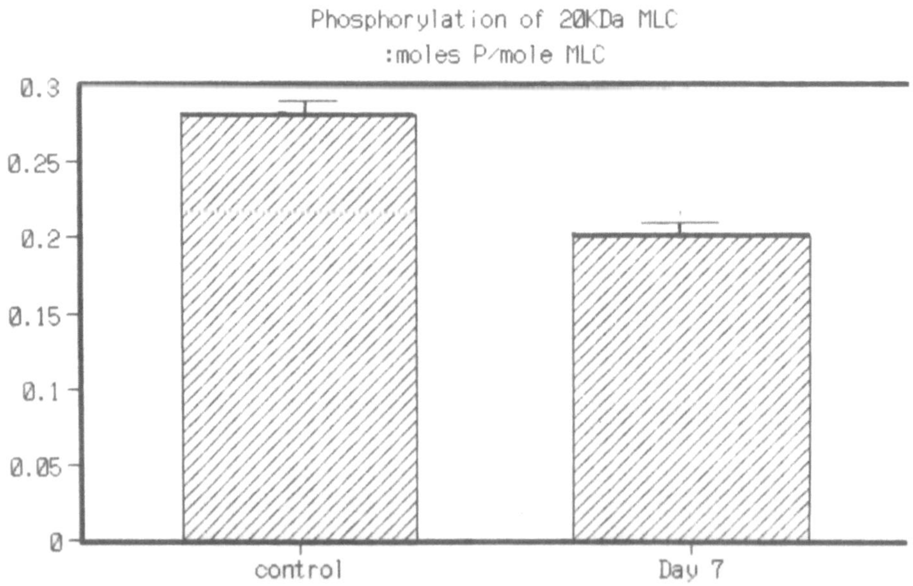

Figure 4. Phosphorylation of 20 kDa myosin light chain of basilar artery on days 0 and 7 of hemorrhage.

Experiment Three

PKC activity of the cytosol fraction in the normal BA was 29.05±7.38 pmol/min/mg wet weight. Values in Figure 5 are represented as a percent of this value. The PKC activity in the cytosol fraction was significantly decreased on days 2 and 7 and that in the membrane fraction was unaltered during the observation period (Figure 5).

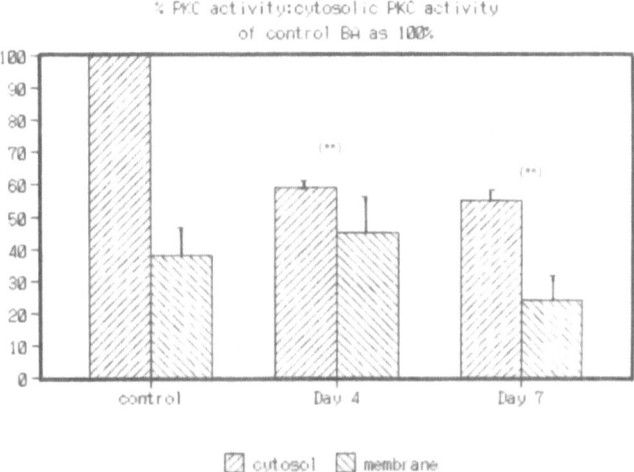

Figure 5. Time course of PKC activity in basilar artery cytosol and membranes. **p<0.01 vs. control.

Experiment Four

The PKC α was a predominant isoform in the canine BA.

DISCUSSION

The phosphorylation level of 20 kDa MLC of spastic BA (day 7) was not significantly increased in comparison to that of control BA. This suggested that a different mechanism other than 20 kDa phosphorylation might be responsible for the occurrence of chronic VS. This finding is consistent with the observation that chronic VA in the canine two-hemorrhage model was not reversed by a topical application of calmodulin inhibitor to the transclivally exposed spastic BA. Our previous study (2-4) led us to the postulation that PKC activation might play an important role in the pathogenetic mechanism underlying chronic VS.

The increased turnover of PC and PE in the spastic BA (day 7) suggested that these phospholipids might serve as a potential source for a prolonged increase in DG content of BAs of beagle dogs subjected to two hemorrhages. It is surmised that cleavage of PC and PE by phospholipase C or D (9) might lead to a DG increase in the spastic BA.

Although the membrane PKC activity was unaltered in spastic arteries (days 4 and 7), the cytosolic PKC activity was down-regulated in those arteries. This phenomenon might have been brought about by the sustained increase in the DG content of spastic BA. These results indicate that the process that affected the PKC activity was in progress in the spastic BA. The data, taken together with our recent observations, suggest that PKC plays a role in the development of chronic cerebral VS.

SUMMARY

The present study aimed to examine the turnover of phospholipids such as PI, PC and PE, the time course of PKC activity and the phosphorylation of 20 kDa MLC in the canine BA undergoing chronic VS. The phosphorylation of 20 kDa MLC was not augmented in the spastic BA. Turnover of PC and PE was detectably stimulated on day 7. The cytosolic PKC activity was down-regulated on days 4 and 7, while the membrane PKC activity remained unchanged during these periods. The present results indicate that a process which affected the membrane lipid metabolism, PKC metabolism and PKC activity occurred in spastic BA.

REFERENCES

1. Rasmussen, H., Takuwa, Y., and Park, S., Protein kinase C in the regulation of smooth muscle contraction, *FASEB J.* 1:177-185, 1987.
2. Sugawa, M., Koide, T., Naitoh, S., Takato, M., Matsui, T., and Asano, T., Phorbol 12,13-diacetate-induced contraction of the canine basilar artery: role of protein kinase C, *J. Cereb. Blood Flow Metab.* 11:135-142, 1991.
3. Matsui, T., Sugawa, M., Johshita, H., Takuwa, Y., and Asano, T., Activation of the protein kinase C-mediated contractile system in canine basilar artery undergoing chronic vasospasm, *Stroke* 22:1183-1187, 1991.
4. Matsui, T., Takuwa, Y., Johshita, H., Yamashita, K., and Asano, T., Possible role of protein kinase C-dependent smooth muscle contraction in the pathogenesis of chronic cerebral vasospasm, *J. Cereb. Blood Flow Metab.* 11:143-149, 1991.
5. Varsos, V.G., Liszczak, T.M., Han, D.H., Kistler, J.P., Vielma, J., Black, P.M., Heros, R.C., and Zervas, N.T., Delayed cerebral vasospasm is not reversed by aminophylline, nifedipine or papaverine in a "two-hemorrhage" canine model, *J. Neurosurg.* 22:492-500, 1988.
6. Bligh, E.G., and Dyer, W.G., A rapid method of total lipid extraction and purification, *Can. J. Biochem. Physiol.* 37:911-917, 1959.
7. Takuwa, Y., Takuwa, N., and Rasmussen, H., Carbachol induces a rapid and sustained hydrolysis of phosphoinositide in bovine tracheal smooth muscle. Measurement of polyphosphoinositides, 1,2-diacylglycerol, and phosphatidic acid, *J. Biol. Chem.* 261:14670-14675, 1986.
8. Hidaka, H., Tanaka, K., Onoda, M., Watanabe, M., Ohta, H., Ito, Y., Tsuru-dome, M., and Yoshida, T., Cell type specific expression of protein kinase C isoenzymes in the rabbit cerebellum, *J. Biol. Chem.* 263:4523-4526, 1988.
9. Exton, J.H., Signaling through phosphatidylcholine breakdown, *J. Biol. Chem.* 265:1-4, 1990.

A NOVEL GENE FAMILY MAY ENCODE ENDOTHELIAL CELL SPECIFIC ADHESION-LIKE MOLECULES: AN EXTRACELLULAR LOOP-REPEAT-LOOP (LRL) MOTIF AND CYTOPLASMIC TYROSINE KINASE DOMAINS

Thomas N. Sato and Ying Qin

Department of Neurosciences
Roche Institute of Molecular Biology
Roche Research Center
Nutley, NJ 07110

Vascular endothelium separates underlying tissues from the blood circulation. Because of this organization, endothelial cells play pivotal roles for signaling between the blood circulation and tissues. The signaling functions of vascular endothelial cells in the central nervous system (CNS) are especially important because of the existence of the blood-brain barrier (BBB) against diffusion-mediated signal transmission between the CNS and the blood circulation. This indicates that any signals have to be integrated into the endothelial signal transduction machinery before they reach either side of the endothelium. The vascular system in the CNS has another unique organization. It is surrounded by a complex and heterologous cellular organization as opposed to the relatively simple and homogeneous tissue organization in peripheral tissues. The developmental program of the endothelium in the CNS may be more complex than in peripheral tissues because development and differentiation of endothelial cells in the CNS may be affected by heterologous cellular interactions.

One way to study the signaling apparatus in endothelial cells that mediate information exchange between the separated cells is to analyze cell surface receptors expressed on either the luminal or abluminal side of endothelial cells. Receptors on the luminal surface can sense factors in the circulation which, by binding to them, can initiate a signal transduction cascade in endothelial cells. Eventually signals are sent to the CNS in different forms. Receptors on the abluminal surface can bind factors in the CNS and these signals can be transduced into the blood circulation.

One of the classes of cell surface receptors involved in this type of signal transduction is a receptor-linked tyrosine kinase. Tyrosine kinases play pivotal roles in many biological systems (1-5). There are a number of growth factors that bind endothelial cell surface receptors linked to cytoplasmic tyrosine kinase domains and promote endothelial cell growth and migration (6-12). There are also substantial numbers of reports indicating that

Frontiers in Cerebral Vascular Biology: Transport and Its Regulation
Edited by L.R. Drewes and A.L. Betz, Plenum Press, New York, 1993

183

brain endothelial cells. The polymerase chain reaction (PCR) was used to amplify tyrosine kinase genes (21) from a bovine brain endothelial cell cDNA library. Among many amplified genes we characterized two novel clones that seemed most interesting because they form a new gene family of receptor-linked tyrosine kinases.

The extracellular domains of both receptors consist of two immunoglobulin (Ig)-like loops flanking three EGF-repeats, and these domains are followed by three fibronectin (FN) type III repeats. The cytoplasmic tyrosine kinase domains are composed of split-type tyrosine kinase domains such as in FGF receptors. The deduced amino acid sequences of both clones revealed intriguing homologies between the two. The sequence identities of the cytoplasmic tyrosine kinase domains are higher than 80%. Although the extracellular regions contain the identical LRL motif structure, the amino acid sequences are more divergent. The highest homology (50-55% identity) is seen in the three EGF-like repeats. The second Ig loop and the first FN type III repeat are about 40% identical. This extracellular LRL motif is unique to this gene family, but closely related to cell adhesion molecules such as the CAM family of proteins and lymphocyte homing receptors. The sequence of one of these genes is similar to another, recently reported, human endothelial cell surface receptor-linked tyrosine kinase (22) and may represent the bovine homolog.

The expression of these genes in adult mice was analyzed histochemically by *in situ* hybridization. In brain, they are expressed only in vascular endothelial cells. The expression of these genes was observed in both major capillaries and in microvessels. In peripheral tissues, expression is also restricted to the vasculature.

Cloning and histochemical analyses have revealed the existence of a novel endothelial receptor-linked tyrosine kinase gene family. It will be interesting to determine how far these endothelial cell specific receptor gene family members extend by PCR analyses using primer pairs specific to these two clones. Identification of putative ligands for these receptors and studying their roles may shed light on as yet unidentified functions of vascular endothelial cells.

REFERENCES

1. Hunter, T., and Cooper, J., Protein-tyrosine kinases, *Annu. Rev. Biochem.*. 54:897-930, 1985.

2. Hunter, T., Gould, K.L., Lindberg, R.A., Meisenhelder, J., Middlemas, D.S., and Thompson, D.P., Protein-tyrosine kinases and their substrates: Old friends and new faces, *in:* "Protein Design and the Development of New Therapeutics and Vaccines," J.B. Hooke, and G. Poste, eds., Plenum Press, New York, 1990.

3. Pawson, T., and Bernstein, A., Receptor tyrosine kinases, genetic evidence for their role in *Drosophila* and mouse development, *Trends Genet.* 6:350-356, 1990.

4. Ullrich, A., and Schlessinger, J., Signal transduction by receptors with tyrosine kinase activity, *Cell* 61:203-212, 1990.

5. Yarden, Y., and Ullrich, A., Growth factor receptor tyrosine kinases, *Annu. Rev. Biochem.* 57:443-478, 1988.

6. Remmers, E.F., Sano, H., and Wilder, R.L., Platelet-derived growth factors and heparin-binding (fibroblast) growth factors in the synovial tissue pathology of rheumatoid arthritis, *Semin. Arthritis Rheum. (U.S.)* 21:191-199, 1991.

7. Blume-Jensen, P., Claesson-Welsh, L., Siegbahn, A., Zsebo, K.M., Westermark, B., and Heldin, C.H., Activation of the human c-kit product by ligand-induced dimerization mediates circular actin reorganization and chemotaxids, *EMBO J.* 10:4121-4128, 1991.

8. Myoken, Y., Kayada, Y., Okamoto, T., Kan, M., Sato, G.H., and Sato, J.D., Vascular endothelial cell growth factor (VEGF) produced by A-431 human epidermoid carcinoma cells and identification of VEGF membrane binding sites, *Proc. Natl. Acad. Sci. USA* 88:5819-5823, 1991.

9. Beitz, J.G., Kim, I.S., Calabresi, P., and Frackelton, Jr., A.R., Human microvascular endothelial cells express receptors for platelet-derived growth factor, *Proc. Natl. Acad. Sci. USA* 88:2021-2025, 1991.

10. Safran, A., Avivi, A., Orr-Urtereger, A., Neufeld, G., Lonai, P., Givol, D., and Yarden, Y., The murine flg gene encodes a receptor for fibroblast growth factor, *Oncogene* 5:635-643, 1990.

11. Gay, C.G., and Winkles, J.A., Heparin-binding growth factor-1 stimulation of human endothelial cells induces platelet-derived growth factor A-chain gene expression, *J. Biol. Chem.* 265:3284-3292, 1990.

12. Ruta, M., Burgess, W., Givol, D., Epstein, J., Neiger, N., Kaplow, J., Crumley, G., Dionne, C., Jaye, M., and Schlessinger, J., Receptor for acidic fibroblast growth factor is related to the tyrosine kinase encoded by the fms-like gene (FLG), *Proc. Natl. Acad. Sci. USA* 86:8722-8726, 1989.

13. Meyer, T., Regenass, U., Fabbro, D., Alteri, E., Rosel, J., Muller, M., Caravatti, G., and Matter, A., A derivative of staurosporine (CGP 41 251) shows selectivity for protein kinase C inhibition and *in vitro* anti-proliferative as well as *in vivo* anti-tumor activity, *Int. J. Cancer* 43:851-856, 1989.

14. Takata, K., and Singer, S.J., Phosphotyrosine-modified proteins are concentrated at the membranes of epithelial and endothelial cells during tissue development in chick embryos, *J. Cell Biol.* 106:1757-1764, 1988.

15. Montesano, R., Pepper, M.S., Belin, D., Vassalli, J.D., and Orci, L., Induction of angiogenesis *in vitro* by vanadate, an inhibitor of phosphotyrosine phosphatases, *J. Cell Physiol.* 134:460-466, 1988.

16. van den Eijnden-van Raaij, A.J., Feijen, A., and Snoek, G.T., EDTA-extractable proteins from calf lens fiber membranes are phosphorylated by Ca2+-phospholipid-dependent protein kinase, *Exp. Eye Res.* 45:215-225, 1987.

17. Kazlauskas, A., and DiCorleto, P.E., Comparison of the phosphorylation events in membranes from proliferating vs. quiescent endothelial cells, *J. Cell Physiol.* 130:228-244, 1987.

18. Mackie, K., Lai, Y., Nairn, A.C., Greengard, P., Pitt, B.R., and Lazo, L.S., Protein phosphorylation in cultured endothelial cells, *J. Cell Physiol.* 128:367-374, 1986.

19. Huang, S.S., and Huang, J.S., Association of bovine brain-derived growth factor receptor with protein tyrosine kinase activity, *J. Biol. Chem.* 261:9568-9571, 1986.

20. Gould, K.L., Cooper, J.A., and Hunter, T., The 46,000-dalton tyrosine protein kinase substrate is widespread, whereas the 36,000-dalton substrate is only expressed at high levels in certain rodent tissues, *J. Cell Biol.* 98:487-497, 1984.

21. Wilks, A.F., Two putative protein-tyrosine kinases identified by application of the polymerase chain reaction, *Proc. Natl. Acad. Sci. USA* 86:1603-1607, 1989.

22. Partanen, J., Armstrong, E., Makela, T.P., Korhonen, J., Sandberg, M., Renkonen, R., Knuutila, S., Huebner, K., and Alitalo, K., A novel endothelial cell surface receptor tyrosine kinase with extracellular epidermal growth factor homology domains, *Mol. Cell. Biol.* 12:1698-1707, 1992.

Interaction of
Brain Endothelial Cells
With Other Cells

PLASMINOGEN ACTIVATION AND ASTROGLIAL-INDUCED NEURAL MICROVESSEL MORPHOGENESIS

John Laterra[1,2,4], Ravi R. Indurti[1], and Gary W. Goldstein[1,3]

Departments of Neurology[1], Neuroscience[2], Pediatrics[3], and Oncology[4]
Kennedy Krieger Research Institute, Johns Hopkins Medical Institutions
707 North Broadway, Baltimore, MD 21205

INTRODUCTION

There are a number of documented and potential influences that perivascular astrocytic cells have on microvascular cells within the central nervous system. The primary experimental focus in this area has been upon the induction, maintenance and regulation by astrocytes of the blood-brain barrier endothelial phenotype (6,20,23). This is a result of the fact that the expression of blood-brain barrier properties most easily distinguishes central nervous system endothelial cells from endothelial cells in peripheral organs. It is likely, however, that astrocytes influence endothelial cells in ways not directly related to blood-brain barrier expression as used in its strictest sense. This paper will focus on the effects of astrocytes and astroglial cells on microvessel morphogenesis. Defining how perivascular astroglia influence microvessel growth and endothelial expression of the tubular phenotype is important to our understanding of the microvascular response to trauma, ischemia and neoplasia within the central nervous system.

Figure 1 depicts the proliferative cycle of the cells comprising the microvascular astrocytic complex. The mature central nervous system microvessel consists of differentiated endothelial cells that express the full complement of blood-brain barrier properties, as well as pericytes. These cells lie within a well-developed extracellular basement membrane and are ensheathed by abluminally located astrocytic foot processes (9). When exposed to an angiogenic stimulus, astrocytes become reactive and proliferate. There is an increased synthesis of proteolytic enzymes by the microvascular cells that result in the partial degradation of basement membrane. This allows microvascular cells to migrate away from the parent microvessel and then proliferate. At some point, the migrating proliferating endothelial cells reorganize into elongated cords that later develop lumens to form primitive immature vessels (5). These immature vessels incompletely express the blood-brain barrier but soon mature into a well-developed, blood-brain barrier vessel (19). It is likely that each phenotype step in this proliferative cycle is regulated to some degree by interactions between astrocytes, pericytes and endothelial cells. In the retina, where the pericyte endothelial ratio is high, pericytes appear to regulate endothelial proliferation through a mechanism in which

Frontiers in Cerebral Vascular Biology: Transport and Its Regulation
Edited by L.R. Drewes and A.L. Betz, Plenum Press, New York, 1993

the formation of pericyte endothelial contacts leads to the activation of TGF beta that directly inhibits endothelial proliferation (22). This paper addresses the role of astroglial-endothelial interactions in the reorganization of proliferating and migrating endothelial cells into tubular structures.

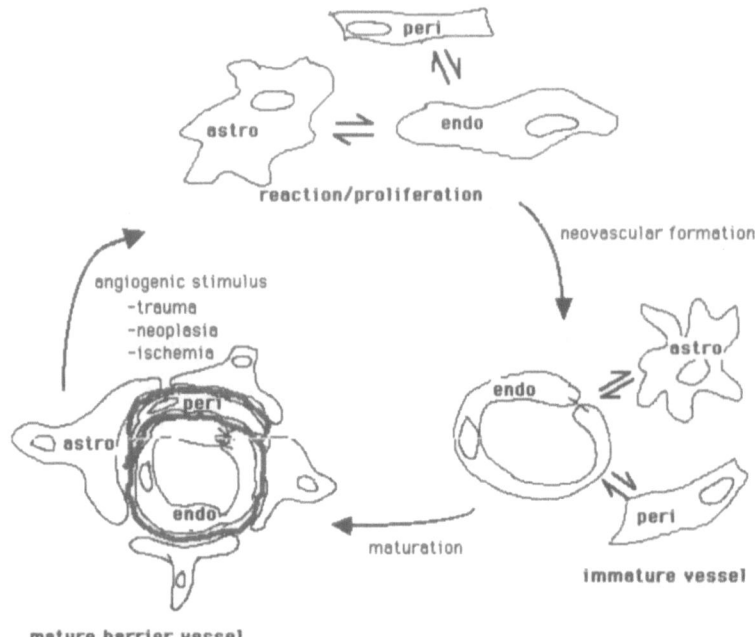

Figure 1. Proliferative cycle of the blood-brain barrier (BBB) microvessel-astrocytic complex. Angiogenic stimuli lead to endothelial (endo), pericytic (peri) and astrocytic (astro) activation, proliferation and migration from parent vessel. Dedifferentiated endothelial cells reorganize into primitive tubular forms that incompletely express BBB properties and later mature to fully re-express the barrier phenotype. These endothelial events are influenced by interactions with periendothelial astrocytes and pericytes.

Although the generation of the blood-brain barrier is not directly addressed in these studies, the formation of endothelial tubes is certainly a prerequisite to blood-brain barrier development, and the biochemical signals that regulate morphogenic astroglial-endothelial interactions might overlap with those that ultimately lead to microvessel maturation and blood-brain barrier expression.

METHODS

Homologous Cell Cultures

Bovine retinal microvessel endothelial (BRE) cells were isolated by the method of Bowman et al. (4), as modified by Laterra et al. (12). BRE cells were cultured on tissue culture plastic pretreated with 20 µg/ml bovine plasma fibronectin (Sigma) in MEM containing D-valine, 20% fetal bovine serum, MEM vitamins, nonessential amino acids, 16 U/ml heparin, 50 µg/ml Endothelial Mitogen (Biomedical Technologies, Inc., Stoughton, MA), 100 U/ml penicillin and 100 µg/ml streptomycin (stock endothelial medium) in humidified 5% CO_2/95% air at 37°C. Endothelial cultures were shown to be essentially 100% endothelial by labeling with 1,1'-dioctadecyl-3,3,3',3'-tetramethyl-indocarbocyanine perchlorate acetylated low density lipoprotein (DiI-acyl-LDL) (26). Cells at passages 3-12 were used in all experiments.

C6 cells (2) were obtained from the American Type Culture Collection (Rockville, MD) and cultured in Dulbecco's Modified Eagles Medium (DMEM) containing 10% bovine calf serum and 50 µg/ml gentamicin sulfate in humidified 5% CO_2/95% air at 37°C.

Neonatal forebrain astrocytes were isolated by the method of Franzakis and Kimelberg (8), cultured in DMEM containing 10% fetal bovine serum and used at first passage.

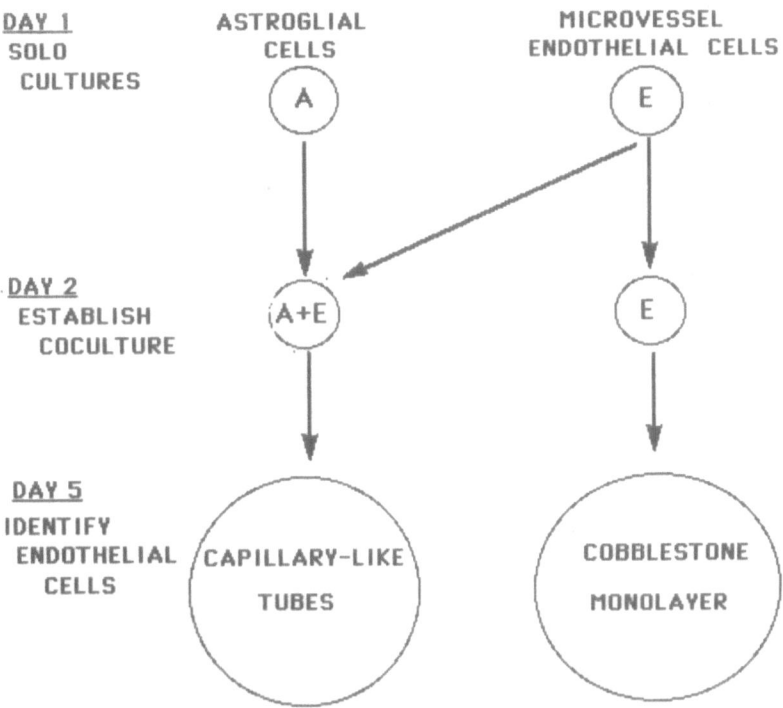

Figure 2. The experimental model of astroglial-induced endothelial differentiation into capillary-like tubes. See Methods for experimental detail.

Cocultures

Astroglial-BRE cell cocultures were established by a modification of the method of Laterra et al. (12). Astroglial cells were trypsinized, resuspended in stock culture medium and plated at 7,500 - 15,000 cells/chamber into eight-chamber Lab-Tek slides pretreated for 1-2 hr with 20 µg/ml of fibronectin in DMEM. After incubation for 24 hr, the astroglial medium was removed and replaced with 40,000 - 80,000 BRE cells suspended in 0.25 ml stock endothelial cell medium containing 50 µg/ml fresh ascorbic acid. Control slides lacked either the astroglial or endothelial cell inoculum. After an additional 72 hr at 37°C, cocultures were rinsed with DMEM, incubated for 3 hr with DiI-acyl-LDL (1:20 in DMEM), rinsed, fixed with 3.7% formaldehyde, and mounted in 90% glycerol. To determine the effects of phorbol esters and protease inhibitor, a subset of cocultures received 10 µl of stock phorbol 12-myristate 13-acetate, 4-alpha-phorbol 12,13-didecanoate, or protease inhibitor 48 hr after plating. These cultures were then labeled with DiI-acyl-LDL and fixed 24 hr later. Photomicrographs were taken on a Leitz Aristoplan microscope equipped for phase contrast and epifluorescence. Capillary-like structure formation was quantitated by computer-assisted image analysis with use of the Microcomputer Imaging Device (MCID) software package of Imaging Research Inc. (Brock University, St. Catherines, Ontario Canada L2S 3A1), a Sierra Scientific High Resolution CCD camera and Compaq DeskPro 386/25 computer (12).

Plasminogen Activator Determinations

Cell layers were rinsed twice with cold phosphate buffered saline and then solubilized by incubating in 100 µl of 0.5% Triton X-100 for 45 min on ice. Solubilized material was centrifuged for 90 sec at 10,000 x g, supernatants collected and 10-µl aliquots used in amidolytic plasminogen activator assays using the S-2251 chromogenic substrate according to the method of Festoff et al. (7).

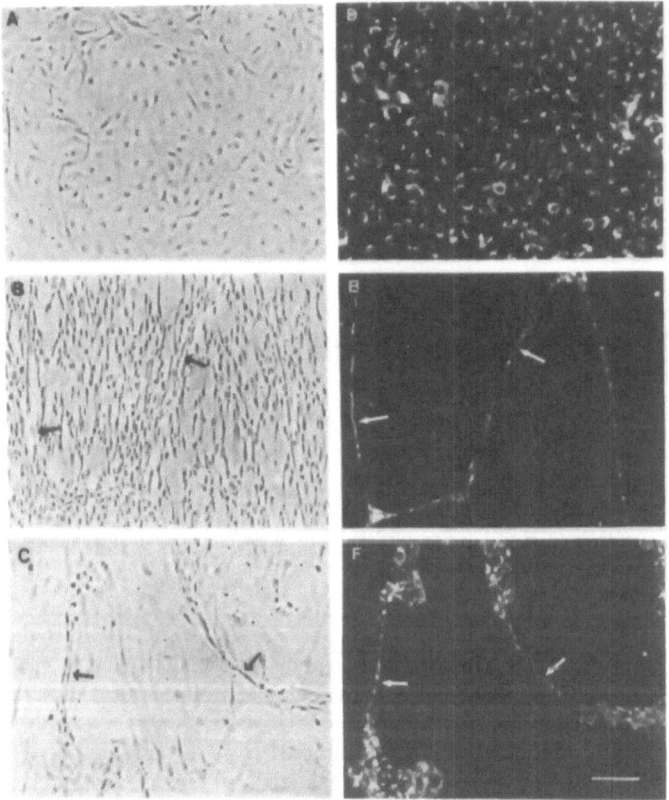

Figure 3. Induction of capillary-like structures by rat brain astrocytes and C6 astroglial cells. Cocultures were established and labeled with the fluorescent endothelial marker DiI-acyl-LDL as outlined in Methods. Cultures were photographed using either phase contrast (A-C) or fluorescence microscopy to specifically image endothelial cells (D-F). Bovine retinal endothelial cells when cultured alone develop a characteristic cobblestoned monolayer (A,D). In response to C6 astroglial cells (B,E) and rat brain astrocytes (C,F), endothelial cells reorganize into capillary-like structures (arrows). Bar = 100 µM. (Taken from 12).

RESULTS AND DISCUSSION

Our laboratory has established a cell culture method by which the morphological response of endothelial cells to astroglial cells can be assessed (12,27). Figure 2 outlines the basic experimental protocol. Cultured astrocytes originally isolated from neonatal rat brains, or glial cells of the C6 line, are plated at specific cell densities on fibronectin-coated tissue culture plastic. Twenty-four hours later, retinal microvascular endothelial cells are added to the astroglial cultures. Within 72 hr of adding the endothelial cells, co-cultures and solo endothelial cultures are labeled with DiI-acyl-LDL, an endothelial cell marker, and then fixed for microscopy and quantitative image analysis. We have previously shown that under these co-culture conditions, the endothelial cells reorganize into capillary-like structures (Figure 3). These structures contain lumen-like openings and accumulate the basement membrane protein laminin (12,12a). In the absence of astroglial cells, the endothelial cells form confluent monolayers. Specificity of this cell-cell interaction has been demonstrated by showing that bovine retinal microvessel endothelial cells fail to organize into capillary-like tubes when co-cultured with the fibroblastic NRK cell line (12). Similarly, adrenal microvascular endothelial cells co-cultured with C6 astroglial cells form markedly fewer capillary-like tubes when compared to the response of bovine retinal endothelial cells under identical conditions (Figure 4). The biological significance of this differential endothelial response is unclear since the isolation and initial culture conditions for the adrenal cells differ from that of the retinal cells (16).

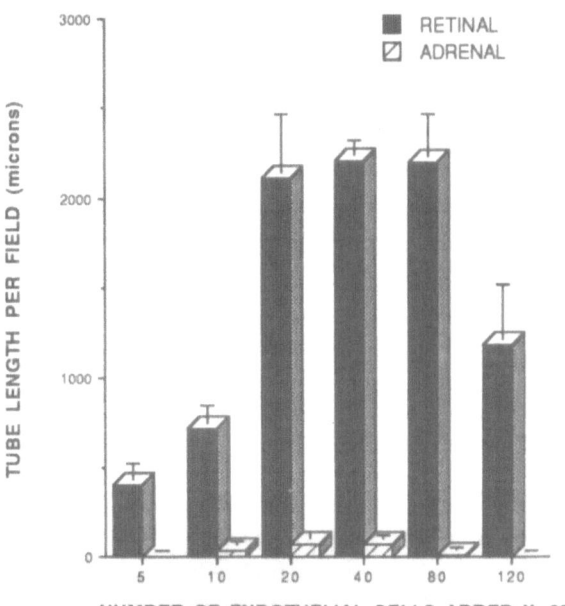

Figure 4. Morphogenic response of bovine retinal and bovine adrenal microvascular endothelial cells to coculture with C6 cells. The indicated numbers of endothelial cells were added to C6 cell cultures established 24 hr earlier as described in Methods. Cocultures were labeled 72 hr later with DiI-acyl-LDL to identify the endothelial cells. Capillary-like structures were quantitated by computer-assisted image analysis.

There are generally three classes of potential molecular mediators of this morphogenic astroglial endothelial interaction. These are extracellular matrix molecules, soluble diffusible factors and membrane-associated molecules that function by direct cell-cell contact.

Precedents exist for the induction of endothelial cell differentiation into capillary-like tubes in response to extracellular matrix and soluble factors (11,14). A series of experiments were performed to determine if soluble factors generated by C6 astroglial cells were sufficient to induce tube formation (12). These experiments are depicted in Figure 5. As shown above, endothelial cells reorganize into capillary-like tubes when cocultured on the same surface as the astroglial cells. This design allows for direct cell-cell contact, although it is unclear whether direct contact is required. When endothelial cells are separated by astroglial cells using a porous membrane, capillary-like structures are not formed. Conditioned media derived from C6 cultures failed to induce tube formation as did conditioned media from C6-endothelial cocultures. These experiments suggest that non-diffusible factors are required for tube formation but do not absolutely rule out the possibility that tube formation is induced by very labile diffusible factors that function over very short intercellular distances.

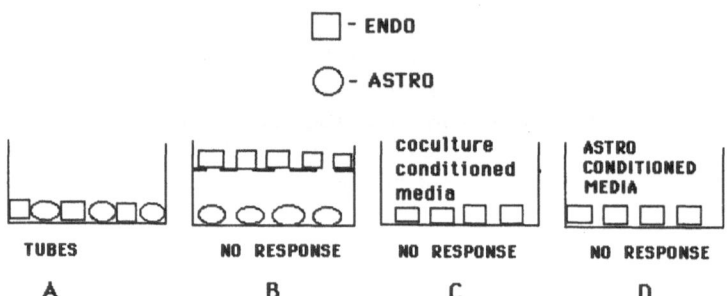

Figure 5. Schematic of experiments designed to determine if soluble factors are sufficient for the induction of capillary-like tubes. Retinal endothelial cells were cultured under conditions that (A) allowed direct astroglial-endothelial contact, (B) separated astroglia from endothelial cells with porous membranes so that interactions occur only by means of diffusible factors, (C) in the presence of astroglial conditioned media, or (D) in the presence of conditioned media obtained from astroglial-endothelial cocultures containing capillary-like structures. Endothelial reorganization into capillary-like tubes was observed only under conditions allowing astroglial-endothelial contact.

Early steps in the cycle of microvascular proliferation require increased expression of proteolytic enzymes. Collagenases and plasminogen activators have been implicated in these early angiogenic steps (10,18). Their requirement is based, in part, on the need to degrade basement membrane prior to endothelial migration from the parent vessel. Later steps in the proliferative cycle that involve the formation of immature endothelial tubes might require a return to baseline in the expression of these proteolytic enzymes. Because of the relative lack of extracellular collagen in brain parenchyma, changes in non-collagenase proteases may be more important than the collagenases during these latter microvascular events. Of the non-collagenous proteases, the plasminogen activator system is most strongly implicated in regulating endothelial angiogenic behavior (1,15,17). The plasminogen activation cascade and its sites of regulation are depicted in Figure 6. The plasminogen activators, urokinase plasminogen activator and tissue plasminogen activator are selective serine proteases that cleave the proenzyme plasminogen to plasmin. Plasmin is a broad spectrum serine protease with multiple functions (21). Once generated within blood, plasmin is able to degrade fibrin resulting in thrombolysis. Tissue plasminogen activator plays a principal role in this cascade. When generated in tissue parenchyma, plasmin modulates the proteolysis of extracellular matrix and cell surface proteins and thereby regulates cell migratory and tissue morphogenic events. This cascade is regulated by specific plasminogen activator inhibitors such as plasminogen activator inhibitor-1, plasminogen activator inhibitor-2 and protease nexins (21). Cells that synthesize plasminogen activators typically also synthesize plasminogen activator inhibitors and it is the balance of activator and inhibitor that determines net pericellular plasmin generation. This proteolytic cascade is also regulated in space by virtue of specific cell surface and extracellular matrix binding sites that sequester and concentrate enzymes of the pathway (3,24). Thus, focal regions within extracellular matrix and on cell surfaces may be modulated by this enzymatic cascade while adjacent sites remain unaffected.

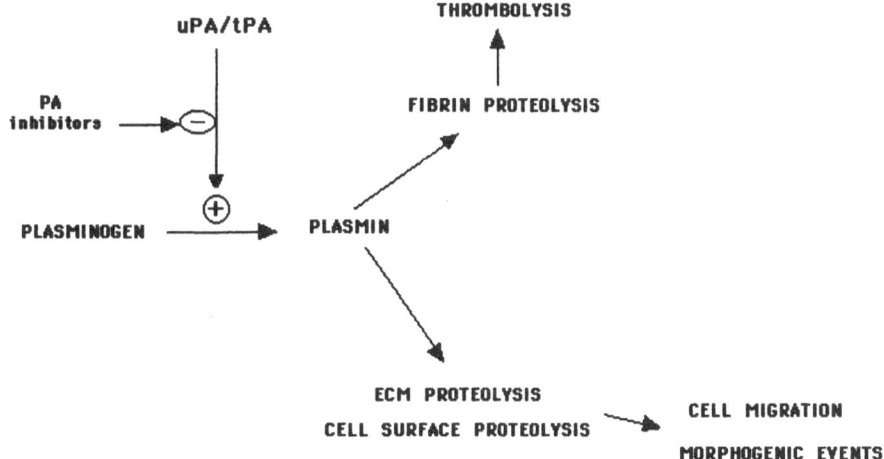

Figure 6. The plasminogen activation cascade and its relationship to thrombolysis and tissue morphogenesis.

We examined the relationship between plasminogen activation and astroglial-induced capillary-like structure formation. Changes in the plasminogen activator activities associated with coculture cell layers during the process of tube formation were quantitated by an amidolytic chromogenic assay (7). Figure 7 demonstrates a nearly 10-fold decrease in cell layer plasminogen activator activity during the 3-day process of C6-induced endothelial differentiation into capillary-like tubes. The sensitivity of retinal endothelial plasminogen activator activity to 0.1 mM amiloride indicates that the predominant form of plasminogen activator in this culture system is urokinase plasminogen activator (25). To determine if the reduction in urokinase activity observed during C6-induced tube formation resulted from or signaled endothelial differentiation into tubes, we examined the effect of inhibitors of plasminogen activation on capillary-like tube formation. Figure 8 indicates that the serine protease inhibitor, aprotinin, increased the extent of tube formation by approximately 100%. Soybean trypsin inhibitor, a structurally distinct protease inhibitor, had similar effects. If the inhibition of periendothelial plasminogen activation is required for tube formation, then inducing endothelial plasminogen activator production was expected to inhibit tube formation. Capillary-like tube formation was quantitated in C6-endothelial cocultures treated with the phorbol ester 12-myristate 13-acetate (PMA), which induces endothelial plasminogen activator production (13) or solvent only as control. PMA greatly inhibited capillary-like tube formation and the concentration dependence of this effect closely correlated with the ability of PMA to increase endothelial plasminogen activator production (Figure 9). The phorbol ester 4-alpha-phorbol 12,13-didecanoate which does not induce endothelial plasminogen activator production, was without effect (13). Treating solo endothelial cultures with aprotinin failed to induce tube formation. These findings suggest that astroglial induction of endothelial tube formation requires the inhibition of periendothelial plasminogen activation. The inhibition of plasminogen activation is not sufficient for tube formation, but requires the concurrent function of other astroglial-derived elements.

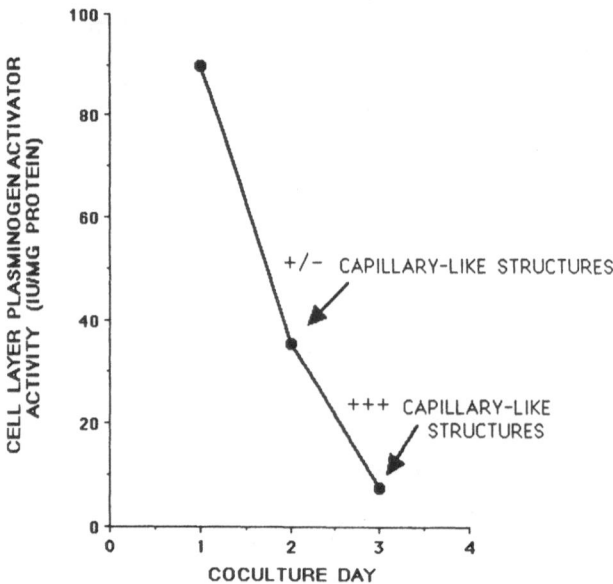

Figure 7. Changes in cell layer plasminogen activator activity during the formation of C6-induced capillary-like structures. Endothelial cells were cultured for 3 days in the presence of C6 astroglial cells under conditions that lead to endothelial differentiation into tubular structures. On each day, cocultures were examined for tubular structures and analyzed for Triton X-100 extractable cell layer plasminogen activator activity by amidolytic chromogenic assay. Capillary-like structure formation correlated with a nearly 10-fold decrease in plasminogen activator activity.

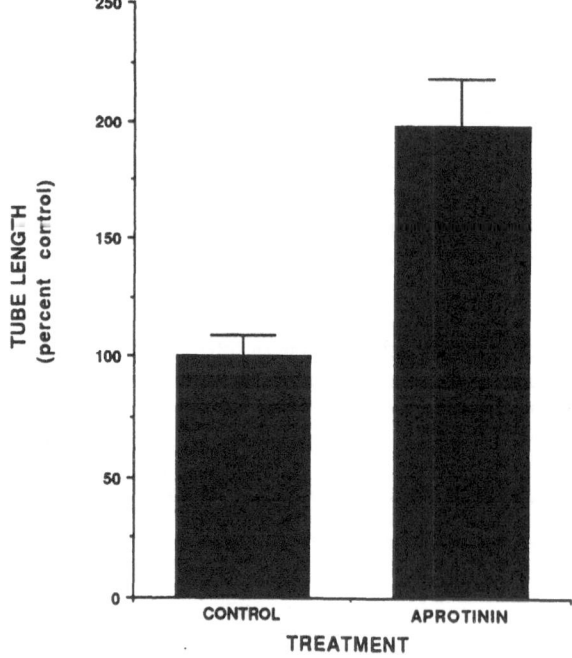

Figure 8. Inhibition of coculture serine proteases induces C6-induced capillary-like structure formation. C6-endothelial cocultures were treated for 24 hr with the serine protease inhibitor aprotinin (1000 IU/ml). These conditions resulted in complete inhibition of cell layer plasminogen activator activity. Addition of solvent only served as control and capillary-like structures were quantitated as described in Methods.

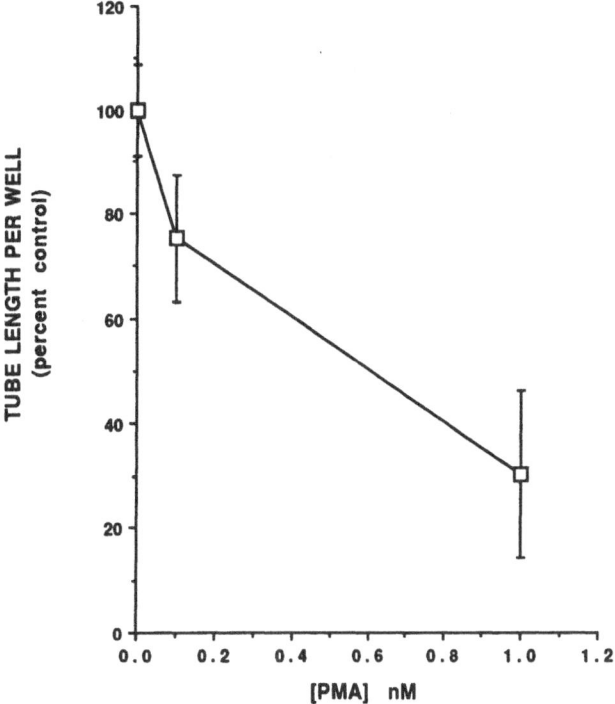

Figure 9. Phorbol 12-myristate 13-acetate (PMA) inhibits capillary-like tube formation in C6-endothelial cocultures. The indicated concentrations of PMA were added to C6 endothelial cocultures, and 24 hr later capillary-like tubes were quantitated as described in Methods. Endothelial cell layers isolated from cultures exposed to 0.1 nM and 1 nM PMA under similar conditions expressed 140% and 200% plasminogen activator activity of controls, respectively.

These findings are consistent with the following hypothesis of how the modulation of periendothelial plasminogen activator activity by perivascular astrocytes may function to induce endothelial differentiation into tubular forms within the central nervous system. Following an angiogenic stimulus, astrocytes and endothelial cells become activated, resulting in a net increase in periendothelial plasminogen activation. In addition to its effects on basement membrane degradation, this increase in proteolysis inactivates either endothelial cell surface receptors or astrocyte-derived ligands that function to maintain endothelial cells in a tubular state. This results in endothelial dedifferentiation and migration. Periendothelial plasminogen activation subsequently decreases by either a decrease in endothelial synthesis of activators or an increase in astroglial- or endothelial-derived inhibitors. This reestablishes functional endothelial cell surface receptors and astroglial-derived ligands that induce the tubular phenotype. Future directions in this work are to determine if the decrease in plasminogen activation seen during astroglial induced capillary-like tube formation results from a decrease in plasminogen activator synthesis or an increase in the production of plasminogen activator inhibitors and to identify the inhibitors involved. In addition, in vivo studies have been designed to test the importance of this mechanism to microvessel growth and differentiation in developing brain and brain tumors.

REFERENCES

1. Ashino-Fuse, H., Takano, Y., Oikawa, T., Shimamura, M., et al., Medroxyprogesterone acetate, an anti-cancer and anti-angiogenic steroid, inhibits the plasminogen activator in bovine endothelial cells, *Int. J. Cancer* 44:859-864, 1989.

2. Benda, P., Lightbody, J., Sato, G., Levine, L., and Sweet, W., Differentiated rat glial cell strain in tissue culture, *Science* 161:370-371, 1968.

3. Blasi, F., Vassalli, J.D., and Dano, K., Urokinase-type plasminogen activator: proenzyme, receptor, and inhibitors, *J. Cell Biol.* 104:801-804, 1987.

4. Bowman, P.D., Betz, A.L., and Goldstein, G.W., Primary culture of microvascular endothelial cells from bovine retina: selective growth using fibronectin-coated substrate and plasma-derived serum, *In Vitro* 18:626-732, 1982.

5. D'Amore, P.A., and Thompson, R.W., Mechanisms of angiogenesis, *Annu. Rev. Physiol.* 49:452-464, 1987.

6. Dehouck, M.P., Meresse, S., Delorme P., et al., An easier, reproducible, and mass-production method to study the blood-brain barrier in vitro, *J. Neurochem.* 54:1798-1801, 1990.

7. Festoff, B.W., Bao, J.S., Maben, C., and Hantai, D., Plasminogen activators and their inhibitors in the neuromuscular system: 1. Developmental regulation of plasminogen activator isoforms during in vitro myogenesis in two cell lines, *J. Cell Physiol.* 144:262-271, 1990.

8. Franzakis, M.V., and Kimelberg, H.K., Dissociation of neonatal rate brain by dispase for preparation of primary astrocyte cultures, *Neurochem. Res.* 9:1689-1698, 1984.

9. Goldstein, G.W., and Betz, A.L., The blood-brain barrier, *Sci. Am.* 254:74-83, 1986.

10. Ingber, D., and Folkman, J., Inhibition of angiogenesis through modulation of collagen metabolism, *Lab. Invest.* 59:44-51, 1988.

11. Kubota, Y., Kleinman, H.K., Martin, G.R., and Lawley, T.J., Role of laminin and basement membrane in the morphological differentiation of human endothelial cells into capillary-like structures, *J. Cell Biol.* 107:1589-1598, 1988.

12. Laterra, J., Guerin, C., and Goldstein, G.W., Astrocytes induce neural microvascular endothelial cells to form capillary-like structures in vitro, *J. Cell. Physiol.* 144:205-215, 1990.

12a. Laterra, J., and Goldstein, G.W., Astroglial-induced in vitro angiogenesis: requirements for RNA and protein synthesis, *J. Neurochem.* 57:1231-1239, 1991.

13. Levin, E.G., and Santell, L., Stimulation and desensitization of tissue plasminogen activator release from human endothelial cells, *J. Biol. Chem.* 263:9360-9365, 1988.

14. Madri, J.A., Pratt, B.M., and Tucker, A.M., Phenotypic modulation of endothelial cells by transforming growth factor-B depends upon the composition and organization of the extracellular matrix, *J. Cell Biol.* 106:1375-1384, 1988.

15. Montesano, R., and Orci, L., Tumor-promoting phorbol esters induce angiogenesis in vitro, *Cell* 42:469-477, 1985.

16. Orlidge, A., and D'Amore, P., Inhibition of capillary endothelial cell growth by pericytes and smooth muscle cells, *J. Cell Biol.,* 105:1455-1462, 1987.

17. Pepper, M.S., Belin, D., Montesano, R., Orci, L., and Vassalli, J.D., Transforming growth factor-beta 1 modulates basic fibroblast growth factor-induced proteolytic and angiogenic properties of endothelial cells in vitro, *J. Cell Biol.* 111:743-755, 1990.

18. Rifkin, D.B., Gross, J.L., Moscatelli, D., and Gabrielides, C., The involvement of proteases and protease inhibitors in neovascularization, *Acta Biol. Med. Germ.* 40:1259-1263, 1981.

19. Rosenstein, J.M., Krum, J.M., Sternberger, L.A., Pulley, M.T., and Sternberger, N.H., Immunocytochemical expression of the endothelial barrier antigen (EBA) during brain angiogenesis, *Develop. Brain Res.,* 66:47-54, 1992.

20. Rubin, L.L., Hall, D.E., Porter, S., Barbu, K., Cannon, C., Horner, H.C., Janatpour, M., Liaw, C.W., Manning, K., Morales, J., Tanner, L.I., Tomaselli, K.J., and Bard, F., A cell culture model of the blood-brain barrier, *J. Cell Biol.* 115:1724-1735, 1991.

21. Saksela, O., and Rifkin, D.B., Cell-associated plasminogen activation: regulation and physiologic functions, *Annu Rev. Cell Biol.* 4:93-126, 1988.

22. Sato, Y., Tsuboi, R., Lyons, R., Moses, H., and Rifkin, D.B., Characterization of the activation of latent TGF-b by co-cultures of endothelial cells and pericytes or smooth muscle cells: a self-regulating system, *J. Cell Biol.* 111:747-763, 1990.

23. Tao-Cheng, J.H., Nagy, Z., and Brightman, M.W., Tight junctions of brain endothelium in vitro are enhanced by astroglia, *J. Neurosci.* 7:3293-3299, 1987.

24. Vassalli, J.D., Baccino, D., and Belin, D., A cellular binding site for the M_r 55,000 form of the human plasminogen activator, urokinase, *J. Cell Biol.* 100:86-92, 1985.

25. Vassalli, J.D., and Belin, D., Amiloride selectively inhibits the urokinase-type plasminogen activator, *FEBS Lett.* 1:187-191, 1987.

26. Voyta, J.C., Via, D.P., Butterfield, C.E., and Zetter, B.R., Identification and isolation of endothelial cells based on their increased uptake of acetylated-low density lipoprotein, *J. Cell Biol.* 99:2034-2040, 1984.

27. Wolff, J.E.A., Laterra, J., and Goldstein, G.W., Steroid inhibition of neural microvessel morphogenesis in vitro: receptor mediation and astroglial dependence, *J. Neurochem.* 58:1023-1032, 1992.

IMMORTALIZED RAT BRAIN MICROVESSEL ENDOTHELIAL CELLS: I - EXPRESSION OF BLOOD-BRAIN BARRIER MARKERS DURING ANGIOGENESIS

Françoise Roux,[1] Odile Durieu-Trautmann,[2] Jean-Marie Bourre,[1]
Arthur Donny Strosberg,[2] and Pierre-Olivier Couraud[2]

[1]INSERM U26, Hôpital F. Widal, 75010 Paris, France
[2]CNRS UPR-0415, ICGM, 22 rue Méchain, 75014 Paris, France

INTRODUCTION

The blood-brain barrier (BBB) controls the exchange of solutes between the blood and the brain parenchyma. It is constituted by the endothelial cells lining the brain microvessels, in close relationship with astrocytes. These endothelial cells display a unique phenotype, characterized essentially by the presence of numerous intercellular tight junctions and the expression of specific enzymes, like γ-glutamyltranspeptidase (γ-GT) and alkaline phosphatase. It is known that the induction of this endothelial phenotype is controlled in vivo by the perivascular astrocytes. The in vitro study of the molecular mechanisms of this induction has been dependent so far on the availability of primary cultures of brain microvessel endothelial cells. In order to get rid of the intrinsic drawbacks of such primary cultures, i.e., contamination by various cell types and rapidly appearing senescence, we immortalized rat brain microvessel endothelial cells by transfection with the E1A-Adenovirus encoding gene. This study describes a phenotypic and pharmacological characterization of the immortalized cellular clone RBE4 and presents evidence for the expression by these cells of BBB enzymatic markers during angiogenesis.

IMMORTALIZATION OF ENDOTHELIAL CELLS

Primary culture of rat brain microvessel endothelial cells were established as described (1). After one passage, cells were transfected, using the calcium/phosphate coprecipitation technique, with the plasmid pE1A-*neo*, containing the adenovirus E1A encoding sequence followed by a neomycin resistance gene. One clone, called RBE4, has been further characterized. These cells exhibit in culture a non-transformed phenotype: contact-inhibited, growth factor- and anchorage-dependent proliferation, expression of endothelial differentiation markers (Factor VIII-related antigen, staining with the lectin *Bandeiraea simplicifolia* I4) (Figure 1), non-tumorigenicity in athymic Nude mice. RBE4 cells are routinely grown on collagen-I-

Frontiers in Cerebral Vascular Biology: Transport and Its Regulation
Edited by L.R. Drewes and A.L. Betz, Plenum Press, New York, 1993

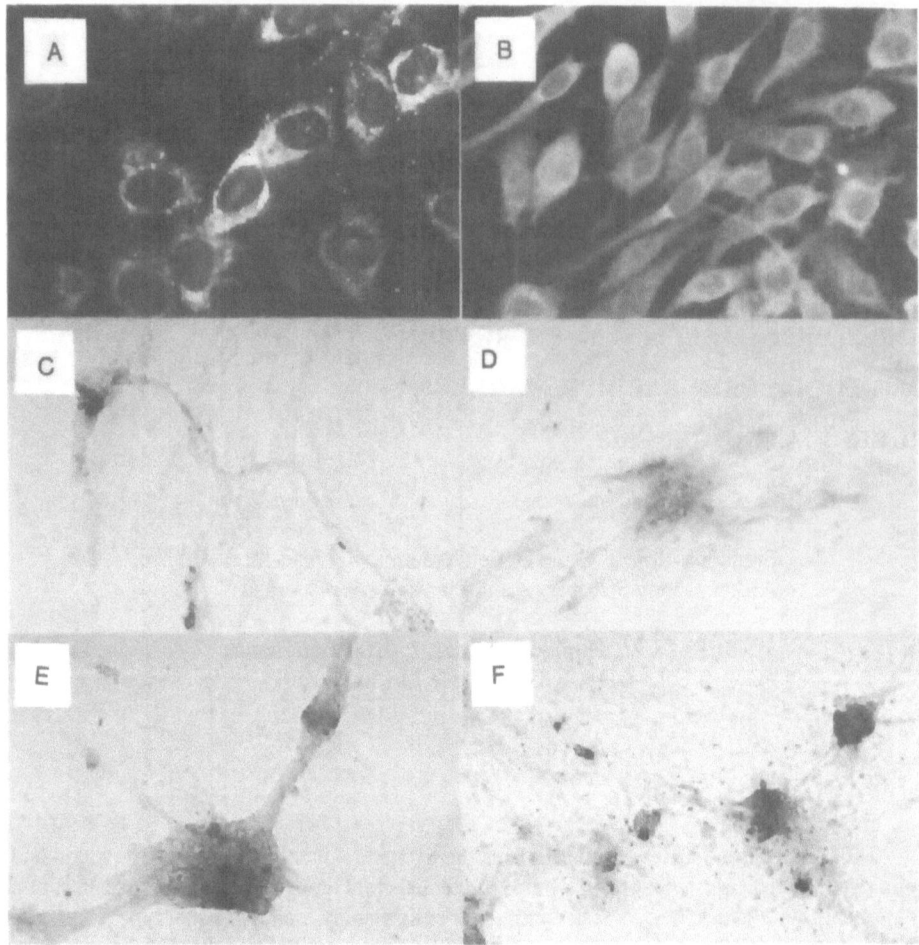

Figure 1. Localization by immunofluorescence of Factor VIII related antigen (A) and by fluorescence of *Bandeiraea simplicifolia* lectin (B) in RBE4 cells. Histochemical localization of γ-GT in presence of bFGF (C-F) C: RBE4 cells at confluence, with a capillary-like structure and some γ-GT positive cells. D: RBE4 cells treated during 6 days by primary astrocyte plasma membranes. E: RBE4 cells treated during 6 days by C6 glioma cell plasma membranes. Cells associate into linear cords of varying thickness or into shapeless aggregates: Numerous cells express γ-GT activity. F: RBE4 cells treated during 6 days by C6 glioma cell plasma membranes and by 0.1 mM 8-bromo cAMP: Most cells are high stained for γ-GT activity. (Fig.1A, x 1500 ; B, x 1000 ; C-F, x 180).

coated dishes in α-MEM/F10 medium, supplemented with 10% fetal calf serum and 1 ng/ml bFGF, and used between the 30th and 60th passages.

EXPRESSION OF BLOOD-BRAIN BARRIER MARKERS DURING ANGIOGENESIS

The activities of two enzymes, which are considered as markers for differentiated brain microvessel endothelial cells, γ-GT and alkaline phosphatase, were demonstrated using histochemical methods (2,3). RBE4 cells were grown on collagen or fibronectin-coated tissue culture plastic in medium containing bFGF. Confluent cultures developed sprouts that extended above the monolayer and organized into capillary-like structures. γ-GT and alkaline phosphatase activities could be detected in some cells of these tubular structures and they were not observed in cells of the surrounding monolayer (Figure 1,2).

RBE4 cells were cocultured with conditioned media or plasma membranes of either primary rat astroglial cells or C6 glioma cells. In monolayers treated with plasma membranes of both type cells, with or without bFGF, tubular structures developed rapidly and formed large shapeless aggregates. Such increased proliferation of cells forming the tubular structures was less obvious after treatment with conditioned media. γ-GT and alkaline phosphatase activities could be shown in numerous cells of these tri-dimensional structures, resulting in a global increase of both BBB marker activities in the cocultures. Addition of the cAMP analogue, 8-bromo cAMP, reduced this angiogenic process and induced the expression of the alkaline phosphatase activity in every cell forming the residual tubular structures or aggregates and in some cells of the monolayer. Effects of this agent on γ-GT activity were less pronounced (Figure 1,2).

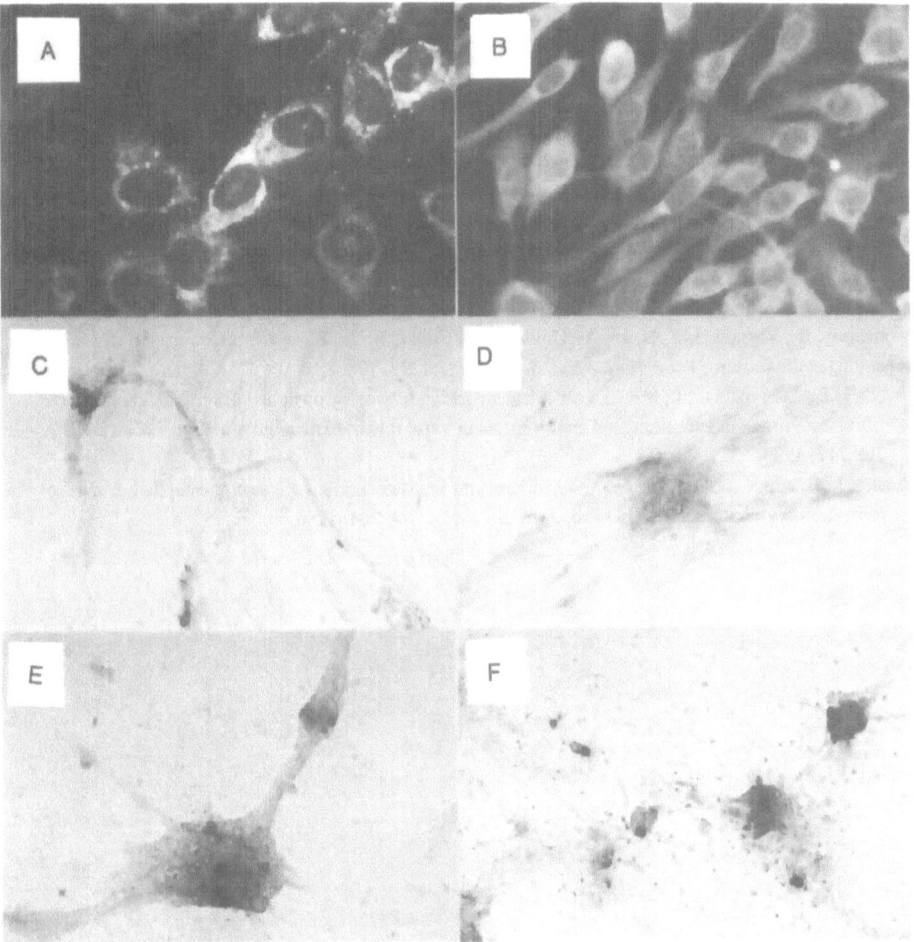

Figure 2. Histochemical localization of alkaline phosphatase, in the absence of bFGF (A-B) or in the presence of bFGF (C-F). A and C: RBE4 cells at confluence. B and D: RBE4 cells treated during 6 days by C6 glioma cell plasma membranes. E: RBE4 cells treated during 10 days by C6 glioma cell-conditioned medium. F: Cells are also treated by 0.1 mM 8-bromo cAMP. Expression of alkaline phosphatase activity shows the same evolution in cocultures and in presence of 8-bromo cAMP, as expression of γ-GT activity. (Fig.2A-F, x 180).

CONCLUSION(S)

These results show that, in endothelial cell monolayers of the clone RBE4, like in primary cultures of microvascular endothelial cells (4,5), capillary tube formation can occur in the presence of a soluble angiogenic mitogen. Moreover such tubular structure formation can be stimulated by astroglial factors, either diffusible or bound to plasma membranes. Astroglial induction of angiogenesis in vitro was first demonstrated by Laterra et al. (6) in primary cultures of bovine retinal microvascular endothelial cells, but the mechanism that leads to this endothelial cell response may require endothelial-astroglial contact. With the immortalized endothelial cells RBE4, some BBB-related enzymes are expressed after the induction of the angiogenic phenotype. This first differentiation step would thus be necessary for the expression of the BBB phenotype. 8-Bromo cAMP blocks this angiogenic response and increases the expression of both studied BBB markers. Further experiments on primary cultures of rat brain microvascular endothelial cells could show whether or not such responses in immortalized cells also occur in non-transformed cerebral endothelial cells.

REFERENCES

1. Roux, F.S., Mokni, R., Hughes, C.C., Clouet, P.M., Lefauconnier, J.M., and Bourre, J.M., Lipid synthesis by rat brain microvessel endothelial cells in tissue culture, *J. Neuropathol. Exper. Neurology* 48:437, 1989.

2. Budi Santoso, A.W., and Bar, T., Postnatal development of gamma-GT activity in rat brain microvessels corresponds to capillary growth and differentiation, *Int. J. Dev. Neuroscience* 4:503, 1986.

3. Ackerman, G.A., Substituted naphthol AS phosphate derivatives for the localization of leucocyte alkaline phosphatase activity, *Lab. Invest.* 11:563, 1962.

4. Montesano, R., Vassalli, J.D., Baird, A., Guillemin, R., and Orci, L., Basic fibroblast growth factor induces angiogenesis in vitro, *Proc. Natl. Acad. Sci. USA* 83:7297, 1986.

5. Ingber, D.E., and Folkman, J., Mechanochemical switching between growth and differentiation during fibroblast growth factor-stimulated angiogenesis in vitro: role of extracellular matrix, *J. Cell Biol.* 109:317, 1989.

6. Laterra, J., Guerin, C., and Goldstein, G.W., Astrocytes induce neural microvascular endothelial cells to form capillary-like structures in vitro, *J. Cell. Physiol.* 144:204, 1990.

IMMORTALIZED RAT BRAIN MICROVESSEL
ENDOTHELIAL CELLS :
II-PHARMACOLOGICAL CHARACTERIZATION

Odile Durieu-Trautmann,[1] Sandrine Bourdoulous,[1] Françoise Roux,[2]
Jean-Marie Bourre,[2] A. Donny Strosberg,[1] and Pierre-Olivier Couraud[1]

[1]CNRS UPR-0415, ICGM, 22 rue Méchain, 75014 Paris, France
[2]INSERM U26, Hôpital F. Widal, 75010 Paris, France

INTRODUCTION

Vascular endothelial cells are known to display a variety of biological functions, including regulation of vascular tone through secretion of vasoactive factors and control of nutrient and cellular trafficking between the blood and the underlying tissue (1,2). Most of the available data concerning these functions have been collected in vitro from studies on macrovascular peripheral endothelial cells. Given the remarkable heterogeneity of endothelia in terms of morphology and function, we intended to assess the ability of brain microvessel endothelial cells, which constitute the blood-brain barrier (BBB), to secrete endothelin (ET-1) and nitric oxide (NO), two potent vasoactive factors, under the control of hormonal stimuli and to express the MHC molecules that may contribute to the adhesion of immune cells to the endothelium. This study was performed with immortalized rat brain microvessel endothelial cells that were isolated after transfection of primary cultures with the E1A-Adenovirus encoding gene. These cells present a non-transformed phenotype and express the blood-brain barrier markers γ-glutamyl transferase and alkaline phosphatase during angiogenesis, as described in the accompanying paper (Roux et al.).

EXPRESSION OF HORMONE RECEPTORS

Isoproterenol-stimulated cAMP Synthesis

RBE4 cells were tested for their ability to produce the cyclic nucleotides cAMP and cGMP under extracellular signal regulation. Figure 1A shows that isoproterenol, a β-adrenergic agonist, stimulated the accumulation of cAMP within RBE4 cells; this stimulation was blocked by the antagonist propranolol (not shown), indicating the involvement of β-adrenergic receptors, positively coupled to the enzyme adenylyl cyclase.

Frontiers in Cerebral Vascular Biology: Transport and Its Regulation
Edited by L.R. Drewes and A.L. Betz, Plenum Press, New York, 1993

205

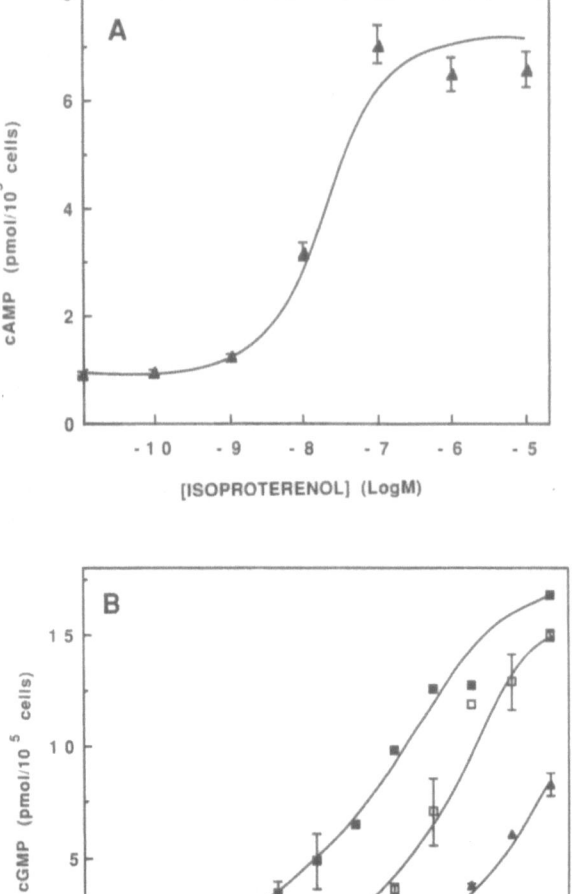

Figure 1. RBE4 cells were incubated for 10 min at 37°C in Hank's salt solution, 20 mM Hepes pH=7.4, 0.5 mM IBMX, with (A) increasing concentrations of (-)isoproterenol or (B) natriuretic peptides, ANP (■), BNP (□), CNP (▲). cAMP and cGMP accumulations were measured using Amersham kits.

Natriuretic Peptide-stimulated cGMP Synthesis

The synthesis of cGMP from GTP is catalyzed by different enzymes, the identity of which has been recently elucidated by molecular cloning: two related natriuretic peptide receptors present a guanylyl cyclase activity in their cytoplasmic domain with only limited homology to the NO-sensitive soluble enzyme (3). Evidence is presented in Figure 1B that the atrial natriuretic peptide (ANP) strongly stimulates the accumulation of cGMP within RBE4 cells. The related brain-derived peptide BNP was also found to activate the enzyme guanylyl cyclase, while the other natural analog CNP appeared much less active. This order of potency (ANP>BNP>CNP) suggests the expression of at least the A-subtype (3). In contrast with this high ANP receptor / guanylyl cyclase activity, no NO-stimulated cGMP formation could be observed in RBE4 cells (not shown).

Our findings that RBE4 cells possess β-adrenergic and ANP receptors are in agreement with previous studies on different types of endothelial cells, including brain microvessel endothelial cells, freshly isolated or in short-term culture. Concerning the absence of soluble (NO-sensitive) guanylyl cyclase activity in these cells, which has been reported in other NO-releasing cells, one might speculate that the insensitivity to NO might constitute, for these cells, a natural protection against NO toxicity (2).

SECRETION OF VASOACTIVE FACTORS

ET-1 Secretion

Vascular endothelial cells secrete ET-1, a potent contracting factor (1). In a specific immuno-enzymatic assay, a constitutive secretion of ET-1 was observed: the accumulation of the peptide in the extracellular medium was linear over 6-8 hr (110 ± 4 pg / 10^6 cells / hr, at confluence), then reached a plateau within about 16 hr. This secretion was stimulated by serum and thrombin, as reported previously with other endothelial cells (not shown); interestingly, although 8Br-cAMP (0.5 mM) strongly induced a release of ET-1, 8Br-cGMP (0.5 mM) and ANP (100 nM) inhibited the secretion (Figure 2).

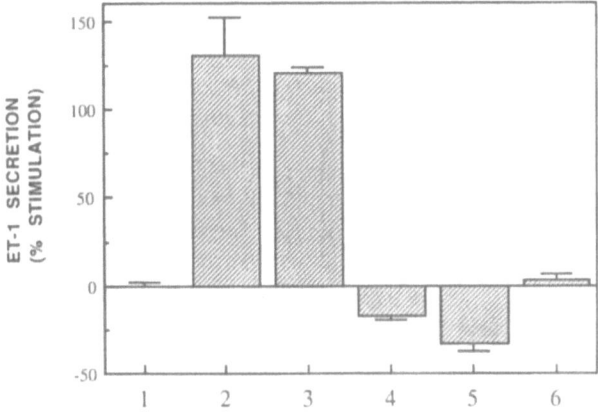

Figure 2. RBE4 cells, grown to confluence in 24-well dishes, were incubated, for 1 hr in serum free medium, supplemented with 0.15% BSA for 1 hr in the presence of the following agent: none(1), 0.5 mM 8Br-cAMP (2), 0.5 mM IBMX and 0.5 mM 8Br-cAMP (3), 0.5 mM 8Br-cGMP (4), 100 nM ANP (5), or 0.5 mM IBMX alone (6). After treatment, supernatants were assayed for ET-1 immunoreactivity. Results are expressed as means ± SE of 8 to 16 determinations; $p < 0.01$ for lanes 2 - 6.

NO Secretion

The endothelium-derived relaxing factor, which has been identified as NO or a closely related compound, is synthesized from L-arginine by at least two different NO synthases (2): one is calcium- and calmodulin-dependent and is constitutively expressed in a few cell types, including some neurons. The other one is inducible by cytokines and was first identified in macrophages. Both enzymes are potentially active in endothelium.

Table 1. Regulation of inducible NO synthase activity in RBE4 cells

Treatment	Nitrite Accumulation (μM)
None	<0.5
IFN	3.1±0.1
TNF	<0.5
TNF + IFN	10.0±0.1
TNF + IFN + NMA (100 μM)	1.7±0.1
TNF + IFN +NoA (10 μM)	7.6±0.4
TNF + IFN + CHX (0.5 μg/ml)	1.2±0.2
IFN + BrcAMP (500 μM)	3.8±0.1
TNF + IFN + BrcAMP (500 μM)	15.0±0.4

Cells grown at confluence in 24-well dishes were treated for 46 hr with 100 U/ml of rat interferon γ (IFN) and/or 50 U/ml of human tumor necrosis factor α (TNF), with or without N-methylarginine (NMA), nitroarginine (NoA), cycloheximide (CHX) or 8-bromo-cAMP (BrcAMP). Results are from a typical experiment performed in triplicate (±S.D.). Similar results were obtained in three other experiments.

Release of NO by RBE4 cells was monitored by a colorimetric determination of NO-derived nitrites. Only the inducible type of NO synthase is detectable in RBE4 cells (Table 1). Nitrite production by RBE4 was induced after treatment with IFN-γ and TNFα potentiated this effect. This activity was blocked by N-methyl-arginine and to a lesser extent by nitro-arginine, and required protein synthesis, as shown by its inhibition by cycloheximide. These properties are quite similar to that of the cytokine-activated NO synthase in macrophages. Our data are in good agreement with a recent study on murine brain endothelial cells in short-term culture (4), suggesting that lack of expression of the constitutive NO synthase might be a special feature of the BBB endothelial cells.

The cell-permeable analog of cAMP, 8Br-cAMP, although exhibiting no effect alone (not shown), was seen to potentiate the inductive effect of cytokines (Table 1).

Prostaglandin Secretion

RBE4 cells were found to constitutively secrete high amounts of PGE_2 (> 1000 pg / ml) but almost no 6-keto-PGF1α, the stable derivative of PGI_2 (< 50 pg / ml). These data are in agreement with a previously reported statement that a high PGE_2 / PGI_2 reflects the microvascular origin of endothelial cells (5).

EXPRESSION OF MHC MOLECULES

A number of disorders that affect the central nervous system, such as the demyelinating diseases multiple sclerosis and its animal model experimental allergic encephalomyelitis (EAE), are characterized by infiltration of immune cells, through the BBB, into the brain parenchyma. During induction of EAE in susceptible animals by transfer of T lymphocytes from animals immunized with myelin basic protein, early steps in the pathogenesis probably involve interactions between cerebral endothelial cells and sensitized lymphocytes (6). For this reason, we investigated MHC class I and class II expression on RBE4 cells. By flow cytometry analysis under basal conditions, RBE4 cells were shown to constitutively express MHC class I, but no class II molecules. Analysis after IFN-γ treatment revealed that MHC class I expression was enhanced and class II expression dramatically induced (Figure 3): maximal class I expression was observed after 16 hr and remained stable for several hours, while 24 hr of treatment were required for optimal class II expression. Further experiments will have to assess the ability of these cells to efficiently present antigens to MHC class I- or class II-restricted T lymphocytes.

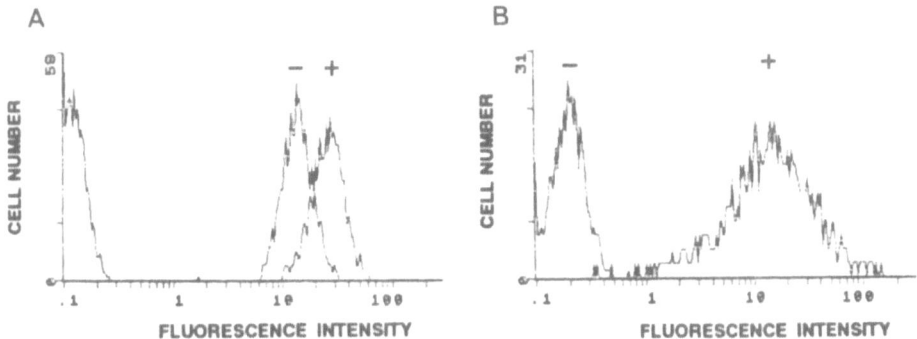

Figure 3. RBE4 cells were either not treated (-) or treated (+) with IFN-γ (50 U/ml) and stained for MHC class I, after 16 hr of treatment (A) or MHC class II, after 48 hr of treatment (B). The cells were then analyzed by flow cytometry.

CONCLUSION

In conclusion, it is shown in this study that the immortalized rat brain microvessel endothelial cells RBE4 have retained the ability to release a number of vasoactive factors, including NO and ET-1, two agents generally considered as local mediators for which astrocytes recently appeared as potential targets. Since astrocytes stimulate the expression, by brain microvessel endothelial cells in vivo and RBE4 cells in vitro, of a BBB-specific phenotype, and since cytokines regulate endothelial function, it is conceivable, from our results, that astrocytes, brain microvessel endothelial cells and leukocytes communicate, at the level of the BBB, through a network of pharmacological as well as immune interactions.

REFERENCES

1. Rubanyi, G.M., and Parker Botelho, L.H., Endothelins, *FASEB J.* 5:2713, 1991.
2. Snyder, S.H., and Bredt, D.S., Nitric oxide as a neuronal messenger, *Trends Pharmacol. Sci.* 12:125, 1991.
3. Koesling, D., Böhme, E., and Schultz, G., Guanylyl cyclases, a growing family of signal-transducing enzymes, *FASEB J.* 5:2785, 1991.
4. Gross, S.S., Jaffe, E.A., Levi, R., and Kilbourn, R.G., Cytokine-activated endothelial cells express an isotype of nitric oxide synthase which is tetrahydrobiopepterin-dependent, calmodulin-independent

and inhibited by arginine analogs with a rank-order of potency characteristic of activated macrophages, *Biochem. Biophys. Res. Commun.* 178:823, 1991.

5. Gerritsen, M.E., Functional heterogeneity of vascular endothelial cells, *Biochem. Pharmacol.* 36:2701, 1987.

6. Wekerle, H., Engelhardt, B., Risau, W., and Meyermann, R., Interaction of T lymphocytes with cerebral endothelial cells in vitro, *Brain Pathology* 1:107, 1991.

EXPRESSION OF THE HT7 GENE

IN BLOOD-BRAIN BARRIER

C.M. Unger, H. Seulberger, G. Breier, U. Albrecht,
M.G. Achen, and W. Risau

Max-Planck-Institut für Psychiatrie
Am Klopferspitz 18A
D-8033 Martinsried
Germany

INTRODUCTION

Homeostasis of the neural microenvironment is important for neuronal function and integrity. Therefore, the neuronal tissue must be protected from the changing metabolite and ion concentrations in the blood. In higher vertebrates, the endothelial cells of the brain capillaries form a selective barrier between the blood and the neuronal cells. Complex tight junctions and a low number of pinocytotic vesicles are the structural basis of this blood-brain barrier (BBB). Specific transport of nutrients, metabolites and ions across this barrier is mediated by carriers, ion channels and receptors located at the cell surface of the BBB endothelium (1). The unique properties of the endothelial cells that form the BBB are reflected by the expression of specific cell surface molecules. Recently we characterized the monoclonal antibody HT7 that was raised against an antigen on chick BBB endothelium (2). Here the localization of the HT7 antigen in the chick and the structure of a cDNA for the mouse homologue of this antigen are described.

RESULTS

The first expression of the HT7 antigen in the brain of the chick embryo was at day 11 (Figure 1). This correlates well with BBB formation (3). The antigen is absent from endothelial cells of the brain circumventricular organs, which lack a functional BBB, and is not detectable in other endothelial cells in the body (Figure 2). The antigen is expressed by other cell types, which form tissue barriers, like the epithelial cells of the choroid plexus (blood-cerebrospinal fluid barrier) (Figure 2a), the pigment epithelial cells of the retina (blood-eye barrier), and the epithelial cells of the kidney and of the liver bile ducts (data not shown). The antigen is also expressed in erythroblasts (Figure 2b), but not in erythrocytes.

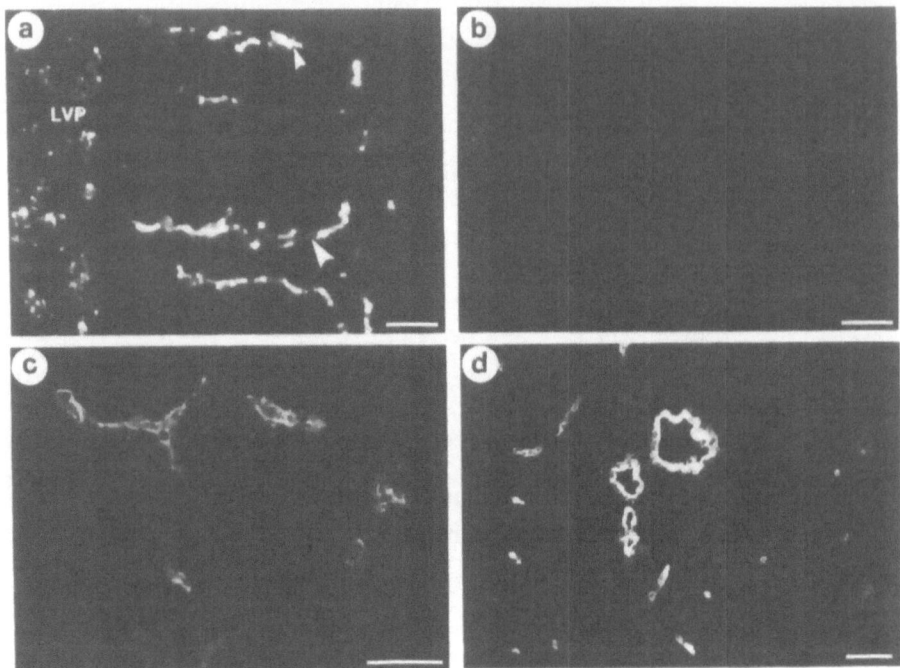

Figure 1. (a) Flourescently labeled brain capillaries of an 8-day chick embryo (after DIO-acetyl-LDL injection). Stem vessels (arrowheads) are seen penetrating the brain from the leptomeningeal vascular plexus (LVP). (b-d) Indirect immunofluorescence using HT7 monoclonal antibodies. (b) Same section as in (a), the HT7 antigen cannot be detected at this early stage. Until day 10 of chick embryonic development the BBB is immature: vessels are permeable to horseradish peroxidase; other BBB markers such as alkaline phosphatase and transferrin receptor cannot be detected. (c) Section of an 11-day embryonic brain. Capillaries in the brain parenchyma are stained. Capillaries at this stage have developed BBB characteristics (impermeable to horseradish peroxidase). (d) Section of an adult chick brain. All vessels in the brain parenchyma are stained. Bars represent 50 mm.

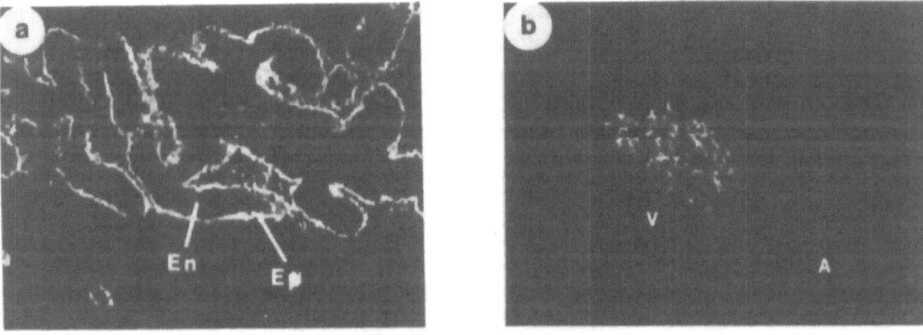

Figure 2. Indirect immunofluorescence using monoclonal HT7 antibodies. (a) Section of an adult chick brain choroid plexus in the region of the lateral ventricle. Ep = choroid plexus epithelium; En = choroid plexus endothelium. The fenestrated endothelial cells of the choroid plexus are negative. The choroid plexus epithelium, which forms the blood-cerebrospinal fluid barrier, expresses the HT7 antigen. (b) Section of an 11-day chick embryo in the thorax region; A = artery, V = vein. Vessels are negative, erythroblasts are stained.

We isolated the antigen from plasma membranes of chick brain, using an immunoaffinity column and reversed phase HPLC. The purified antigen is a highly glycosylated transmembrane protein with an apparent molecular weight of 43-52 kDa (4). Partial amino acid sequence of the purified protein was determined. By using degenerate oligonucleotides and the polymerase chain reaction, we cloned the corresponding cDNA. The sequence revealed that HT7 is a novel member of the immunoglobulin superfamily with two immunoglobulin-like domains (4). In addition, the HT7 protein has a hydrophobic signal sequence, one membrane-spanning domain and a cytoplasmic tail. The murine HT7 cDNA was cloned by screening of a mouse kidney lambda gt11 cDNA library under low stringency hybridization conditions. The nucleotide sequence of the murine HT7 cDNA is shown in Figure 3.

```
                                                            ACGAGGCGAC   10

      ATG GCG GCG GCG CTG CTG CTG GCG CTG GCC TTC ACG CTC TTG AGC GGC CAA GGC GCC TGC   70
   1  Met Ala Ala Ala Leu Leu Leu Ala Leu Ala Phe Thr Leu Leu Ser Gly Gln Gly Ala Cys

      GCG GCG GCG GGC ACC ATC CAA ACC TCT GTC CAG GAA GTC AAC TCC AAA ACA CAG CTT ACC  130
  21  Ala Ala Ala Gly Thr Ile Gln Thr Ser Val Gln Glu Val Asn Ser Lys Thr Gln Leu Thr

      TGC TCT TTG AAC AGC AGT GGC GTT GAC ATC GTT GGC CAC CGC TGG ATG AGA GGT GGC AAG  190
  41  Cys Ser Leu Asn Ser Ser Gly Val Asp Ile Val Gly His Arg Trp Met Arg Gly Gly Lys
              *** *** ***

      GTA CTG CAG GAG GAC ACT CTG CCC GAC CTG CAT ACG AAG TAC ATA GTG GAC GCA GAT GAC  250
  61  Val Leu Gln Glu Asp Thr Leu Pro Asp Leu His Thr Lys Tyr Ile Val Asp Ala Asp Asp

      CGC TCT GGG GAA TAT TCC TGC ATC TTC CTT CCT GAG CCT GTG GGC AGA AGC GAG ATC AAT  310
  81  Arg Ser Gly Glu Tyr Ser Cys Ile Phe Leu Pro Glu Pro Val Gly Arg Ser Glu Ile Asn

      GTG GAA GGG CCA CCC AGG ATC AAG GTC GGA AAG AAA TCA GAG CAT TCC AGT GAG GGA GAG  370
 101  Val Glu Gly Pro Pro Arg Ile Lys Val Gly Lys Lys Ser Glu His Ser Ser Glu Gly Glu

      CTT GCG AAA CTG GTC TGC AAG TCC GAT GCA TCC TAC CCT CCT ATT ACA GAT TGG TTC TGG  430
 121  Leu Ala Lys Leu Val Cys Lys Ser Asp Ala Ser Tyr Pro Pro Ile Thr Asp Trp Phe Trp

      TTT AAG ACC TCT GAC ACT GGG GAA GAA GAG GCA ATC ACC AAT AGC ACT GAA GCC AAT GGC  490
 141  Phe Lys Thr Ser Asp Thr Gly Glu Glu Glu Ala Ile Thr Asn Ser Thr Glu Ala Asn Gly
                                                         *** *** ***

      AAG TAT GTG GTG GTA TCC ACG CCT GAG AAG TCA CAG CTG ACC ATC AGC AAC CTT GAC GTA  550
 161  Lys Tyr Val Val Val Ser Thr Pro Glu Lys Ser Gln Leu Thr Ile Ser Asn Leu Asp Val

      AAT GTT GAC CCT GGC ACC TAC GTG TGT AAT GCC ACC AAC GCC CAG GGC ACT ACT CGG GAA  610
 181  Asn Val Asp Pro Gly Thr Thr Val Cys Asn Ala Thr Asn Ala Gln Gly Thr Thr Arg Glu
                                           *** *** ***

      ACC ATC TCA CTG CGT GTG CGG AGC CGC ATG GCA GCC CTC TGG CCC TTC CTA GGC ATC GTG  670
 201  Thr Ile Ser Leu Arg Val Arg Ser Arg Met Ala Ala Leu Trp Pro Phe Leu Gly Ile Val

      GCT GAG GTC CTG GTG TTG GTT ACC ATC ATC TTT ATC TAT GAG AAG AGG CGG AAG CCA GAC  730
 221  Ala Glu Val Leu Val Leu Val Thr Ile Ile Phe Ile Tyr Glu Lys Arg Arg Lys Pro Asp

      CAG ACC CTG GAC GAG GAT GAC CCT GGC GCC GCC CCA CTG AAG GGC AGT GGA ACT CAC ATG  790
 241  Gln Thr Leu Asp Glu Asp Asp Pro Gly Ala Ala Pro Leu Lys Gly Ser Gly Thr His Met

      AAT GAC AAG GAC AAG AAT GTA CGC CAG AGG AAC GCC ACC TGA GTGGTGGGCAGGCGGGGGAGGG    855
 261  Asn Asp Lys Asp Lys Asn Val Arg Gln Arg Asn Ala Thr  *

      AGGTGCCCAGGGTGCCTGACCCCAGGCCAGCGTCTACCTCCACTCCAGTATCCCATCCTGTCCCGATTTGAACCTACCC    934
      AACCCAACCTATCCCAACCCAAGTGAAGACAGAGCCTTACCTTACAGAAAACCCACCTGGAAGAAGCAAGCCACTTGCA  1013
      GCCCCTGTTTCTAATTTAAACTAAATGAGGTTTCTATGCAGACAATCCATTCCTTAGGGGTTTATGTTTTTATTTTTCC  1092
      TTCCCTTCTGAAGTGTTGTCACTACAGCCCTGTGGAGTGGGGGAATGGGCCTTGTCCTTGGTCAGGAGGGAAGGCCAG  1171
      TGCATGCTCTGACTTACTGTTGGAGGGGGCTGGGCCTGCTGGAAACCCCCCAAATAAAACCTAACCCACCAAAAAAn    1247
```

Figure 3. Nucleotide and deduced protein sequence of a cDNA encoding the mouse HT7 protein. The putative signal sequence is underlined, the conserved cysteine residues are shown in italics and bold, the potential N-linked glycosylation sites are marked with asterisks, the putative transmembrane region is boxed and the polyadenylation signal is shown in bold. Independently, two groups have isolated cDNA clones with an almost identical nucleotide sequence (5,6).

The amino acid sequence identity between the HT7 antigen homologues from chick and mouse is 100% in the transmembrane region, 70% in the cytoplasmic tail, and 40% in the extracellular domains. The high degree of conservation in the transmembrane and cytoplasmic regions suggests that these structural domains have an important functional role. Schematic representations of possible structures of the mouse and chick HT7 proteins are shown in Figure 4. Comparison of the immunoglobulin-like domains of the extracellular domain of the chick HT7 protein with those of other members of the immunoglobulin superfamily indicated that they are of the C2 type according to the criteria of Williams and Barclay (7). In contrast to the chick protein, the extracellular domain of the murine HT7 protein contains one C2 type and one V type immunoglobulin domain. This is the result of an insertion of 6 amino acids in the second immunoglobulin domain, which leads to the formation of an additional C'' beta-sheet structure.

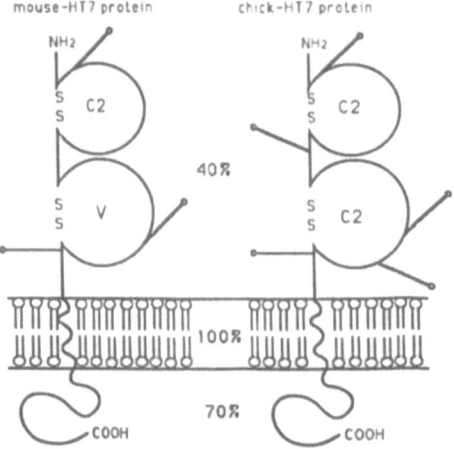

Figure 4. Schematic model of the possible structure of the mouse and chick HT7 proteins. The large circles represent immunoglobulin domains. Putative N-glycosylation sites are shown as small circles. The sequence identity between the mouse and the chick protein is 100% in the transmembrane region, 70% in the cytoplasmic tail and 40% in the two immunoglobulin-like domains.

CONCLUSIONS

The HT7 antigen is a novel constituent of BBB endothelium. Therefore, it may be an important tool for the investigation of the mechanism and regulation of BBB formation in vivo and in vitro and of the loss of this regulation under pathological conditions such as in brain tumors. Based on the expression of the HT7 protein on the luminal surface of BBB endothelium, together with the observation that, like HT7, many cell adhesion proteins and receptors belong to the immunoglobulin protein superfamily, we suppose that the HT7 protein might be involved in cell adhesion or transport processes at the BBB.

ACKNOWLEDGMENTS

Financial support from the Alexander von Humboldt Foundation in the form of a research fellowship to MGA is gratefully acknowledged.

REFERENCES

1. Risau, W., and Wolburg, H., Development of the blood-brain barrier, *Trends in Neurosci.* 13:174-178, 1990.
2. Risau, W., Hallmann, R., Albrecht, U., and Henke-Fahle, S., Brain induces the expression of an early cell surface marker for blood-brain barrier-specific endothelium, *EMBO J.* 5:3179-3183, 1986.
3. Risau, W., Hallmann, R., and Albrecht, U., Differentiation-dependent expression of proteins in brain endothelium during development of the blood-brain barrier, *Dev. Biol.* 117:537-545, 1986.
4. Seulberger, H., Lottspeich, F., and Risau, W., The inducible blood-brain barrier specific molecule HT7 is a novel immunoglobulin-like cell surface glycoprotein, *EMBO J.* 9:2151-2158, 1990.
5. Altruda, F., Cervella, P., Gaeta, M.L., Daniele, A., Giancotti, F., Tarone, G., Stefanuto, G., and Silengo, L., Cloning of cDNA for a novel mouse membrane glycoprotein (gp42): shared identity to histocompatibility antigens, immunoglobulins and neural-cell adhesion molecules, *Gene* 85:445-452, 1989.
6. Miyauchi, T., Kanekura, T., Yamaoka, A., Ozawa, M., Miyazawa, S., and Muramatsu, T., Basigin, a new, broadly distributed member of the immunoglobulin superfamily, has strong homology with both the immunoglobulin V domain and the beta-chain of major histocompatibility complex class II antigen, *J. Biochem.* 107:316-323, 1990.
7. Williams, A., and Barclay A. N., The immunoglobulin superfamily--domains for cell surface recognition, *Annu. Rev. Immunol.* 6:381-405, 1988.

ASTROCYTES SECRETE A FACTOR INDUCING THE EXPRESSION OF HT7-PROTEIN AND NEUROTHELIN IN ENDOTHELIAL CELLS OF CHORIOALLANTOIC VESSELS

Robert C. Janzer, Johannes A. Lobrinus, Pushpa Darekar, and Lucienne Juillerat

Institute of Pathology, Division of Neuropathology
University of Lausanne, Switzerland

SUMMARY

Among the most specific markers of the blood-brain barrier phenotype of endothelial cells are the well-characterized immunoglobulin-like surface glycoprotein HT7 (19) and the probably related or identical glycoprotein neurothelin (18). Both can be induced in chorioallantoic vessels by transplants of embryonic mouse brain. Other blood-brain barrier markers have been shown to be inducible by type-1 astrocytes in endothelial cells of non-neural origin (4). In the present work we tested the hypothesis that this cellular interaction between astrocytes and endothelial cells is mediated by a soluble factor(s).

Chorioallantoic vessels of embryonic day 9 chick embryos were exposed for 4 or 10 days to a constant and localized delivery of astrocyte-conditioned medium by using a piece of gelfoam posed onto the chorioallantoic membrane, as a localized reservoir, which was connected to a miniosmotic pump system delivering astrocyte-conditioned medium at a steady rate. We found that in a significant number of chorioallantoic vessels located near the gelfoam, endothelial cells exposed to astrocyte-conditioned medium for a period of 4 or 10 days, but not to glioma-, fibroblast- or endothelial cell-derived conditioned medium, expressed the HT7 antigen and neurothelin.

These results provide evidence that type-1-astrocytes are capable of inducing blood-brain barrier related properties in endothelial cells of non-neural origin through a soluble factor(s).

INTRODUCTION

Increasing knowledge on the specific characteristics of the cerebral microvascular endothelial cells representing the site of the blood-brain barrier has led to an impressive list of unique properties not shared with microvascular endothelial cells of non-neural origin (for review see 5,13,14). These properties are not intrinsic to microvascular endothelial

Frontiers in Cerebral Vascular Biology: Transport and Its Regulation
Edited by L.R. Drewes and A.L. Betz, Plenum Press, New York, 1993

217

cells, but are induced by the surrounding brain tissue, as shown in "in vivo" transplantation models (12,18,21). More specifically, astrocytes have been shown to induce some of these properties in endothelial cells of non-neural origin "in vivo" (4) and "in vitro" (3,20 22,23). Among the markers of the blood-brain barrier phenotype actually considered to be strongly correlated with the permeability barrier function are the well characterized immunoglobulin-like cell surface glycoprotein HT7 (1,12,19) and the probably related or identical glycoprotein neurothelin (17,18). Both of these markers have been shown to be inducible in endothelial cells of chorioallantoic vessels by transplants of embryonic mouse brain (12,18). The exact mechanism of this induction is, however, not known. With the presented experiments we wanted to test the hypothesis that an astrocyte-derived soluble factor is the mediator of this cellular interaction between brain tissue and endothelial cells of non-neural origin.

MATERIALS AND METHODS

In Ovo Experiments

Commercially available fertilized chicken eggs were obtained from a local breeder. After incubation at 37.8°C and 60% humidity for 9-10 days, the eggs were fenestrated. A drop of sterile PBS put on the hole in the eggshell allowed the detachment of the chorioallantoic membrane. Small pieces (~75 mm³) of gelfoam (Upjohn, Zürich) were soaked with conditioned medium from type-1 astrocytes, C6 glioma cells, END-p23 cells (pulmonary endothelial cell line), L929 (fibroblast cell line), with unconditioned medium or with PBS, and the gelfoam was deposited onto the chorioallantoic membrane. At E13 or E14 the gelfoam was connected to a polyethylene tube, with an inside diameter of 0.76 mm, attached on the other end outside of the egg to a miniosomitic pump (Alzet model 2002). The pump, containing 227 µl of the same medium as applied on E9 or E10 and placed in a saline-filled container, allowed the constant delivery of its contents (astrocyte-conditioned medium or controls) during 5 days, after which it was removed and the eggs examined at E18 or E19.

From a total of 116 treated eggs, 33 (28%) survived. Eighteen eggs were exposed to astrocyte-conditioned medium for 10 days. Three eggs were already examined at E13 after 4 days of exposure to astrocyte-conditioned medium. Twelve eggs were used for the control experiments using conditioned medium from C6-glioma, a fibroblast- and an endothelial cell line, unconditioned medium and PBS.

Cell Culture

All the cells used were grown in uncoated tissue culture flasks (Costar, Tecnorama, Wallisellen) in DMEM containing 10% fetal calf serum (Seromed, Kakola, Basel), 90 U/ml penicillin and 90 µg/ml streptomycin. In most experiments fetal calf serum was omitted for the last 4 days of culturing before collecting the supernatant. Supernatants were pooled and stored at -20°C for a maximum of 2 months. Primary astrocytes were purified from neonatal Sprague-Dawley rat brain cortex as described earlier (4,11). C6 glioma cells were a gift from D. Monard (Basel), L929 fibroblast cells from S. Corradin-Betz (Epalinges) and the medium-T transformed mouse pulmonary endothelial cell line END-p23 from A. Aguzzi (Vienna) (24).

Tissue Preparation and Immunohistochemistry

The embryos were sacrificed 4 or 10 days after the beginning of the continuous exposure to astrocyte-conditioned medium. The chorioallantoic membrane with the adherent gelfoam was dissected out, the specimen cryoprotected by immersion in saturated sucrose solution containing 0.8% paraformaldehyde, and cryofixed by immersion in isopentane precooled in liquid nitrogen. Serial step cryostat sections (7 μm thickness) mounted on gelatinized glass slides were exposed to primary antibodies to the HT7-protein (a gift from W. Risau, Munich) and to neurothelin (a gift from B. Schlosshauer, Tübingen) overnight at 4°C at a dilution of 1:50 and 1:10,000, respectively. After several washing steps, the sections were stained by standard immunoperoxidase methods (PAP or ABC). The secondary antibodies were obtained from DAKO (Instrumentengesellschaft, Zürich) and used at a dilution of 1:300 for 45 min at room temperature. Visualization was performed with diaminobenzidine as the chromogen. Most sections were counterstained with hematoxylin. Normal chick brain was used as the positive control. Negative controls included the omission of the first incubation step with the specific antibodies and the use of non-immune serum as the first antibody.

RESULTS AND DISCUSSION

In all surviving chick embryos, the gelfoam itself was well-tolerated and did not induce a significant inflammatory response. In all cases treated with conditioned or unconditioned medium, an angiogenic response was observed in the chorioallantoic membrane. The control cases receiving only PBS showed no significant angiogenic response, indicating that the gelfoam by itself is not angiogenic.

In all 18 eggs exposed to astrocyte-conditioned medium for 10 days, the immunohistochemical staining gave superposable results for the HT7-protein and neurothelin, adding evidence to previous data (19), indicating that these proteins are probably identical or closely related. The two antibodies may detect different epitopes on the same molecule. The positive vessels were capillaries and venules showing a homogeneous HT7/neurothelin expression over their entire luminal surface. Sometimes the entire cellular cut surface of the endothelial cells was positive, especially for the HT7-antibody. This distribution corresponds to the localization of HT7 reported in chick brain microvessels by immuno-electron microscopy (1). Most of the HT7/neurothelin-positive small and medium sized vessels were found close to the gelfoam. However, not all of these vessels exposed most intensely to astrocyte-conditioned medium were positive, and some negative vessels were immediate neighbors of positive vessels. The number and distribution of positive vessels varied from one experiment to another, but in 12 out of 18 cases, more than 20% of the vessels close to the gelfoam expressed HT7 and/or neurothelin. The heterogeneous response of endothelial cells to the astrocyte-derived factor(s) may reflect the presence of endothelial cell subgroups, especially since there is increasing evidence of brain microvascular endothelial cell heterogeneity. Brain microvessels have regionally heterogeneous γ-glutamyl-transpeptidase (25) and monoamine oxidase activities (8) and variable electrical resistance and macromolecular permeability (16). Variable expression of endothelial markers and differences in growth behavior were noted in brain endothelial cell lines (15). In two brain endothelial cell lines obtained by a standardized immortalization protocol, we noted remarkable differences in morphology, enzyme activities, and in their response to extracellular matrix and astrocytic factors (6,7). These observations suggest that there is either a mosaic like or segmental microanatomic distribution of different subtypes of brain endothelial cells. It is therefore reasonable to postulate chorioallantoic vessel endothelial cell heterogeneity even if there is only indirect evidence to support this hypothesis.

With increasing distance from the gelfoam, the number of positive vessels decreased. This might be a result of limited access and/or stability of the glial factor(s). Alternatively, it may be speculated that there is concomitant secretion of an antagonizing factor by astrocytes related to vascular endothelial growth factor (2,10), which is identical to the vascular permeability factor (9), or that the secreted inducing factor is present in an inactive form and has to be activated by factors enriched in or around the gelfoam.

Chorioallantoic membranes exposed to astrocyte-conditioned medium for 4 days showed positive immunoreactivity for HT7-protein and neurothelin in a similar way as did those exposed to the astrocytic factor for 10 days. The number of positive vessels was comparable in both series; however, the staining intensity was less intense in the group treated only 4 days. We did not observe HT7/neurothelin induction in control cases.

These results provide evidence for the concept of the induction of blood-brain barrier related properties in endothelial cells of non-neural origin by a soluble factor(s) secreted by type-1 astrocytes.

ACKNOWLEDGMENTS

This work was supported by a grant from the Swiss National Fund for Scientific Research (Grant No. 31.9132.87). We thank A. Aguzzi, S. Corradin-Betz, D. Monard, W. Risau and B. Schlosshauer for kindly providing cell lines or antibodies.

REFERENCES

1. Albrecht, U., Seulberger, H., Schwarz, H. and Risau, W., Correlation of blood-brain barrier function and HT7 protein distribution in chick brain circumventricular organs, *Brain Res.* 535:49-61, 1990.

2. Breier, G., Albrecht, U., Sterrer, S., and Risau, W., Expression of vascular endothelial growth factor during embryonic angiogenesis and endothelial cell differentiation, *Development* 114:521-532, 1992.

3. Dehouk, M.P., Meresse, S., Delorme, P., Fruchart, J.C. and Cecchelli, R., An easier, reproducible and mass-production method to study the blood-brain barrier in vitro, *J. Neurochem.* 51:1798-1801, 1990.

4, Janzer, R.C. and Raff, M.C., Astrocytes induce blood-brain barrier properties in endothelial cells, *Nature* 325:253-257, 1987.

5. Joó, F., The cerebral microvessel in culture, an update, *J. Neurochem.* 58:1-17, 1992.

6. Juillerat-Jeanneret, L., Aguzzi, A., Wiestler, O.D., Darekar, P., and Janzer, R.C., Dexamethasone selectively regulates the activity of enzymatic markers of cerebral endothelial cell lines, *In Vitro Cell. Dev. Biol.*, 1992 (in press).

7 Juillerat-Jeanneret, L., Darekar, P., and Janzer, R.C., Heterogeneity of microvascular endothelial cells of the brain: a comparison of the effects of extracellular matrix and soluble astrocytic factors, 1992 (submitted).

8. Lasbennes, F., Sercombe, R., Verrechia, C., and Seylaz, J., Vascular monoamine oxidase activity in the rat brain: variation with the substrate and the vascular segment, *Life Sci* 36:2263-2268, 1985.

9. Olander, J.V., Connolly, D.T., and DeLarro, J.E., Specific binding of vascular permeability factor to endothelial cells, *Biochem. Biophys. Res. Comm.* 175:68-76, 1991.

10. Pepper, M.S., Ferrara, N., Orci, L., and Montesano, R., Vascular endothelial growth factor (VEGF) induces plasminogen activators and plasminogen activator inhibitor-1 in microvascular endothelial cells, *Biochem. Biophys. Res. Comm.* 181:902-906, 1991.

11. Raff, M.C., Abney, E.R., and Fok-Seang, J., Reconstruction of a developmental clock in vitro: a critical role for astrocytes in the timing of oligodendrocyte differentiation, *Cell* 42:61-69, 1985.

12. Risau, W., Hallmann, R., Albrecht, U., and Henke-Fahle, S., Brain induces the expression of an early cell surface marker for blood-brain barrier specific endothelium, *EMBO J.* 5:3179-3183, 1986.

13. Risau, W., and Wolburg, H., Development of the blood-brain barrier, *Trends in Neurosciences* 13:174-179, 1990.

14. Rubin, L.L., The blood-brain barrier in and out of cell culture, *Curr. Top. Neurobiol.* 1:360-364, 1991.

15. Rupnik, M.A., Carey, A., and Williams, S.K., Phenotypic diversity in cultured cerebral microvascular endothelial cells, *In Vitro Cell. Dev. Biol.* 24:435-444, 1988.

16. Rutten, M.J., Hoover, R.L., and Karnovsky, M.J., Electrical resistance and macromolecular permeability of brain endothelial monolayer cultures, *Brain Res.* 425:301-310, 1987.

17. Schlosshauer, B., Neurothelin: molecular characteristics and developmental regulation in the chick CNS, *Development* 113:129-140, 1991.

18. Schlosshauer, B., and Herzog, K.-H., Neurothelin: an inducible cell surface glycoprotein of blood-brain barrier specific endothelial cells and distinct neurons, *J. Cell. Biol.* 110:1261-1274, 1990.

19. Seulberger, H., Lottspeich, F., and Risau, W., The inducible blood-brain barrier specific molecule HT7 is a novel immunoglobulin-like cell surface glycoprotein, *EMBO J.* 110:2151-2158, 1990.

20. Shivers, R.R., Arthur, F.E., and Bowman, P.D., Induction of gap junctions and brain endothelium-like tight junctions in cultured bovine endothelial cells: local control of cell specialization, *J. Submicroscop. Cytol. Pathol.* 20:1-14, 1988.

21. Stewart, P., and Wiley, M.J., Developing nervous tissue induces formation of blood-brain barrier characteristics in invading endothelial cells: a study using quail-chick transplantation chimeras, *Dev. Biol.* 84:183-192, 1981.

22. Tao-Cheng, J.H., Nagy, Z., and Brightman, M.W., Tight junctions of brain endothelium in vitro are enhanced by astroglia, *J. Neurosci.* 7:3293-3299, 1987.

23. Tio, U., Deenen, M., and Marani, E., Astrocyte-mediated induction of alkaline phosphatase activity in human umbilical cord vein endothelium: an in vitro model, *Eur. J. Morphol.* 28:289-300, 1990.

24. Williams, L., Courtneidge, S.A., and Wagner, E.F., Embryonic lethalities and endothelial tumors in chimeric mice expressing polyoma middle-T oncogene, *Cell* 52:121-131, 1988.

25. Wolff, J.E.A., Belloni-Olivi, L., Bressier, J.P., and Goldstein, G.W., γ-glutamyltranspeptidase activity in brain microvessels exhibits regional heterogeneity, *J. Neurochem.* 58:909-915, 1992.

CHORIOALLANTOIC MEMBRANE (CAM)
VESSELS DO NOT RESPOND TO
BLOOD-BRAIN BARRIER (BBB) INDUCTION

J.A. Holash and P.A. Stewart

Department of Anatomy and Cell Biology
University of Toronto, Toronto, Ontario
Canada, M5S 1A8

INTRODUCTION

In early development, the brain and spinal cord develop from the neural tube, an avascular derivative of the embryonic ectoderm. Primitive endothelial cells in the surrounding mesoderm form a perineurial vascular plexus from which penetrating vessels grow into the neural tube to form the definitive cerebral vasculature. It has been shown that the neural tissue environment induces formation of blood-brain barrier features in the penetrating vessels (1). Numerous in vitro studies (2-6) suggested that astrocytes, whose processes ensheath the cerebral capillaries, may be the cells responsible for this induction.

Recently Janzer and Raff (7) have transplanted purified astrocytes to the anterior chamber of the eye and to the chick chorioallantoic membrane (CAM) and found that, in both sites, the vessels that are associated with the grafts are impermeable to Evan's blue, a vascular tracer that does not pass through BBB vessels. Their control grafts consisted of meningeal- or skin-derived fibroblasts. Vessels associated with the fibroblast grafts allowed visible leakage of Evan's blue. These authors concluded that the astrocytes had induced barrier features in the exogenous vessels. We feel that the results from the anterior chamber grafts are inconclusive since it is well known that iridial vessels that supply the graft already have barrier features (8-10). Furthermore, preliminary evidence from our laboratory shows that in the anterior chamber, astrocyte grafts tend to form a layer on the surface of the iris that is poorly vascularized, whereas fibroblasts form an invasive mass that is well vascularized and associated with significant inflammation. Both of these factors would be associated with low tracer accumulation in the astrocyte grafts and high tracer accumulation in the fibroblast grafts.

Results from Janzer and Raff's chorioallantoic grafts are also difficult to interpret because no evidence regarding relative vascular supply was presented since CAM vessels

Frontiers in Cerebral Vascular Biology: Transport and Its Regulation
Edited by L.R. Drewes and A.L. Betz, Plenum Press, New York, 1993

223

have a highly sensitive vasomotor system that makes delivery of tracer to different areas of CAM problematic, and since foreign bodies, such as the millipore filter disc that they included elicits the ingrowth of inflammatory cells (11) that release permeability-altering substances.

We have repeated the CAM grafting experiments of Janzer and Raff in order to investigate the relationship between grafted cells, inflammation, and barrier features in the CAM vessels. We grafted purified astrocytes isolated from newborn rat brain, primary fibroblasts isolated from newborn rat skin, and early neural tube from two-three day chick embryos. The vessels in or associated with the resulting grafts were examined for two histochemical barrier markers: glucose transporter (GT) (12-14) and HT7 (15,16) (also known as neurothelin and a variety of other names (17)) and for ultrastructural BBB characteristics.

MATERIALS AND METHODS

Astrocytes were isolated and purified according to the method of McCarthy and De Vellis (18). Fibroblasts were cultured from fragments of new-born rat skin, grown in F12/DMEM medium, and sub-passaged no more than 5 times. Spinal cords were isolated from 2-3 day old chicken embryos and cleared of adherent mesenchyme in 0.5% trypsin in Hank's balanced salt solution. Varying amounts of mesoderm were left adhering to some of the spinal cords. The cultured cells, astrocytes or fibroblasts were removed from the culture dishes with trypsin, incubated in DiI (19) for 1 hr to label the cells, washed 4-5 times in fresh medium and spun into a pellet. Meanwhile, a sterile ring of millipore filter (0.45 μm pore size) was placed on the CAM to help us relocate the grafted cells. A portion of the pellet was drawn into a micropipette and placed on the CAM in the center of the ring using gentle mouth suction. In some cases grafted cells were placed on the CAM without a filter ring.

Either chick or quail embryos 6-7 days in ovo were used as hosts for the CAM grafts. An opening in the shell was made over the air space and the shell membrane removed. Grafts were placed on the CAM and the eggs were sealed with cellophane tape and re-incubated until the hosts were 13-18 days in ovo. Grafts were harvested and frozen for histochemistry or fixed by immersion in 2.5% glutaraldehyde in 1M phosphate buffer or Bouin's fixative.

Frozen sections of unfixed tissue, 10 μm thick, were cut and collected on polylysine-coated slides. For indirect immunoperoxidase staining, sections were incubated in the appropriate dilution of one of the following: rabbit polyclonal antisera to GFAP (to label astrocytes), rabbit polyclonal antisera to the glucose transporter, or mouse monoclonal antisera to HT7, at 4°C overnight. Subsequently, sections labelled with antibodies to GFAP or GT were incubated in biotinylated goat-antirabbit IgG (Amersham), and HT7-labelled sections were incubated in biotinylated rabbit-anti mouse IgG (H + L) (ICN) for 30 min at room temperature. Antibody binding was visualized using the avidin biotin complex reagent (Vector) and by reacting the sections for HRP using the cobalt-oxidase method (20). In all cases control slides in which primary antibodies were omitted showed that labelling was not a result of non-specific binding of the secondary antibody.

For immunofluorescent double labelling with the quail endothelial marker QH1 and GT, slides were incubated in a monoclonal antibody to QH1 combined with the rabbit polyclonal to GT overnight at 4°C. Subsequently slides were incubated in donkey-anti rabbit (FITC) IgG (H+L) and rabbit-anti mouse (TRITC) IgG F(ab')2.

Glutaraldehyde-fixed tissue was processed for routine electron microscopy using *en bloc* staining with uranyl acetate to enhance interendothelial junctions, and the graft vessels were examined in a Hitachi 7000 electron microscope. Bouin's-fixed tissue was embedded in paraffin, sectioned and stained using hematoxylin and eosin and evaluated for degree of inflammation.

RESULTS

Astrocyte and Fibroblast Grafts

When astrocytes or fibroblasts were cultured with a millipore filter ring, a florid inflammatory response developed. Millipore filter rings alone also induced inflammation with a pronounced foreign body response. When the ring was omitted, however, grafts were often poorly vascularized and became necrotic. Serendipitously, the inflammation-induced angiogenesis promoted graft vascularization, and graft survival was much improved. Treatment with 2-10 µg of methylprednisolone, administered in a slow-release agent (Depo-Medrol®, Upjohn) 3 days after the cells and ring were placed on the CAM, reduced the inflammation and further improved graft cell survival. We were not able to resolve the inflammation completely, as higher doses of steroid resulted in the deaths of most of the hosts.

We found that astrocytes did not induce GT, or high levels of HT7 (Figure 1A) in any of the grafts. Vessels in the host brain, however expressed normal levels of both GT and HT7 (Figure 1B), showing that their expression was not affected by the steroid treatment. Ultrastructurally, the few vessels that we were able to find in grafts of either type did not have the ultrastructural features of BBB vessels. Their junctions were short and the occluding areas sparse. Numerous vesicles were found in the endothelial cytoplasm. We could not distinguish astrocyte graft vessels from fibroblast graft vessels. We conclude that astrocytes do not induce barrier features in CAM vessels.

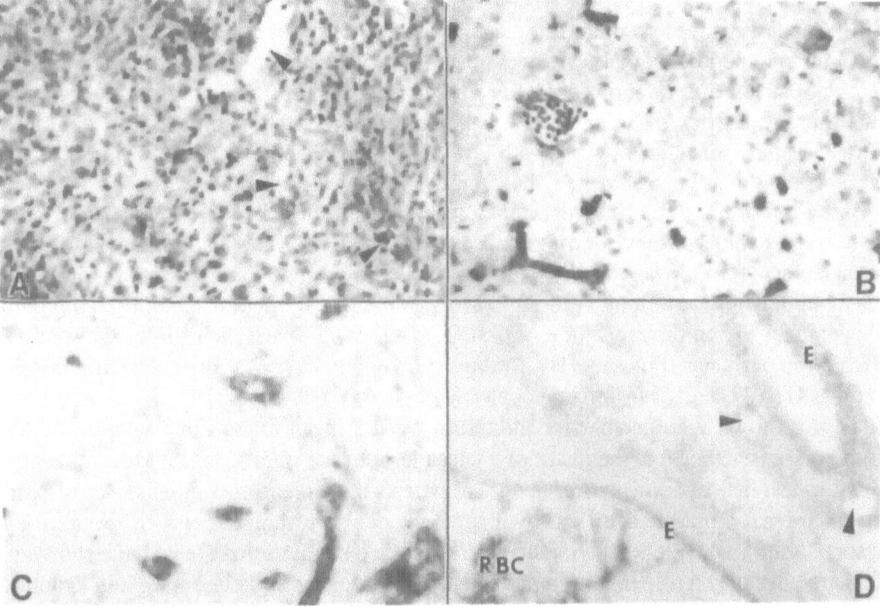

Figure 1 A. Astrocyte graft treated with anti-inflammatory agent and reacted for HT7. No positive staining is seen in the vessels (arrow heads). Counter-stained with cresyl violet. X 190 **B.** Host brain of similarly treated embryo reacted for HT7 shows that steroid treatment does not affect expression of HT7. Counter-stained with cresyl violet. X 190 **C.** Untreated fibroblast graft showing strong HT7 reactivity in its vessels. X 152 **D.** Normal CAM showing mildly positive HT7 staining in the epithelium (E) and red blood cells (RBC) but not in CAM endothelium (arrow heads). X 152

Interestingly, most grafts induced a low level of HT7 expression. In fact, the HT7 staining was distinct in 5/7 fibroblast grafts that were not treated with anti-inflammatory agents (Figure 1C) and was not because of non-specific reaction with the secondary antibody. Positive HT7 staining was also seen in the macrophages surrounding the filter rings.

Neural Tube Grafts

To determine whether the lack of BBB features in CAM vessels in response to astrocyte grafts was a result of a failure in the inductive capabilities of astrocytes or a failure of the CAM vessels to respond, we transplanted two-day neural tube to the CAM, since early embryonic neuroepithelium is known to induce BBB features in non-barrier vessels (1). For these experiments we used chick neural tube and quail CAM so that we could take advantage of the quail endothelial antibody QH1 to determine unequivocally the source of the graft vessels. In grafts that were thoroughly cleared of mesoderm before explanting, all of the vessels were of quail origin. In cases in which some mesoderm was left adhering to the spinal cords the vessels were chimeric, with some endothelial cells expressing the quail marker and others not. When we double-stained the slides for GT and QH1, we found that the chick endothelial cells express this barrier marker, but that the quail endothelial cells did not. These preliminary results suggest that developing neural tube induces formation of at least one BBB feature in ingrowing endothelium, but that vascular endothelium of CAM origin is not competent to respond to this induction.

DISCUSSION

Until recently it was commonly held that astrocytes induce the development of BBB features in brain endothelial cells. Astrocytic endfeet ensheath most of the abluminal surface of brain capillaries, and in choroid plexus stroma and circumventricular organs, where the barrier is absent, astrocyte endfeet do not have an intimate relationship with the capillaries, but are separated from them by a sizeable extracellular space. This intriguing structural association, and the in vitro evidence cited above strongly suggests that astrocytes play a role in the formation of the BBB. However, in at least two sites, the iris (8,10,21) and peripheral nerves (22-25), capillaries have barrier features even in the absence of an intimate association with astrocytes. Furthermore, recent studies have shown that some barrier features form very early in development, long before the astrocytic ensheathment occurs (26). Beginning at embryonic day 12 in the rat (26), fenestrated, sinusoid-like vessels begin to grow into the brain from the perineurial plexus. By 15 days the vessels have lost their fenestrations (26) and express GT (27) and transferrin receptor (28).

It is not known whether barrier induction occurs in an all-or-none fashion early in development or whether it is the result of a cascade of tissue interactions at different stages. The early appearance of some barrier features suggests that at least the initial stages of barrier induction are accomplished either by primitive neuroepithelial cells or by astrocyte precursors. Later in development, the numerous in vitro studies cited above have shown that mature astrocytes play a role in up-regulating barrier-associated proteins, at least in endothelia that are already committed to barrier formation (3,6). It seems reasonable to suggest, therefore, that mature astrocytes may play a maintenance role in the BBB.

The results reported here show that mature astrocytes do not induce barrier levels of GT or HT7 in CAM vessels, and, furthermore, that CAM vessels cannot be induced to express these markers even by primitive neuroepithelial tissue, although vessels derived from mesoderm of embryonic origin can be induced to express GT by neuroepithelium. HT7 shows homology to the immunoglobulin superfamily of cell surface receptors, which transport immunoglobulins across epithelial interfaces (15). Expression of these receptors has been shown to be up-regulated by factors that are released during the inflammatory process (29). Because of this, we suspect that up-regulation of HT7 in graft vessels that we saw may have been a response to inflammation. The distinct reaction in fibroblast grafts that were not treated with anti-inflammatory agents shows that this is likely, and that the expression of HT7 under these circumstances cannot be taken as evidence for BBB induction.

ACKNOWLEDGMENTS

This work was supported by the Medical Research Council of Canada through a Grant to PAS and a Studentship to JAH. The authors are grateful to the following for their generous gifts of antibodies: Werner Risau, Max-Planck Institüt, Martinsried, Germany for HT7 antibody, Dr. W.M. Pardridge, Department of Medicine, UCLA, for glucose transporter antibody and Drew Noden, Department of Anatomy, Cornell University for QH1 antibody.

REFERENCES

1. Stewart, P.A., and Wiley, M.J., Developing nervous tissue induces formation of blood-brain barrier characteristics in invading endothelial cells: A study using quail-chick transplantation chimeras, *Dev. Biol.* 84:183, 1981.

2. Beck, D.W., Vinters, H.V., Hart, M.N., and Cancilla, P.A., Glial cells influence polarity of the blood-brain barrier, *J. Neuropathol. Exp. Neurol.* 43:219, 1984.

3. Maxwell, K., Berliner, J.A., and Cancilla, P.A., Induction of gamma-glutamyl transpeptidase in cultured cerebral endothelial cells by a product released by astrocytes, *Brain Res.* 410:309, 1987.

4. Arthur, F.E., Shivers, R.R., and Bowman, P.D., Astrocyte-mediated induction of tight junctions in brain capillary endothelium: an efficient in vitro model, *Dev. Brain Res.* 36:155, 1987.

5. Tao-Cheng, J.-H., Nagy, Z., and Brightman, M.W., Tight junctions of brain endothelium in vitro are enhanced by astroglia, *J. Neurosci.* 7:3293, 1987.

6. Maxwell, K., Berliner, J.A., and Cancilla, P.A., Stimulation of glucose analogue uptake by cerebral microvessel endothelial cells by a product released by astrocytes, *J. Neuropathol. Exp. Neurol.* 48:69, 1989.

7. Janzer, R.C., and Raff, M.C., Astrocytes induce blood-brain barrier properties in endothelial cells, *Nature* 325:253, 1987.

8. Freddo, T.F., and Raviola, G., Freeze-fracture analysis of the interendothelial junctions in the blood vessels of the iris in *Macaca mulatta*, *Invest. Ophthalmol. Vis. Sci.* 23:154, 1982.

9. Dehouck, M.-P., Méresse, S., Delorme, P., Fruchart, J.-C., and Cecchelli, R., An easier, reproducible, and mass-production method to study the blood-brain barrier in vitro, *J. Neurochem.* 54:1798, 1990.

10. Townes-Anderson, E., and Raviola, G., Morphology and permeability of blood vessels in the prenatal rhesus monkey eye: How plasma components diffuse into the intraocular fluids during development, *Exp. Eye Res.* 35:203, 1992.

11. D'Arcy, P.F., and Howard, E.M., A new anti-inflammatory test, utilizing the chorio-allantoic membrane of the chick embryo, *Br. J. Pharmacol. Chemother.* 29:378, 1967.

12. Kasanicki, M.A., Jessen, K.R., Baldwin, S.A., Boyle, J.M., Davies, A., and Gardiner, R.M., Immunocytochemical localization of the glucose transport protein in mammalian brain capillaries, *Histochem. J.* 21:47, 1989.

13. Harik, S.I., Kalaria, R.N., Andersson, L., Lundahl, P., and Perry, G., Immunocytochemical localization of the erythroid glucose transporter: abundance in tissues with barrier functions, *J. Neurosci.* 10:3862, 1990.

14. Farrell, C.L., Yang, J., and Pardridge, W.M., GLUT-1 glucose transporter is present within apical and basolateral membranes of brain epithelial interfaces and in microvascular endothelia with and without tight junctions, *J. Histochem. Cytochem.* 40:193, 1992.

15. Seulberger, H., Lottspeich, F., and Risau, W., The inducible blood-brain barrier specific molecule HT7 is a novel immunoglobulin-like cell surface glycoprotein, *EMBO J.* 9:2151, 1990.

16. Albrecht, U., Seulberger, H., Schwarz, H., and Risau, W., Correlation of blood-brain barrier function and HT7 protein distribution in chick brain circumventricular organs, *Brain Res.* 535:49, 1990.

17. Seulberger, H., Unger, C.M., and Risau, W., HT7, neurothelin, basigin, gp42 and OX-47 - many names for one developmentally regulated immunoglobulin-like surface glycoprotein on blood-brain barrier endothelium, epithelial tissue barriers and neurons, *Neurosci. Lett.* 140:93, 1992.

18. McCarthy, K.D., and de Vellis, J., Preparation of separate astroglial and oligodendroglal cell cultures from rat cerebral tissue, *J. Cell Biol.* 85:890, 1980.

19. Honig, M.G., and Hume, R.I., DiI and DiO: versatile fluorescent dyes for neuronal labelling and pathway tracing, *TINS* 12:333, 1989.

20. Itoh, K., Konishi, A., Nomura, S., Mizuno, N, Nakamura, A., and Sugimoto, T., Application of coupled oxidation reaction to electron microscopic demonstration of horseradish peroxidase: cobalt-glucose oxidase method, *Brain Res.* 175:341, 1979.

21. Harik, S.I., Kalaria, R.N., Whitney, P.M., Andersson, L., Lundahl, P., Ledbetter, S.R., and Perry, G., Glucose transporters are abundant in cells with "occluding" junctions at the blood-eye barriers, *Proc. Natl. Acad. Sci. USA* 87:4261, 1990.

22. Bradbury, M.W.B., and Crowder, J., Compartments and barriers in the sciatic nerve of the rabbit, *Brain Res.* 103:515, 1976.

23. Rechthand, E., and Rapoport, S.I., Regulation of the Microenvironment of peripheral nerve: role of the blood-nerve barrier, *Prog. Neurobiol.* 28:303, 1987.

24. Latker, C.H., Shinowara, N.L., Miller, J.C., and Rapoport, S.I., Differential localization of alkaline phosphatase in barrier tissues of the frog and rat nervous systems: A cytochemical and biochemical study, *J. Comp. Neurol.* 264:291, 1987.

25. Froehner, S.C., Davies, A., Baldwin, S.A., and Lienhard, G.E., The blood-nerve barrier is rich in glucose transporter, *J. Neurocytol.* 17:173, 1988.

26. Bar, T., and Wolff, J.R., The formation of capillary basement membranes during internal vascularization of the rat's cerebral cortex, *Z. Zellforsch.* 133:231, 1972.

27. Dermietzel, R., and Krause, D., Molecular anatomy of the blood-brain barrier as defined by immunocytochemistry, *Int. Rev. Cytol.* 127:57, 1991.

28. Risau, W., Hallmann, R., and Albrecht, U., Differentiation-dependent expression of proteins in brain endothelium during development of the blood-brain barrier, *Dev. Biol.* 117:537, 1986.

29. Kvale, D., Løvhaug, D., Sollid, L.M., and Brandtzaeg, P., Tumor necrosis factor-α up-regulates expression of secretory component, the epithelial receptor for polymeric Ig[1], *The Journal of Immunol.* 140:3086, 1988.

CEREBRAL ENDOTHELIUM AND ASTROCYTES COOPERATE IN SUPPLYING DOCOSAHEXAENOIC ACID TO NEURONS

Steven A. Moore

Department of Pathology
The University of Iowa
Iowa City, IA 52242

INTRODUCTION

Elongated, more highly polyunsaturated derivatives of linolenic acid, especially docosahexaenoic acid (DHA), accumulate in neurons of the brain (1). We have previously shown that neurons themselves are incapable of significant essential fatty acid elongation and desaturation and, thus, are dependent upon a supply of preformed DHA (2). However, astrocytes and cerebral endothelium actively elongate and desaturate the ω-3 fatty acid precursors, linolenic acid and eicosapentaenoic acid (EPA) (2,3). By doing so they may together represent an important source of DHA for the central nervous system.

In the present work, potential interactions among cerebral endothelial cells, astrocytes, and neurons in the elongation, desaturation and subsequent accumulation of ω-3 fatty acids in neurons were investigated. Cultures of astrocytes, neurons and cerebral microvascular endothelium were studied individually and in co-culture combinations with one another. The findings suggest that endothelial cells of the blood-brain barrier may cooperate with astrocytes in providing DHA to neurons.

METHODS

Cerebromicrovascular endothelial cells were isolated from mouse or rat brains as described previously (3). The cultures were maintained in Lewis medium containing 10% fetal bovine serum (FBS). The endothelial cell cultures were characterized and determined to be >95% pure by (a) light and electron microscopic appearance, (b) the presence of γ-glutamyltranspeptidase activity, (c) the uptake of DiI-Ac-LDL, and (d) Griffonia simplicifolia agglutinin histochemical staining followed by flow cytofluorimetric analysis (4). Contamination of cultures by smooth muscle and astroglia was determined by immunohistochemical staining with an anti-muscle-actin antibody (5) and anti-glial fibrillary acidic protein antibodies. Cells were subcultured weekly and cultures between passages 8 and 18 were used in each experiment.

Primary cultures of astrocytes and neurons were prepared from whole cerebellum of 7- to 10-day-old rat pups by a trypsinization, trituration technique that disaggregates brain perikarya (6,7). Highly enriched astrocyte and neuronal cultures were obtained from the same starting cell suspension. Astrocyte cultures were maintained in Eagle's MEM with Earle's salts supplemented with 33 mM glucose, 2 mM glutamine, 180 μM gentamicin and 10% FBS. Neuronal cultures were maintained in Eagle's MEM with Earle's salts supplemented with 33 mM glucose, 1 mM glutamine, 180 μM gentamicin, 20 mM KCl, 80 μM 5-fluorodeoxy-2'-uridine, 2.5% chick embryo extract and 10% FBS. Astrocyte cultures were routinely ≥95% type I astrocytes by immunohistochemical characterization. Neuronal cultures from cerebellum contained approximately 90% glutamatergic granule neurons and 5-7% GABAergic inhibitory interneurons, with the majority of contaminating cells type I astrocytes.

Cultures of astrocytes or endothelium were established on 25-mm filter chambers (Falcon), and cultures of cerebellar granule cell neurons or astrocytes were established in 6-well tissue culture plates as depicted in Figure 1. The cells were incubated alone or in co-culture with either [1-14C]linolenic acid or [1-14C]EPA. Elongation and desaturation products in the top and bottom medium and in cell glycerolipids were determined by reversed phase HPLC after chloroform:methanol lipid extraction and subsequent fatty acid methylation with BF$_3$ (2,3).

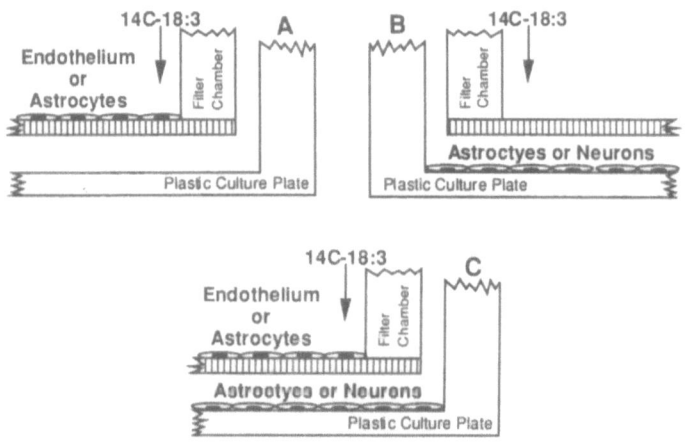

Figure 1. Cell cultures were established with (A) cells only on the filter of Falcon 25 mm filter chambers, (B) cells only on the plastic of 6-well tissue culture plates, or (C) cells on both the filter and plastic surfaces in a co-culture. Endothelial cells were passaged to filters, while astrocytes and neurons were seeded as primary cultures directly onto either the filter or the plastic well.

RESULTS AND DISCUSSION

Although the essential polyunsaturated fatty acid docosahexaenoic acid (DHA) is highly enriched in brain neurons (1), it cannot be synthesized de novo by animal tissues and must ultimately be obtained from the diet. In addition, the ω-3 fatty acid precursors linolenic acid and eicosapentaenoic acid (EPA) can be converted to DHA through a process of fatty acid chain elongation and desaturation. The liver is a major site where dietary ω-3 fatty acid is converted to DHA, and the main source of DHA for the central nervous system is considered to be liver-derived DHA (8).

In addition to DHA formed in the liver, plasma also ordinarily contains ω-3 fatty acid precursors (9), and circulating linolenic acid and EPA may be additional sources of DHA for the brain. This pathway would require concomitant or sequential transfer of ω-3 fatty acid precursors across the blood-brain barrier and conversion to DHA in the brain. It has been known for many years that brain tissue is capable of forming DHA from appropriate precursors (10), and recent work has shown that cerebral microvascular endothelium and astrocytes are capable of active elongation and desaturation of linolenic acid and EPA (2,3). Cerebral endothelia produce and release primarily EPA, but astrocytes produce and release large amounts of DHA (2,3). Neurons, on the other hand, are incapable of significant ω-3 fatty acid elongation and desaturation and are thus dependent upon a source of preformed DHA (2). Interactions among these three cell types during the conversion of ω-3 fatty acid precursors to DHA and the transfer of DHA to neurons were the subject of the present studies.

Endothelial Cell Targeting of ω-3 Fatty Acids

In order for circulating linolenic acid or EPA to be sources of DHA for the brain, they must first cross the blood-brain barrier. In the present study cerebral endothelial cells were cultured on filters as an in vitro model of the blood-brain barrier (Figure 1). In this model both linolenic acid and EPA passed from the apical to the basal surface of the endothelium. Not only were these fatty acids transferred to the bottom medium, they were enriched in the medium facing the endothelial basolateral surface by 3- to 9-fold over the apical medium (Figure 2). This suggests that the cerebral endothelium may actively target ω-3 fatty acids for release from its basal surface where these fatty acids can be enriched in the brain.

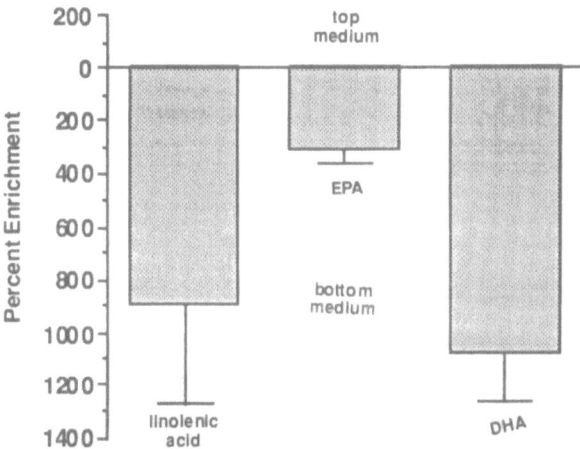

Figure 2. Cell cultures were incubated with either [1-¹⁴C]linolenic acid or [1-¹⁴C]EPA placed in the top medium of the filter chamber. After 96 hr, the bottom medium of endothelial cell cultures was highly enriched with the fatty acid originally placed in the top medium. In co-cultures of endothelium and astrocytes, the bottom medium was highly enriched with DHA produced by the astrocytes. Bars represent means ± SE of triplicate cultures.

Cooperation of Endothelial Cells and Astrocytes

Since cerebral endothelial cells appear capable of targeting ω-3 fatty acids into the brain, the next step to test was the interaction between endothelium and astrocytes in producing DHA. In co-cultures, the ω-3 fatty acids targeted by endothelium to the bottom medium were readily taken up, elongated and desaturated by astrocytes to form DHA that in turn was released into the bottom medium. This astrocyte-derived DHA was enriched in the bottom medium to a degree similar to the linolenic acid and EPA targeted into the bottom medium by endothelial cells cultured alone (Figure 2). In addition these endothelial cell/astrocyte co-cultures produced greater total amounts of elongation and desaturation products than either cell type alone, as much as 50% more. These studies suggest that endothelial cells of the blood-brain barrier may cooperate with astrocytes by (a) providing astrocytes with ω-3 fatty acid precursors, (b) increasing the degree of elongation and desaturation of ω-3 fatty acids in the brain, and (c) blocking the egress of DHA from the brain.

Enrichment of Neuronal Cell Lipids With Docosahexaenoic Acid

The final portion of this pathway to be tested was the transfer of DHA from astrocytes to neurons. In the present studies, astrocyte-derived DHA was readily incorporated by co-cultured neurons. In fact, only those neurons co-cultured with astrocytes were found to contain radiolabeled DHA (Figure 3). In these astrocyte/neuron co-cultures as much as 90% of the DHA produced from radiolabeled precursors was found either in the medium or in neuronal cell lipids. In addition to providing DHA for neurons, the total radiolabeled elongation and desaturation products of linolenic acid found in neuronal lipids was increased by 5- to 10-fold when neurons were co-cultured with astrocytes (Figure 3).

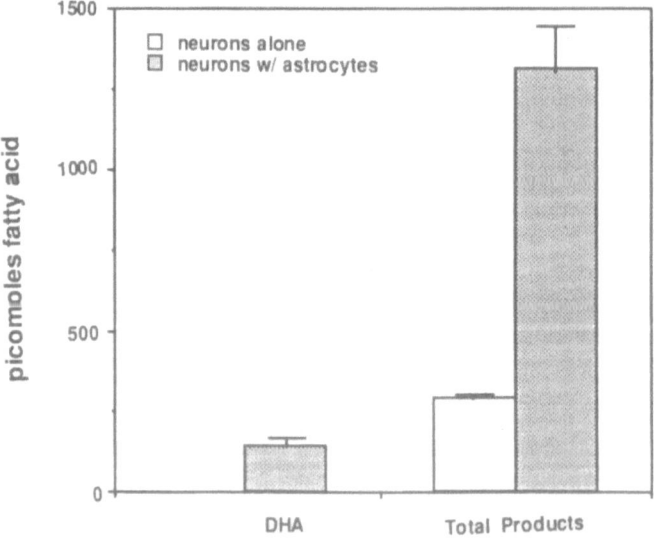

Figure 3. Cell cultures were incubated with [1-^{14}C]linolenic acid placed in the top medium of the filter chamber. After 96 hr the neuronal cell lipids contained no radiolabeled DHA when cultured alone. Neurons co-cultured with astrocytes did, however, contain DHA. Likewise, neuronal cell lipids contained greater amounts of total elongation/desaturation products when co-cultured with astrocytes. Bars represent means ± SE of triplicate cultures.

SUMMARY

These findings confirm that astrocytes and cerebral endothelium, not neurons, are the cells primarily responsible for the desaturation of essential fatty acids in the brain. They suggest that cerebral endothelial cells can target ω-3 fatty acids for release from their basolateral surface and thus aid in the enrichment of ω-3 fatty acids observed in the brain. The studies further suggest that endothelium and astrocytes play an important supportive role in the brain by cooperating in the elongation and desaturation of ω-3 essential fatty acid precursors in the brain and in the transfer of DHA to neurons. In doing so, endothelium and astrocytes may contribute positively to the high level of fatty acid desaturation necessary for normal neuronal function.

In composite, the present studies and previously published work (2,3,8) support a model for supplying DHA to central nervous system neurons that could utilize either DHA or its ω-3 fatty acid precursors circulating in the blood (Figure 4). If preformed DHA were available, the cerebral endothelium would take it up and transfer it into the brain. An additional sequential pathway would utilize circulating linolenic acid or EPA. In this pathway cerebral endothelium would take up ω-3 fatty acid precursors and target them preferentially into the brain, performing some elongation and desaturation in the process. Astrocytes would subsequently complete the conversion of precursors to DHA, releasing it for uptake by neurons.

ACKNOWLEDGMENTS

This work was supported by National Institutes of Health Grants NS-01096, NS-27914, and NS-24621. Linda Gorman, Elizabeth Yoder and Erin Hurt performed the technical aspects of these studies.

REFERENCES

1. Salem, Jr., N., Kim, H.-Y., and Yergey, J.A., Docosahexaenoic acid: Membrane function and metabolism, *in:* "Health Effects of Polyunsaturated Fatty Acids in Seafoods," A.P. Simopoulos, ed., pp. 263-317, Academic Press, New York, 1986.
2. Moore, S.A., Yoder, E., and Spector, A.A., Role of the blood-brain barrier in the formation of long-chain ω-3 and ω-6 fatty acids from essential fatty acid precursors, *J. Neurochem.* 55:391-402, 1990.
3. Moore, S.A., Yoder, E., Murphy, S., Dutton, G.R., and Spector, A.A., Astrocytes, not neurons, produce docosahexaenoic acid (22:6ω-3) and arachidonic acid (20:4ω-6), *J. Neurochem.* 56:518-524, 1991.
4. Sahagun, G., Moore, S.A., and Hart, M.N., Permeability of neutral versus anionic dextrans in cultured brain microvascular endothelium, *Am. J. Physiol.* 259:H162-H166, 1990.
5. Skalli, O., Ropraz, P., Trzeciak, A., Benzonana, G., Gillessen, D., Gabbiani, G., A monoclonal antibody against a smooth muscle actin: a new probe for smooth muscle differentiation, *J. Cell Biol.* 103:2787-2796, 1986.
6. Murphy, S., Generation of astrocyte cultures form normal and neoplastic central nervous system, *in:* "Methods in Neuroscience," P.M. Conn, ed., vol. 2, pp. 33-47, Academic Press, New York, 1990.
7. Dutton, G.R., Isolation, culture, and use of viable central nervous system perikarya, *in* : "Methods in Neuroscience," P.M. Conn, ed., vol. 2, pp. 87-102, Academic Press, New York, 1990.
8. Scott, B.L., and Bazan, N.G., Membrane docosahexaenoate is supplied to the developing brain and retina by the liver, *Proc. Nat. Acad. Sci. USA* 86:2903-2907, 1989.
9. Edelstein, C., Biochemistry and Biology of the Plasma Lipoproteins, A.M. Scanu and A.A. Spector, eds., Marcel Dekker, New York, pp. 495-505, 1986.
10. Dhopeshwarkar, G.A., and Subramanian, C., Biosynthesis of polyunsaturated fatty acids in the developing brain: I. Metabolic transformations of intracranially administered 1-^{14}C linolenic acid, *Lipids* 11:67-71, 1976.

Immunologic Functions

EFFECT OF CYTOKINES ON ICAM EXPRESSION AND T CELL ADHESION TO CEREBROVASCULAR ENDOTHELIAL CELLS

R.M. McCarron[1], L. Wang[1], M. K. Racke[2],
D. E. McFarlin[2], and M. Spatz[1]

[1] Stroke Branch and [2] Neuroimmunology Branch, NINDS
National Institutes of Health, Bethesda, MD 20892

INTRODUCTION

The transvascular migration of lymphocytes and other leukocytes from blood to peripheral tissues is preceded by adherence of these cells to endothelial cells (EC) lining the vasculature. These adhesive interactions utilize specific molecules (e.g., integrins) on the surfaces of leukocytes and EC. In many cases, the expression of these molecules has been linked to organ-specific homing of lymphocytes (1). Many central nervous system (CNS) disorders such as experimental allergic encephalomyelitis (EAE) and multiple sclerosis (MS) are characterized by the increased infiltration of peripheral blood leukocytes into the CNS. The enhanced adhesive interactions between leukocytes and cerebrovascular EC, which constitute the blood-brain barrier (BBB), may be a mechanism partially responsible for the enhanced transmigration of the BBB observed in these disorders. For example, various adhesion molecules have been indicated to play a role in the pathogenesis of EAE and MS (2,3).

Because of the highly restrictive nature of the BBB, the amount of leukocyte traffic into the CNS under normal conditions is low or absent (4). One reason for this observation could be the low levels of intercellular adhesion molecule-1 (ICAM-1; CD54) expression by cerebrovascular EC in situ. Relatively low levels of ICAM-1 have been shown to be expressed in vitro on unstimulated cerebrovascular EC cultures (4-6). ICAM-1 is an inducible glycoprotein adhesion molecule expressed by a variety of cells. The potential pathophysiological role of ICAM-1 is indicated by its increased expression on cerebrovascular EC in vivo in EAE models (2-4,7). This increase in ICAM-1 expression correlated with both attachment and extravasation of lymphocytes and onset of clinical disease (8). Immunohistochemical studies indicated that ICAM-1 expression on human cerebrovascular EC was also increased in MS and other neurological diseases and may contribute to the pathogenesis of these disorders (9).

In the present study ICAM-1 expression on cultured cerebrovascular EC and its possible role in adhesion of encephalitogenic lymphocytes were studied. The regulation of both ICAM expression and adhesion were also concomitantly examined in the presence of cytokines that are believed to play relevant roles in EAE.

METHODS

Mice

Female SJL/J mice obtained from The Jackson Laboratory (Bar Harbor, ME) at 3 to 4 days of age were utilized to isolate cerebrovascular EC used in this study. All MBP-specific T cell lines originated from SJL/J mice immunized at 8 to 12 weeks of age.

Cerebrovascular EC Cultures

The isolation and characterization of cerebrovascular EC from SJL mice have been previously described (10). EC were cultivated to confluency on 96-well flat bottom microtiter plates at 37°C in M-199 medium (containing 25 mM Hepes buffer, Earle's salts and L-glutamine) with 20% heat-inactivated FCS, 1% BME amino acids, 1% BME vitamins, 100 gm/ml streptomycin, 100 U/ml penicillin G (all from Gibco; Long Island, NY) and 2 μg/ml endothelial cell growth supplement (Collaborative Research; Bedford, MA). EC cultures contained >95% EC as determined by staining with EC-specific anti-Factor VIII-related antigen (Accurate Chemical and Scientific Corp.; Westbury, NY) as previously described (11). EC cultures were incubated in media alone or with indicated concentrations of recombinant murine interferon-α (IFN), recombinant murine interleukin-1α (IL-1), recombinant human tumor necrosis factor-α (TNF)(all from Genzyme Corp.; Boston, MA) for 3 days at 37°C, or human transforming growth factor-β1 (TGFβ)(R & D Systems; Minneapolis, MN) for 24 hr; all treatments were performed in quadruplicate wells.

Lymphocyte Preparation

MBP-specific T cell lines were generated from SJL/J mice as previously described (12). Prior to use in adhesion assays, T cells were incubated (1 x 10^6 cells/ml) for 4 days in fresh media containing 5 x 10^6 irradiated splenocytes/ml and 25 μg/ml MBP. All cultures were >95% LFA-1- and L3T4-positive as determined by FACS analysis and passively transferred EAE as previously described (13).

Assessment of T Lymphocyte Adhesion

T cell lines (1 x 10^7 cells/ml) were labeled with Na$_2$^{51}CrO$_4$ (0.2 mCi/ml; New England Nuclear; Boston, MA), washed twice with HBSS/4% FCS and resuspended (4 x 10^5/ml) in RPMI 1640 media supplemented with 10% FCS, 100 U/ml penicillin, 100 μg/ml streptomycin and 2 mM glutamine. T cells were added (2 x 10^4/50 μl/well) to microtiter plates containing EC culture monolayers and plates were incubated at 37°C for 30 min. After gently removing nonadherent cells and washing (3 times with HBSS), adherent cells were lysed (2% Triton-X solution, 6 hr, 37°C) and counted on a LKB Compugamma 1282 gamma counter. The percent adhesion was calculated by: [Triton-X cpm/(Triton-X cpm + washes cpm)] x 100%. The data are expressed as mean cpm ± SE.

ICAM Expression

ICAM-1 expression on EC was assessed with monoclonal rat antibody to murine lymphocyte activation antigen, Mala-2 (clone YN1/1.7.4; ATCC, Rockville, MD), which represents the murine homologue of human adhesion molecule ICAM-1 (14). EC cultures were fixed (2% formaldehyde; 2 min), incubated at 37°C with 1:10 Mala-2 (45 min) and 1:100 rhodamine-conjugated F(ab')$_2$ goat anti-rat IgG (Sigma Chem. Corp.; St. Louis, MO) (30 min) and examined by indirect immunofluorescence microscopy. Mala-2 expression was quantitated by ELISA using aggregated rabbit IgG (Sigma) to block nonspecific binding sites, Mala-2 (1:200), biotin-conjugated anti-rat IgG, and avidin-horseradish peroxidase (Sigma) as previously described (15). The optical density (O.D.) was immediately read at 492 nm (reference 630 nm).

RESULTS

T Lymphocyte-EC Adhesion

MBP-specific, encephalitogenic T cells demonstrated significant levels of adhesion to untreated cerebrovascular EC cultures (Table 1). Pretreatment of EC for 72 hr with IFN, TNF and/or IL-1 significantly increased the level of adherence of T cells (Table 1). The effects of various concentrations of these cytokines (i.e., 10-400 U/ml IFN, 0.1-100 U/ml TNF, and 10-100 U/ml, IL-1) demonstrated that these increases in the levels of T cell adhesion were dose-dependent on the exogenous cytokine. No additive effects were observed in EC treated with various combinations of these cytokines (results not shown).

Table 1. Effect of cytokines on T cell adhesion to cerebrovascular EC

Treatment [a]	Pre-treatment of EC Cultures [b]		Inhibition [c] (%)
	None	+ TGFβ	
None	25.6 ± 0.6	16.8 ± 0.4	8.8
IFN (100 U/ml)	34.8 ± 0.6	20.7 ± 0.5	14.1
IL-1 (50 U/ml)	38.2 ± 0.7	21.9 ± 0.7	16.3
TNF (100 U/ml)	52.2 ± 0.8	31.8 ± 0.4	20.4

[a] Cerebrovascular EC were incubated for 3 days in the presence of indicated cytokines
[b] prior to the assay for T cell adhesion as described in the Materials and Methods section. Cerebrovascular EC were untreated (None) or preincubated (24 hr) with 10 ng/ml TGF β prior to the addition of indicated cytokines. Data are expressed as the mean percent ± SE
[c] of T cells adhering to quadruplicate EC cultures. Inhibition is calculated as the difference between TGF β-treated and untreated cultures.

EC cultures treated with TGFβ (10 ng/ml) exhibited significantly decreased basal levels of T cell adhesion (Table 1). Pretreatment with TGFβ also partially inhibited the levels of T cell adhesion to EC treated with IFN, IL-1 or TNF (Table 1). The relative degree of inhibition caused by TGFβ varied with each cytokine and was correlated with the degree of enhanced adhesion (i.e., TGFβ inhibited adhesion to TNF-treated EC to a greater degree than untreated, IFN- or IL-1-treated EC). The TGFβ-induced inhibition was dose-dependent with less inhibition in cultures pretreated with 2 ng/ml TGFβ (results not shown). No inhibition was observed if TGFβ was added at the same time or after the addition of IFN, IL-1 or TNF.

ICAM Expression by EC

The expression of ICAM on EC as measured by ELISA indicated that untreated EC expressed basal levels of ICAM and that treatment of EC with cytokines enhanced ICAM expression (Table 2). The capacity of all cytokines to enhance ICAM expression was dose-dependent similar to that previously observed for T cell adhesion. Treatment of EC cultures with the combination of IFN and TNF considerably enhanced ICAM expression to levels greater than were observed in the presence of either cytokine alone (results not shown).

Table 2. Effect of cytokines on ICAM expression by cerebrovascular EC

Treatment [a]	Pre-treatment of EC Cultures [b]		Inhibition [c] (%)
	None	+ TGFβ	
None	0.286 ± 0.023	0.181 ± 0.011	10.5
IFN (100 U/ml)	0.467 ± 0.036	0.352 ± 0.021	11.5
IL-1 (50 U/ml)	0.549 ± 0.041	0.446 ± 0.028	10.3
TNF (100 U/ml)	0.620 ± 0.049	0.503 ± 0.039	11.7

[a] Cerebrovascular EC were incubated for 3 days in the presence of indicated cytokines prior to the assay for ICAM expression as described in the Materials and Methods section.

[b] Cerebrovascular EC were untreated (None) or preincubated (24 hr) with 10 ng/ml TGF β prior to the addition of indicated cytokines. Data for ICAM expression are expressed as the mean O.D. 490 nm ± SE of quadruplicate EC cultures.

[c] Inhibition is calculated as the O.D. difference between TGF β-treated and untreated cultures.

As shown in Table 2, pretreatment of EC with TGFβ inhibited the basal level of ICAM expression on EC cultures. Addition of cytokines to TGFβ-pretreated EC still enhanced ICAM expression; the levels of ICAM expression in these cultures were inhibited at levels quantitatively similar to cultures not preincubated with TGFβ (Table 2).

DISCUSSION

The results presented here demonstrate ICAM expression on cultures of murine cerebrovascular EC and adhesive interactions between these cells and LFA-1-positive, MBP-specific encephalitogenic murine T cells. The adhesion of lymphocytes to cultured EC has been previously shown to involve the integrin LFA-1 which is expressed by these cells and that binds to ICAM (1,5). It is also demonstrated here that the cytokines IFN, IL-1 and TNF increase the level of ICAM expression by cerebrovascular EC (Table 2). These cytokines also increase the level of T cell adhesion to cerebrovascular EC to a degree very similar to their effects on ICAM expression.

It was shown that treatment of EC with transforming growth factor-β1 (TGFβ) down-regulated the level of T cell adhesion on untreated EC and inhibited the up-regulation of adhesion induced by IFN, IL-1 and/or TNF (Table 1). However, TGFβ had no effect on the ability of IFN, IL-1 and/or TNF to up-regulate ICAM expression (Table 2). These results indicate that T cell adhesion and ICAM expression are not totally dependent phenomena. This lack of a direct correlation between ICAM expression and T cell adhesion was also seen in experiments using EC simultaneously treated with both IFN and TNF, which resulted in only slight increases in the level of T cell adhesion compared with TNF alone, but significant increases in ICAM expression. These observations suggest that factors other than just ICAM may also be involved in the adhesion of T cells to cerebrovascular EC observed here.

The potential clinical relevance of IFN, IL-1 and TNF in both EAE and MS has been indicated by a number of investigations. These include the detection of these cytokines in lesions and direct relationships between the presence of these cytokines and disease severity (7,16-19). In addition to these cytokines, TGFβ may also be involved in the clinical expression of these disorders. Unlike the aforementioned cytokines, TGFβ appears to be predominantly involved in disease-limiting mechanisms. It has been observed to suppress both the onset and occurrence of relapses of EAE, and antibodies to TGFβ have been shown to exacerbate disease (13,20). One of the possible mechanisms for these observations may be related to the ability of TGFβ to inhibit lymphocyte attachment to EC (21).

The capacity of IFN, IL-1 and TNF to up-regulate adhesive interactions between encephalitogenic T cells and cerebrovascular EC and the opposing effects induced by TGFβ implicate the adhesion phenomenon as a potential mechanism for the effects of these cytokines that have been previously described. Their ability to modulate interactions between lymphocytes and cerebrovascular EC implicate them, as well as the adhesion phenomenon itself, as important factors involved in the migration of cells across the BBB and development of inflammatory lesions in the CNS.

REFERENCES

1. Cavender, D.E., Interactions between endothelial cells and the cells of the immune system, *Int. Rev. Exp. Pathol.* 32:57, 1991.

2. O'Neill, J.K., Butter, C., Baker, D., Gschmeissner, S.E., Kraal, C., Butcher, E.C., and Turk, J.L., Expression of vascular addressins and ICAM-1 by endothelial cells in the spinal cord during chronic relapsing experimental allergic encephalomyelitis in the Biozzi AB/H mouse, *Immunol.* 72:520, 1991.

3. Cannella, B., Cross, A.H., and Raine, C.S., Upregulation and coexpression of adhesion molecules correlate with relapsing autoimmune demyelination in the central nervous system, *J. Exp. Med.* 172:1521, 1990.

4. Wilcox, C.E., Ward, A.M.V., Evans, A., Baker, D., Rothlein, R., and Turk, J.L., Endothelial cell expression of the intercellular adhesion molecule-1 (ICAM-1) in the central nervous system of guinea pigs during acute and chronic relapsing experimental allergic encephalomyelitis, *J. Neuroimmunol.* 30:43, 1990.

5. Dustin, M.L., Rothlein, R., Bhan, A.K., Dinarello, C.A., and Springer, T.A., Induction by IL 1 and interferon-γ: tissue distribution, biochemistry, and function of a natural adherence molecule (ICAM-1), *J. Immunol.* 137:245, 1986.

6. Cross, A.H., Cannella, B., Brosnan, C.F., and Raine, C.S., Homing to central nervous system vasculature by antigen-specific lymphocytes. I. Localization of ^{14}C-labeled cells during acute, chronic, and relapsing experimental allergic encephalomyelitis, *Lab. Invest.* 63:162, 1990.

7. Cannella, B., Cross, A.H., and Raine, C.S., Adhesion-related molecules in the central nervous system, *Lab. Invest.* 65:23, 1991.

8. Raine, C.S., Cannella, B., Dujvestijn, A.M., and Cross, A.H., Homing to central nervous system vasculature by antigen-specific lymphocytes. II. Lymphocyte/endothelial cell adhesion during the initial stages of autoimmune demyelination, *Lab. Invest.* 63:476, 1990.

9. Sobel, R.A., Mitchell, M.E., and Fondren, G., Intercellular adhesion molecule-1 (ICAM-1) in cellular immune reactions in the human central nervous system, *Am. J. Pathol.* 136:1309, 1990.

10. McCarron, R.M., Kempski, O., Spatz, M., and McFarlin, D.E., Presentation of myelin basic protein by murine cerebral vascular endothelial cells, *J. Immunol.* 134:3100, 1985.

11. McCarron, R.M., Spatz, M., Kempski, O., Hogan, R.N., Muehl, L., and McFarlin, D.E., Interaction between myelin basic protein-sensitized T lymphocytes and murine cerebral vascular endothelial cells, *J. Immunol.* 137:3428, 1986.

12. McCarron, R.M., Racke, M.K., Spatz, M., and McFarlin, D.E., Cerebral vascular endothelial cells are effective targets for in vitro lysis by encephalitogenic T lymphocytes, *J. Immunol.* 147:5039, 1991.

13. Racke, M.K., Dhib-Jalbut, S., Cannella, B., Albert, P.S., Raine, C.S., and McFarlin, D.E., Prevention and treatment of chronic relapsing experimental allergic encephalomyelitis by transforming growth factor-β_1, *J. Immunol.* 146:3012, 1991.

14. Prieto, J., Takai, F., Gendelman, R., Christenson, B., Biberfeld, P., and Patarroyo, M., MALA-2, mouse homologue of adhesion molecule ICAM-1 (CD54), *Eur. J. Immunol.* 19:1551, 1989.

15. Tanaka, M., and McCarron, R.M., The inhibitory effect of tumor necrosis factor and interleukin-1 on Ia induction by interferon-γ on endothelial cells from murine central nervous system microvessels, *J. Neuroimmunol.* 27:209, 1990.

16. Kuroda, Y., and Shimamoto, Y., Human tumor necrosis factor-α augments experimental allergic encephalomyelitis in rats, *J. Neuroimmunol.* 34:159, 1991.

17. Selmaj, K., Raine, C.S., Cannella, B., and Brosnan, C.F., Identification of lymphotoxin and tumor necrosis factor in multiple sclerosis lesions, *J. Clin. Invest.* 87:949, 1991.

18. Beck, J., Rondot, P., Catinot, L., Falcoff, E., Kirchner, H., and Wietzerbin, J., Increased production of interferon gamma and tumor necrosis factor precedes clinical manifestation in multiple sclerosis: do cytokines trigger off exacerbations? *Acta Neurol. Scand.* 78:318, 1988.

19. Maimone, D., Gregory, S., Arnason, B.G., and Reder, A.T., Cytokine levels in the cerebrospinal fluid and serum of patients with multiple sclerosis, *J. Neuroimmunol.* 32:67, 1991.

20. Kuruvilla, A.P., Shah, R., Hochwald, G.M., Liggitt, H.D., Palladino, M.A., and Thorbecke, G.J., Protective effect of transforming growth factor β_1 on experimental autoimmune diseases in mice, *Proc. Natl. Acad. Sci.* 88:2918, 1991.

21. Gamble, J.R., and Vadas, M.A., Endothelial cell adhesiveness for human T lymphocytes is inhibited by transforming growth factor-β, *J. Immunol.* 146:1149, 1991.

EXPRESSION OF ENDOTHELIAL CELL

ACTIVATION ANTIGENS IN MICROVESSELS

FROM PATIENTS WITH MULTIPLE SCLEROSIS

Paula Dore-Duffy[1], Ruth Washington[1], and Ljubisa Dragovic[2]

[1]Wayne State University Multiple Sclerosis Clinical
Research Center, Department of Neurology
Division of Neuroimmunology
Wayne State University School of Medicine
Detroit, Michigan 48201
[2]Office of the Oakland County Medical Examiner
Pontiac, Michigan 48341

INTRODUCTION

Postcapillary endothelium, at sites of inflammation, undergoes a number of changes referred to as "activation" (1). Activated endothelial cells (EC) are characterized by: 1) increased cell surface expression of immunorelevant proteins (class I and class II MHC antigens (Ags), adhesion molecules and transferrin receptors), 2) increased leukocyte adherence, 3) prominence of biosynthetic organelles, 4) synthesis and release of interleukin-1 (IL-1), IL-6, platelet-derived growth factor (PDGF), platelet activating factor (PAF), heparin-binding fibroblast growth factors and eicosanoids (2), and 5) upregulation of cFos and cMyc (2). Although the mechanisms and sequence of events governing EC activation have not been conclusively described, evidence does point to an immunoregulatory role for released cytokines from activated leukocytes (3).

Multiple sclerosis (MS) is a chronic progressive demyelinating disease of the central nervous system (CNS) that is characterized by an infiltration of peripheral blood mononuclear leukocytes. The leukocyte infiltration seen in MS shares common characteristics with that seen in acute/chronic inflammation. Locally released lymphokines have been identified as possible mediators of tissue damage in this disease (4). We hypothesize that activation of the CNS endothelium and expression of EC activation antigens are important in the pathogenesis of MS.

Much of what is known about the expression of EC activation Ags has been performed in vitro using human umbilical cord EC (HUVEC). A fundamental flaw with such studies is that cultured EC are likely to be de-differentiated. Further, with culturing, there is a loss of microvascular cells such as smooth muscles cells and pericytes, which may have an

Frontiers in Cerebral Vascular Biology: Transport and Its Regulation
Edited by L.R. Drewes and A.L. Betz, Plenum Press, New York, 1993

243

important role in EC responses. We developed a technique that allows us to measure the expression of EC activation markers in the intact CNS microvessel using laser cytometry. We used this technique to evaluate antigenic expression in CNS microvessels from patients with MS.

METHODS

Autopsy Material

Autopsy material was obtained through the collaboration of the Oakland County Medical Examiner's Office and through individual donation to the Wayne State University MS Clinical Research Center. Additional samples of MS brain were kindly donated by Drs. Jack Antel, Montreal Neurological Institute, and Donald Gilden, University of Colorado School of Medicine. The diagnosis of MS was confirmed by neuropathologic evaluation. Samples for microvessel preparation were obtained from periventricular white matter regions showing plaque involvement.

Preparation of Human Microvessels[5]

Microvessels were prepared from autopsy material and from prefrozen brain by a modification of the method described previously (5). Tissue was homogenized in 10 volumes of Ringer's solution with Hepes and 10% fetal bovine serum (adjusted to pH 7.4) using a glass-teflon homogenizer. After 20 up and down strokes at 420 rpm, the homogenate was centrifuged and the pellet resuspended in Ringer's plus 15% Dextran 70. The suspension was centrifuged at 5,000 x g for 20 min. The pellet was filtered through a nylon mesh and then passed through glass beads and washed. Yields equalled about 0.1 - 0.2 mg protein per gram of starting material. Purity of the microvessel preparations was determined by visual examination and measurement of gamma-glutamyl-transpeptidase activity (Sigma diagnostic kit 545).

Staining of Microvessels

Microvessels were allowed to dry on coverslips and then fixed with 3% paraformaldehyde and permeabilized with 0.01% triton X-100 for 15 min at room temperature. Coverslips were washed in PBS + 1% BSA and stained with monoclonal antibodies (mAb) at saturation density for 30 min, washed, and then incubated with a second antibody, if appropriate.

Monoclonal Antibodies

Antibody directed against human intercellular adhesion molecule-1 (ICAM-1) was kindly provided by Dr. R. Rothlein, Boehringer Mannheim, Ridgefield, CT. Antibody recognizing vascular cell adhesion molecule-l (VCAM-1; INCAM 110) (IgG) was kindly provided by Dr. M. Bevilacqua, Harvard University (6). Mo3f (IgG2) which recognizes the human urokinase plasminogen activator receptor (uPA-R) was kindly provided by Dr. R. Todd III, University of Michigan (7,8). Antibody directed against human transferrin receptor (tfR) was purchased from Biosource International, Camarillo, CA. Antibody directed against human class II antigen (HLA-DR) and fluorochrome-conjugated second antibodies were purchased from Becton Dickinson, Mountain View, CA. Isotype similar control antibodies include M45/q (IgG2) kindly provided by Dr. R. Todd III, University of Michigan (9), and MSIgG purchased from Coulter Immunology, Hialeah, FL.

Analysis of Relative Fluorescence Intensity by Laser Cytometry

Stained coverslips were inverted and mounted on standard glass slides and then analyzed on the Meridian ACAS 470 laser cytometer. An attenuated laser beam focused at about 1 mm PMT (set to read negative for isotope control antibody) excites the fluorochrome. The emission is recorded by a 16 bit microcomputer to produce false color images of fluorescent intensity (FI) and distribution. A minimum of 20 randomly selected microvessels per slide from two to three slides per determination were analyzed. Results are expressed as average FI. Transmission-light and fluorescence-light images were recorded using the 40 oil objective.

RESULTS

MS microvessels did not vary significantly from control preparations with regard to size or staining patterns. Morphologic appearance on light microscopy was similar, although it was more difficult to prepare clean microvessels from prefrozen tissue. However, microvessels from MS brains exhibited less factor VIII than control microvessels upon laser cytometric analysis of relative FI (Table 1).

TABLE 1. Factor VIII in MS microvessels

	Average FI/microvessel (mean ± SD)
MS BRAINS	2,008 ± 550 (50)*
CONTROL	2,793 ± 599 (50)

*Significantly different (p < 0.02) from control vessels
(N)= Number of microvessels analyzed

Microvessel preparations were examined for expression of VCAM-1, ICAM-1, HLA-DR Ags, tfr, uPA-R, and factor VIII using indirect immunofluorescent staining techniques (Table 2). Normal human microvessels exhibited no significant staining with mAb to any EC activation antigen. However, normal microvessels stained marginally (low intensity) with anti-tfR, and expressed detectable factor VIII. Expression of activation markers was observed only on MS microvessels (Table 2).

TABLE 2. Expression of endothelial cell activation Ags on MS microvessels

	Average FI/microvessel (mean ± SD)				
No Stain	Class II Ags	ICAM	VCAM	tfr	uPA-R
MS (N= 5)					
<120	1,404 ± 171*	1,455 ± 400*	1,531 ± 155*	1,598 ± 109*	1,568 ± 398*
CONTROL (N = 3)					
<200	386 ± 25	334 ± 6	440 ± 102	1,008 ± 90	467 ± 96

*Significantly different (p < 0.01) from normal microvessels. Greater than 50 microvessels per brain were analyzed on the Meridian Laser Cytometer.

Microvessels prepared from MS brain expressed significant levels of EC activation markers. Greater than 80% of MS microvessels exhibited detectable VCAM-1 at very high density. MS microvessels from periplaque areas also exhibited high levels of ICAM-1. Approximately 50-55% of microvessels isolated expressed ICAM-1 at high density. Control brain did not express ICAM-1 or expressed this antigen at barely detectable levels. A similar increase in expression on MS microvessels was seen for uPA-R, which was found on 45% of MS microvessels isolated and on less than 5% of normal human microvessels. MS microvessels expressed MHC class II antigen on an average of 45% of microvessels examined as assessed by positive staining of EC with antibody directed against HLA-DR antigens (Table 2). The receptor for transferrin (tfR) was constitutively expressed on normal microvessels. In MS, the percentage of microvessels expressing tfr was similar to controls, but MS microvessels stained much brighter.

Microvessels were examined for dual expression of activation markers. HLA-DR antigens tended to co-express most frequently with VCAM-1 (50%). Only 7% of HLA-DR+ microvessels were VCAM-1-, and 5% of DR- microvessels were VCAM-1+. uPA-R was expressed on 70% of MS microvessels, and 28% of these co-expressed with HLA-DR antigens. HLA-DR positive microvessels co-expressed ICAM-1 on 43% of the vessels examined. uPA-R positive microvessels co-expressed ICAM-1 on 31% of those examined; of those expressing VCAM-1, 28% co-expressed ICAM-1.

DISCUSSION

It is generally accepted that EC are fundamentally altered during trauma or during acute and chronic inflammation (2). One of the earliest alterations to take place reflects microvascular activation. At the microscopic level, this may include changes in EC morphology, but at the molecular level, changes included enhanced surface expression of EC activation markers such as class I, class II MHC antigens, ICAM-1, VCAM-1, ELAM-1, transferrin receptors and uPA-R (human EC) (2,10). There is increased phosphotyrosine, increased phosphorylation of P29 proteins, increased metallaproteinases, eicosanoids, stress proteins, and altered Factor VIII concentration (11). Autocrine and paracrine factors such as IL-1, PDGF, TGF-β and heparin binding factors are released and are thought to be important regulatory substances (2).

Our results show that microvessels from MS patients express EC activation markers. Many of the microvessels expressed lower levels of factor VIII, supporting previous studies that this substance is released from activated EC (11). In areas with visible and palpable lesions, roughly 3-10% of microvessels express none of the activation markers that we measured.

Our results imply that EC activation is an integral step in the pathogenesis of MS. Ia positive EC have been reported by Traugott (12) and Hayashi et al. (13) using immunofluorescence techniques. The distribution of Ia positive EC was found to be random throughout the CNS (12), and this was estimated to be about 10% of EC. In a later paper, Raine and associates (14) reported that no EC were class II positive. Differential expression of activation markers supports the hypothesis that EC activation may involve a sequence or series of steps (15). The capacity of EC to respond to a given signal will depend on its previous interactions and "state of activation." For example, the capacity to present antigen to class II restricted T-cells may be augmented by previous interactions with CD8+, CD4+ leukocytes, monocytes and their respective cytokines. Alternatively, primed CD4+ cells may not require class II induction on EC. Once primed cells have accumulated at the BBB or perivascularly, they may redirect EC to exhibit other molecules necessary for nonantigen driven recruitment of leukocytes into the CNS.

SUMMARY

Although the mechanisms governing EC activation are not well understood, evidence points to a role for locally released cytokines from activated leukocytes. We propose that the sequence of events that result in EC activation are important in perivascular leukocyte infiltration into the CNS seen in MS. In the present study we examined expression of EC activation antigens on cerebral microvessels from patients with MS using immunofluorescence staining and quantitation by laser cytometry. Normal human microvessels do not express MHC class II antigens (Ags), intercellular adhesion molecule-1 (ICAM-1), vascular cell adhesion molecule-1 (VCAM-1), or the urokinase plasminogen activator receptor (uPA-R). They express low levels of transferrin receptors and express factor VIII. Microvessels prepared from MS brain with plaque involvement expressed decreased factor VIII and increased transferrin receptors (tfR). Expression of the adhesion molecules VCAM-1, and ICAM-1 were found on 80% of isolated microvessels. HLA-DR Ags were expressed on 40-60% of microvessels, and the uPA-R was expressed on 50% of MS microvessels examined. MHC class II Ags co-express with VCAM-1 and ICAM-1 more frequently than with the uPA-R. Results indicate that activation of EC in MS is likely to be an important factor in disease pathology.

ACKNOWLEDGMENTS

This work was supported, in part, by a grant from the National Multiple Sclerosis Society, the generous support of the DreaMS group, and a grant from the National Institutes of Health, CA-42246. The authors wish to thank Dr. Sami Harik for helpful suggestions regarding this study, and for the technical expertise in the preparation of microvessels. We also thank Mr. Chuck Booth and Dr. Robert Skoff for helpful suggestions and technical expertise involving laser cytometry.

REFERENCES

1   Pober, J.S., Cytokine-mediated activation of vascular endothelium: Physiology and pathology, *Am. J. Pathol.* 133:426-436, 1988.

2. Wilder, R.L., Case, J.P., Crofford, L.J., Kukumian, G.K., Lafyatis, R., Remmers, Sano, H., Sternberg, E.M., and Yocum, D.E., Endothelial cells and the pathogenesis of rheumatoid arthritis in Lewis rats, *J.Cell Biochem.* 45:162,1990.

3. Pober, J.S., Gimbrone, M.A., Cotran, R.S., Reiss, C.S., Burakoff, S.J., Fiers, W., and Ault, K.A., Ia expression by vascular endothelium is inducible by activated T-cells and by human gamma interferon, *J. Exp. Med.* 157:1339-1348, 1983.

4. Selmaj, K., Raine, C.S., and Cross, A.H., Anti-tumor necrosis factor therapy abrogates autoimmune demyelination, *Ann. Neurol.* 30:694-700, 1991.

5. Betz, A.L., Sodium transport in capillaries isolated from rat brain, *J. Neurochem.* 41:1150-1157, 1983.

6. Bevilacqua, M., Pober, J.S., Wheeler, M.E., Cotran, R.S., and Gimbrone, M.J., Interleukin 1 acts on cultured human vascular endothelium to increase the adhesion of polymorphonuclear leukocytes, monocytes, and related leukocyte cell lines, *J. Clin. Invest.* 76:2003, 1985.

7. Todd, R.F., Mizukami, I., Vinjamuri, S.D., Trochelman, R.D., Hancock, W.W., and Liu, D.Y., Human mononuclear phagocytes activation antigens, *Blood Cells* 16:167-168, 1990.

8. Min, H.Y., Semnani, R., Mizukami, I.F., Watt, K., Todd, R.F. III., and Liu, D.Y., cDNA for Mo3, a monocyte activation antigen, encodes the human receptor for urokinase plasminogen activator, *J. Immunol.* 148:3636-3642, 1992.

9. Mizukami, I.F., Vinjamuri, S.D., Trochelman, R.D., and Todd, R.F., A structural characterization of the Mo3 activation antigen expressed on the plasma membrane of human mononuclear phagocytes, *J. Immunol.* 144:1891-1848, 1990.

10. Kahaleh, M.B., The role of vascular endothelium in the pathogenesis of connective tissue disease: Endothelial injury, activation, participation and response, *Clin. Exp. Rheumatol.* 8:595-601, 1990.

11. Tannebaum, S.H., and Gralick, H.R., γ-Interferon modulates von Willebrand factor release by cultured human endothelial cells, *Blood* 75:2177-2184, 1990.

12. Traugott, V., Scheinberg, L.C., and Raine, C.S., On the presence of Ia-positive endothelial cells and astrocytes in multiple sclerosis lesions and its relevance to antigen presentation, *J. Neuroimmunol.* 8:1-14, 1985.

13. Hayashi, T., Morimoto, C., Burks, J.S., Kerr, S., and Hauser, S.L., Dual-label immunohistochemistry of the active multiple sclerosis lesion: Major histocompatibility complex and activation antigens, *Ann. Neurol.* 24:523-531, 1988.

14. Raine, C.S., Lee, S.C., Scheinberg, L.C., Duijvestin, A.M., and Cross, A.H., Adhesion molecules on endothelial cells in the central nervous system: An emerging area in neuroimmunology of multiple sclerosis, *Clin. Immunol. Immunopathol.* 57:173-187, 1990.

15. Doukas, J., and Pober, J.S., Lymphocyte mediated activated of cultured endothelial cells (EC). CD4+ T cells inhibit EC class II MHC expression despite secreting IFNγ and increasing EC class I MHC and intracellular adhesion molecule-1 expression, *J. Immunol.* 145:1088-1098, 1990.

GENERATION OF AN ANTI-MOUSE BRAIN ENDOTHELIAL CELL MONOCLONAL ANTIBODY THAT RECOGNIZES 84-110 kDa AND 36 kDa DETERMINANTS THAT ARE UPREGULATED BY CYTOKINES

Jeymohan Joseph[1], Marissa Nunez[1], James L. Grun[2]
Fred D. Lublin[1], Michael N. Hart[3] and Robert L. Knobler[1]

[1]Division of Neuroimmunology and [2]Department of Biochemistry
Jefferson Medical College, Philadelphia, PA, and
[3]Department of Neuropathology, University of Iowa
Iowa City, IA 52242

INTRODUCTION

Experimental allergic encephalomyelitis (EAE) is an organ-specific disease, mediated by T cells trafficking into the central nervous system (CNS) (1). EAE has served as a model system for the human demyelinating disease, multiple sclerosis (MS) (2). The mechanisms regulating the tissue tropism in organ-specific diseases such as MS are under intense investigation. Local antigen presentation (i.e., myelin basic protein), by blood-brain barrier-derived cells (endothelial cells, perivascular microglia, astrocytes), can contribute to a selective localization of T cells into the CNS (3). Expression of CNS-specific adhesion molecules, or localized CNS upregulation of systemic adhesion molecules, could also facilitate T cell trafficking into the CNS.

Organ-specific endothelial cell determinants have been identified that influence the homing of recirculating lymphocytes (4-6). Since endothelial cells are heterogenous, with unique characteristics in different organs, the tissue-specific expression of such endothelial cell molecules could be a critical factor in local immune reactions (8). These determinants may orchestrate the interaction of inflammatory cells with cerebral endothelial cells. Therefore, identification of these molecules is essential to an improved understanding of their role, which may also serve as a potential site for therapeutic intervention in CNS inflammatory diseases.

In an effort to define novel CNS endothelial cell-specific determinants involved in interactions with encephalitogenic T cells, we have been generating monoclonal antibodies to cytokine-activated mouse brain endothelial cells from the EAE-susceptible SJL strain. We report here on the characterization of one of several antibodies generated. This antibody

Frontiers in Cerebral Vascular Biology: Transport and Its Regulation
Edited by L.R. Drewes and A.L. Betz, Plenum Press, New York, 1993

249

recognizes a murine panendothelial cell determinant that is upregulated by the cytokines tumor necrosis factor (TNF) and gamma interferon (γ-IFN).

MATERIALS AND METHODS

Endothelial Cell Cultures

Cerebral microvascular endothelial cells (CEC) were isolated from the brains of BALB/c, (BALB/c X SJL)F1, B10.S and SJL strains of mice using the method described by Rupnick et al. (8). Briefly, cerebral microvessels were isolated by digestion of cerebral white matter in 0.5% collagenase (Sigma, St. Louis, MO) and density centrifugation of homogenized material in 15% dextran. The vascular pellet was further purified by gradient centrifugation on Percoll (45%). Capillaries were plated onto 0.1% gelatin-coated plates, and endothelial cells grew out in about 10 days. Endothelial cell lines were established from these initial outgrowths and were cultured in Medium 199 (GIBCO, Grand Island, NY) supplemented with 20% fetal bovine serum, 2 mM L-glutamine, 90 μg/ml heparin (porcine, Sigma, St. Louis, MO) and 20 μg/ml endothelial cell growth factor (ECGF, Collaborative Research, Cambridge, MA).

Fat pad-derived endothelial cells (FEC), were isolated as described by Wagner and Matthews (9). Mouse epididymal fat pads were treated with 0.2% collagenase for 40 min at 37°C. The resulting slurry was then centrifuged at 100 x g for 10 min, and the adipocytes separated from the capillaries during centrifugation, floating to the top. Further purification was achieved by gradient centrifugation on Percoll (45%). Capillaries were plated and cultured as described above for the CECs.

Endothelial cell identity was established by the uptake of DiI-Ac-LDL (acetylated low density lipoprotein, Biomedical Technologies, Stoughton, MA) and specific binding of *Bandeiraea simplicifolia* BSI-B$_4$. The endothelial cell lines used in these studies were between passages 12 and 16.

Immunization and Monoclonal Antibody Generation

SJL mouse strain-derived CECs were treated with 200 U/ml recombinant mouse γ-IFN (Amgen, Thousand Oaks, CA) and 100 U/ml recombinant mouse TNF (Genzyme, Boston, MA) for 72 hr. Wistar rats were injected intraperitoneally with 3.6 x 10^7 cytokine treated endothelial cells in precipitated alum (used in the ratio of 3 parts cells and 2 parts alum). The animals were given a booster injection six weeks later using the same number of endothelial cells. Three days after the booster injection, spleen cells from the rat were fused with a Sp2/0-Ag14 non-secreting mouse myeloma cell line (ATCC). Antibody-secreting colonies were initially screened by ELISA, using cytokine-activated glutaraldehyde-fixed SJL CECs. Several colonies selected by ELISA screening were further studied by flow cytometry and Western blot analysis.

Antibody Labeling and Flow Cytometry

Endothelial cells were removed from the flasks using 0.25% trypsin and 0.1% EDTA for 1 min at 37°C and labeled with undiluted hybridoma supernatants. Fluorescein-conjugated goat anti-rat IgG (Organon-Teknika Cappel, Cochranville, PA) was used as a secondary reagent. The percentage of positive cells and mean fluorescence intensities (MFI) were determined by duplicate sample analysis on a flow cytometer (EPICS C, Coulter Diagnostics, Hialeah, FL), equipped with an argon laser tuned to 488 nm. Gate windows on the flow cytometer were chosen to exclude dead cells and cellular debris.

Western Blot Analysis

Untreated- and cytokine-treated endothelial cells were lysed using a single detergent lysis buffer (Triton X-100, NaN$_3$, PMSF, Aprotinin). The cell lysates were denatured by heating in sample buffer containing DTT. SDS gel electrophoresis was run with pre-stained protein markers(24,000-110,400) using Tris-glycine electrophoresis buffer. A Western transfer was performed using Tris-glycine buffer with SDS. The hybridoma supernatants were used as the primary antibodies for staining the nitrocellulose filters. Alkaline phosphatase-conjugated goat anti-rat IgG (H + L, Promega Biotech, Madison, WI) was used as a secondary reagent.

RESULTS AND DISCUSSION

CECs derived from SJL mice, which were used for antibody generation, as well as BALB/c, (BALB/c X SJL)F1, and B10.S mice, were labeled equally well with the antibody, suggesting that it does not detect a strain-specific determinant (Figure 1A). However, the antibody did label endothelial cells from other sites, i.e., HECs or FECs, so that it appears to detect a panendothelial marker (Figure 1B). Cytokine-activated (TNF + γ-IFN), splenic mononuclear cells (MNCs), from either SJL or BALB/c mice, were not labeled with this antibody. This suggests that the antibody neither recognized a common murine determinant nor the MHC class I and II antigens (Figure 1C). Other cell populations have not yet been screened with the antibody.

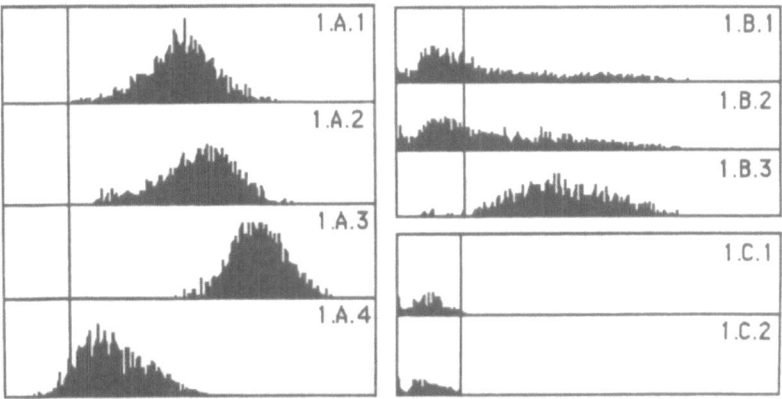

Figure 1. A-C. Fluorescence histograms of the reactivity of the rat anti-mouse endothelial cell hybridoma supernatant with cerebral endothelial cells derived from A.1. SJL, A.2. BALB/C, A.3.(BALB/C X SJL)F1 and A.4. B10.S strains of mice. B.1. SJL epididymal fat pad endothelial cells, B.2. BALB/c epididymal fat pad endothelial cells B.3. BALB/c hepatic endothelial cells. C.1. SJL lymphocytes, C.2. BALB/c lymphocytes. X-axis, fluorescence intensity; Y-axis, cell number.

Western blot analysis demonstrated antibody binding to both an 84-110 kDa and a 36 kDa determinant. This was established using SJL and B10.S CECs (Figure 2), as well as BALB/c CECs and HECs, along with (BALB/c x SJL)F1 CECs (Figure 3, Lane A). Exposure to TNF + γ-IFN can upregulate the expression of both the 84-110 kD and 36 kDa determinants (Figures 2-3, Lane B). In contrast, splenic MNC lysates exposed to the same cytokine cocktail remained unlabeled (Figure 4).

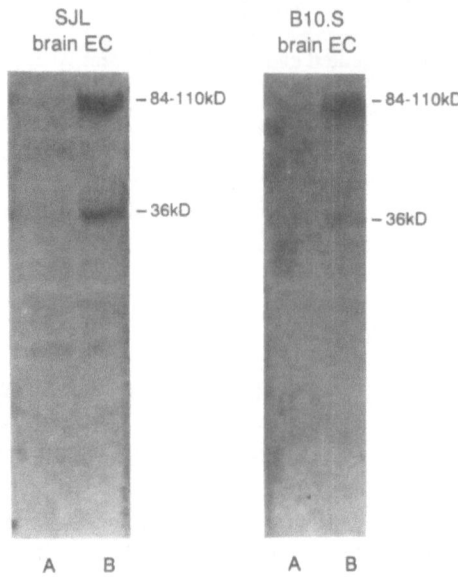

Figure 2. Western blot analysis of rat anti-mouse endothelial cell hybridoma supernatant binding to SJL and B10.S cerebral endothelial cells. Lane A. Untreated endothelial cell lysates, Lane B. Cytokine-(TNF, 100 U/ml + γ-IFN, 200 U/ml for 48 hr) treated endothelial cell lysates.

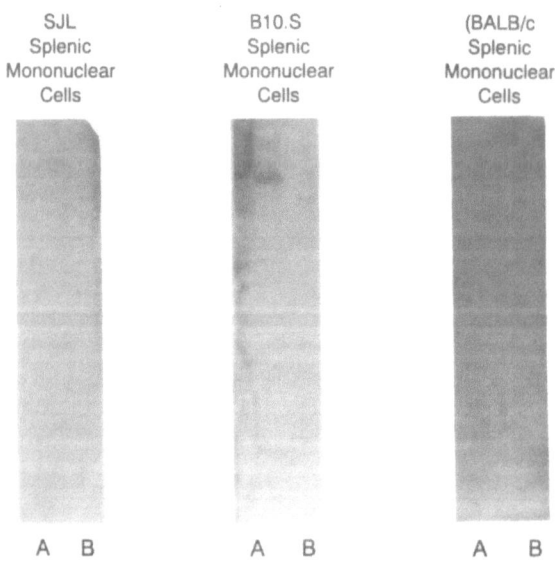

Figure 3. Western blot analysis of rat anti-mouse endothelial cell hybridoma supernatant binding to BALB/c cerebral and hepatic endothelial cells and (BALB/c x SJL)F1 cerebral endothelial cells Lane A. Untreated endothelial cell lysates, Lane B. Cytokine-(TNF, 100 U/ml + γ-IFN, 200 U/ml for 48 hr) treated endothelial cell lysates.

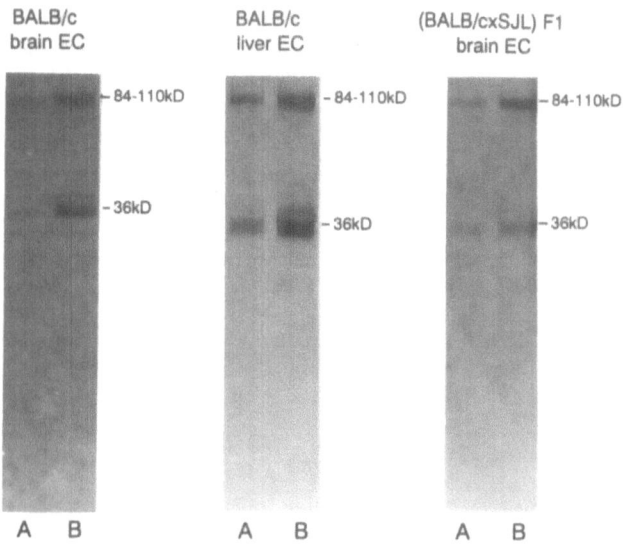

Figure 4. Western blot analysis of rat anti-mouse endothelial cell hybridoma supernatant binding to SJL, B10.S and BALB/c splenic mononuclear cells. Lane A. Untreated cell lysates, Lane B. Cytokine-(TNF, 100 U/ml + γ-IFN, 200 U/ml for 48 hr) treated cell lysates.

In conclusion, we have generated an antibody that recognizes panendothelial specific 84-110 kDa and 36 kDa determinants whose expression is upregulated by exposure to the cytokines TNF and γ-IFN. We are currently analyzing an endothelial cell derived cDNA library to further characterize the determinants recognized by this antibody. In addition, we are presently testing the ability of this antibody to block the adhesion of specific populations of inflammatory cells to tissue-specific endothelial cells.

ACKNOWLEDGMENTS

This study was supported by research grants RG-2072 and RG-1722 from the National Multiple Sclerosis Society and NS 22145 and NS 24621 from the National Institutes of Health. The excellent technical assistance of Ms. Concetta D'Imperio is acknowledged.

REFERENCES

1. Raine, C.S., Canella, B., Duijvestijn, A.M., and Cross, A.H., Homing to central nervous system vasculature by antigen specific lymphocytes. II. Lymphocyte/endothelial cell adhesion during the initial stages of autoimmune demyelination, *Lab. Invest.* 63:476, 1990.
2. Alvord, E.C., Kies, M.W., and Suckling, A.J., Experimental allergic encephalomyelitis: a useful model for multiple sclerosis, *Prog. Clin. Biol. Res.* 146:1, 1984.
3. McCarron, R.M., Kempski, O., Spatz, M., and McFarlin, D.E., Interaction of myelin basic protein sensitized T lymphocytes and murine cerebral vascular endothelial cells, *J. Immunol.* 134:310, 1984.
4. Streeter, P.R., Berg, E.L., Rouse, B.T.N., Bargatze, R.F., and Butcher, E.C., A tissue specific endothelial cell molecule involved in lymphocyte homing, *Nature* 331:41, 1988.

5. Berg, E.L., Yoshino, T., Rott, L.S., Robinson, M.K., Warnock, A., Kishimoto, T.K., Picker, L.J., and Butcher, E.C. The cutaneous antigen is a skin lymphocyte homing receptor for the vascular lectin endothelial cell-leukocyte adhesion molecule 1, *J. Exp. Med.* 174:1461, 1991.

6. Springer, T.A., Adhesion receptors of the immune system, *Nature* 346:425, 1990.

7. Auerbach, R., Alby, L., Morrissey, L.W., Tu, M., and Joseph, J., Expression of organ specific antigens on capillary endothelial cells, *Microvasc. Res.* 29:401, 1985.

8. Rupnick, M.A., Carey, A., and Williams, S.K., Phenotypic diversity in cultured cerebral microvascular endothelial cells, *In Vitro* 24:435, 1988.

9. Wagner, R.C., and Matthews, M.A., The isolation and culture of capillary endothelium from epididymal fat, *Microvasc. Res.* 20:280, 1975.

Barrier Function and Characteristics

THE USE OF INTRACEREBRAL MICRODIALYSIS TO STUDY BLOOD-BRAIN BARRIER TRANSPORT IN HEALTH, AFTER MODIFICATION AND IN DISEASE

Elizabeth C.M. de Lange, Meindert Danhof, Albertus G. de Boer, and Douwe D. Breimer

Center for Bio-Pharmaceutical Sciences, Division of Pharmacology
P.O. Box 9503, 2300 RA, Leiden, The Netherlands

INTRODUCTION

Microdialysis is a relatively new in vivo technique that allows for the monitoring of concentrations of endogenous and exogenous compounds in brain tissue. Briefly, the technique involves the stereotactic implantation of a semi-permeable dialysis probe into a selected tissue. A physiological solution is pumped through the probe and, by means of the dialysis principle, small molecular weight compounds will diffuse into the dialysis fluid. The concentration of a compound in the dialysate, which can be determined by any suitable analysis, supposedly reflects its extracellular concentration. This technique has been applied mostly to estimate the extracellular concentrations of endogenously produced substances. Recently, its application has been extended in the field of pharmacokinetics.

Of great interest is the measurement of intracerebral drug concentrations because it allows the study of blood-brain barrier (BBB) transport. It offers the opportunity to obtain concentration-time profiles of the drug in individual animals.

Intracerebral microdialysis is an invasive technique. The brain tissue reactions to the implantation of microdialysis probes have been investigated. It has been shown that there is an initial decrease in glucose metabolism immediately after insertion of the probe, and glial reactions usually begin after 2-3 days. Furthermore, it appeared that despite some degree of tissue trauma, the BBB is intact shortly after implantation (2,3). The extent of implantation damage is usually considered small and dependent on several factors, including the size and shape of the probe, its chemical composition and the duration of its implantation

An important question is whether the microdialysis procedure as used in our laboratory alters blood-brain barrier (BBB) functionality. Therefore, it is of importance to study its ability to reflect BBB transport under several experimental conditions.

In this paper some results on intracerebral microdialysis studies with acetaminophen, atenolol and the anticancer drug methotrexate are presented. In these studies the emphasis has been on the effects of composition and temperature of the perfusion solution. BBB transport characteristics were also determined following hyperosmotic BBB disruption and experimentally induced tumor.

Frontiers in Cerebral Vascular Biology: Transport and Its Regulation
Edited by L.R. Drewes and A.L. Betz, Plenum Press, New York, 1993

257

INTRACEREBRAL MICRODIALYSIS METHOD

Male SPF Wistar rats (Sylvius Laboratory Breeding Facility, Leiden, The Netherlands) weighing 180-200 g were maintained on a standard laboratory rat diet (RMH-TH, Hope Farms, Woerden, The Netherlands) and used throughout the studies. The rats were anaesthetized with an intramuscular injection of 150 µl of Hypnorm® (Janssen Pharmaceutica, Goirle, The Netherlands) and placed in a stereotactic apparatus. Incisions were made to expose the skull, which was thereafter locally anaesthetized with a 0.6% solution of lidocaine. Holes were drilled in the lateral plane of the skull to allow for the horizontal introduction of the microdialysis probe through the cortex at 2mm below the bregma. The dialysis fibre (O.D. 0.29 mm) consisted of saponified cellulose ester that had been previously covered with silicon glue except for the area of the dialysis zone. Stainless steel needles, glued to both ends of the dialysis fibre, were secured with dental cement on the top of the skull. The animals were allowed to recover from probe implantation and anaesthesia for 24 hr before the start of the first experiment.

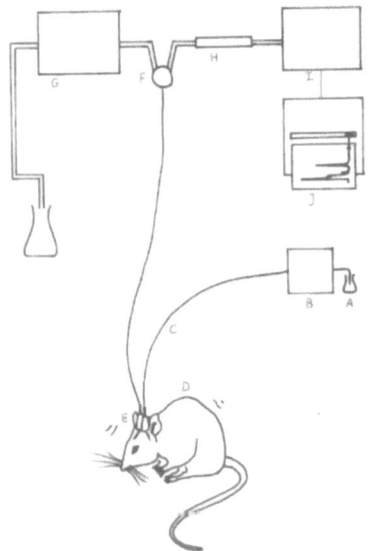

Figure 1. a) physiological solution, b) Gilson pump, c) polyethylene tubing, d) subcutaneous cannula, e) probe and stainless steel needles, f) HPLC loop, g) HPLC pump, h) HPLC column, i) detector, j) recorder.

A dialysate temperature of 38°C was achieved using a subcutaneous cannula (polyethylene tube, I.D. 0.58 mm) at the back of the rat through which the dialysis fluid passed before entering the microdialysis probe.

For intravenous (IV) drug administration and serial blood sampling, polyethylene cannulas were implanted in the femoral vein and artery, respectively, under ether anaesthesia. The recovery period was at least 2 hr before the start of the experiment.

For the microdialysis experiments, the stainless steel needles at both sides of the microdialysis probes were connected by means of polyethylene tubing (O.D. 0.51 mm, I.D. 0.08 mm, 80 cm) to a perfusion pump (Gilson) and to the sample loop connected to HPLC. The dialysis probe was perfused at 7 µl/min with an artificial ECF solution at pH7.4 according to Moghaddam and Bunney (6).

The rat was dialyzed for 30 min to obtain dialysis equilibrium and blank data. Drugs were administered as a single intravenous bolus injection. Thereafter, drug concentrations were measured for 120 min. The doses of the drugs used in these studies were 825 μg of acetaminophen and 10 mg of atenolol. At regular intervals blood samples were drawn from the femoral artery and heparinized. Plasma was obtained by centrifugation and stored at -20°C until analysis.

The drugs were analyzed by a reversed high-pressure liquid chromatography (HPLC) with phase column (Spherisorb, 10 cm* 4.6 or 2 mm I.D., S3 ODS2, Phase Separations, Waddinxveen, The Netherlands) with an appropriate buffer solution and electrochemical detection for acetaminophen and fluorescence detection for atenolol.

RESULTS

After IV bolus injection of the two model drugs, acetaminophen and atenolol (5), reproducible concentration versus time profiles were obtained. Concentrations in brain were considerably lower compared to those in plasma, which is reflected in the ratio AUC(cortex): AUC (plasma) that were found to be 0.18 and 0.043 for acetaminophen and atenolol, respectively. This is in accordance with the general idea that BBB transport is restricted, especially for hydrophilic drugs (4,8).

Effect of Dialysate Composition and Temperature

In this study our purpose was to examine the influences of experimental conditions on the outcome of the transport experiments. In intracerebral microdialysis experiments, many dialysate compositions have been used with more or less the same composition as the ECF. Deviations may have effects on function characteristics of the tissue under examination (6,10). We used a hypotonic dialysis solution (with a tenfold decrease in sodium chloride content) to achieve a periprobe disturbance in ion concentrations that could affect BBB permeability. Table 1 shows the influence of composition of the dialysis solution on the AUC values of atenolol and acetaminophen when studies were conducted according to a longitudinal design with repeated experiments on three consecutive days. The use of the physiological solution resulted in quite reproducible values of the AUC; only for acetaminophen was there a slight decrease in the AUC in the third experiment. However, the use of a hypotonic perfusion solution appeared to increase brain AUC values dramatically.

Table 1. Effect of composition of the dialysis solution (38°C) on cortical AUC ± SEM values expressed as a percentage of the control values (Day 1) in repeated experiments.

Dialysis solution	Drug	Day 1	Day 2	Day 3
isotonic	acetaminophen	100±4	98±5	76±5*
hypotonic	acetaminophen	100±9	154±21*	114±13*
isotonic	atenolol	100±21	103±16	98±18
hypotonic	atenolol	100±25	378±24*	427±24*

*= significance p<0.05

This increase is more pronounced for the hydrophilic drug atenolol, of which transport is expected to be more sensitive to BBB changes. It is suggested that the composition of the dialysate solution can indeed alter BBB function characteristics and the isotonic dialysate solution as used in our experiments is suitable to perform studies on BBB transport.

Until recently most microdialysis experiments have been performed with the bulk dialysate solution at room temperature. Therefore, a temperature gradient may exist between the dialysis fluid and the periprobe tissue. The effect of dialysate temperature (24°C and 38°C) on pharmacokinetics of atenolol and acetaminophen in the brain was assessed, and no temperature effect on cortical AUC values could be demonstrated (Table 2). For acetaminophen the experiment was also performed using the hypotonic perfusion solution. In this case a larger AUC was obtained at a dialysate temperature of 24°C.

Table 2. Effect of dialysis solution temperature on cortical AUC ± SEM values as a percentage of the control values obtained at 38°C. (*= significantat $p<0.05$)

Dialysis solution	Drug	38°C	24°C
isotonic	acetaminophen	100±11	103±8
hypotonic	acetaminophen	100±20	192±21*
isotonic	atenolol	100±10	95±17

*= significance $p<0.05$

It seems that the periprobe tissue loses its capability to render temperature differences ineffectual when the tissue under examination is already "stressed." Therefore, a dialysis solution of 38°C rather than room temperature has to be considered as more physiological.

Hyperosmotic BBB Disruption

The changes in BBB transport after the intracarotid infusion of a hypertonic solution of mannitol has been reported frequently (7). In this study we used this experimental model for the opening of the BBB to investigate the ability of intracerebral microdialysis to detect changes in BBB transport of atenolol.

The hypertonic mannitol solution (25% in 1 mM phosphate buffer, pH7.4, at 38°C)was infused via the external into the left internal carotid artery (3.6 ml/30 sec) of male Wistar rats (230-250 gm). As a control, we used a saline solution in 1 mM phosphate buffer, pH7.4. After 0.5 min, a bolus dose of 10 mg of atenolol was administered intravenously. BBB transport of atenolol was increased 3.4fold after mannitol infusion (n=11) as compared to the saline infusion (n=9). This effect appeared to occur only at the infusion site of mannitol because the AUC values obtained in the contralateral hemisphere (n=9) were comparable to those in the saline controls. Therefore, both regional differences in BBB transport as well as changes in BBB transport are reflected in the application of this technique.

Experimental Brain Tumor Model

The uptake of methotrexate (MTX) was studied in an experimental brain tumor model in male Wag/rij rats (Harlan B.V., Zeist, The Netherlands) weighing 180-200 gr. For the intracortical implantation of the experimental brain tumor, vital parts of a subcutaneous transplanted rhabdomyo-sarcoma were cut into pieces of 1 mm². After an intramuscular injection of 200 μl Hypnorm, an incision was made to expose the skull. A hole was drilled at 2 mm ipsilateral to the Bregma point. A guide cannula (Microlance needle, 21 G 2, Becton Dickinson B.V., Etten-Leur, The Netherlands) and an inner cannula (a closed Microlance needle, 22 G 1 1/4) were used to deliver the tumor fragment into the left cortex at a depth of 2 mm. The skull was then covered with dental cement and the skin was securely stitched. After 10 days, the animals received the transversal microdialysis probe also at a depth of 2 mm below the Bregma point. Before the start of the experiment the animals were allowed to recover for 24 hr.

In our experimental design, we used 4 groups of 6 rats each. The first group was used to determine the penetration of MTX into the cortex under standard microdialysis conditions ("normal"). The second group contained the animals that had a sham tumor implantation procedure ("sham"). In the third and fourth groups a piece of tumor was implanted in the left cortex, and the profiles of MTX were measured respectively "inside" (left cortex) and "outside" (right cortex) the tumor. Also plasma concentrations were determined. The AUC ratio of brain dialysate versus plasma of "normal": "sham": "inside": "outside" appeared to be 100:96:232:96.

Although a 2.3-fold increase was found in transport of MTX across the BBB in the tumor as compared to the normal brain, the sham surgery and the tumor-bearing state as such appeared to be of no influence. This suggests that within the experimental brain tumor the BBB may be partially absent.

CONCLUSION

In conclusion, these experiments show the potential use of intracerebral microdialysis to study BBB transport. It has been shown to detect drug-dependent changes in the extent of drug transport into the cortex under various experimental microdialysis conditions (of which the use of the physiological dialysis solution at 38°C is expected to be the best posssible condition). Also differences in BBB transport could be shown after BBB modification and in an implanted experimental brain tumor.

We expect that the use of intracerebral microdialysis can be an important additional technique to obtain information about drug transport across the BBB in health and disease.

ACKNOWLEDGMENTS

Investigations were supported in part by the Dutch Cancer Society.

REFERENCES

1. Agon, P., Goethals, P., Van Haver, D., and Kaufman, J.-M., Permeability of the blood-brain barrier for atenolol studied by positron emission tomography, *J. Pharm. Pharmacol.* 43:597-600, 1991.
2. Benveniste, H., Drejer, A., Shoesboe, A., and Diemer, N.H., Regional cerebral glucose phosphorylation and blood flow after insertion of a microdialysis fiber through the dorsal hippocampus in the rat, *J. Neurochem.* 49:729-734, 1987.
3. Benveniste, H., and Diemer, N.H., Cellular reactions to implantation of a microdialysis tube in the rat hippocampus, *Acta Neuropath.* (Berl) 74:234-238, 1987.
4. Bradbury, M.W., Transport across the blood-brain barrier, *In:* "Implications of the Blood-Brain Barrier and Its Manipulation," Vol I, E.A. Neuwelt, ed., Plenum Medical Book Company, New York and London, pp. 119-126, 1989.

5. Van Bree, J.B.M.M., Drug transport across the blood-brain barrier, Thesis, University of Leiden, Leiden, 1990.

6. Moghaddam, B., and Bunney, B.S., Ionic composition of microdialysis perfusing solution alters the pharmacological responsiveness and basal outflow of striatal dopamine, *J. Neurochem.* 53:652-654, 1989.

7. Neuwelt, E.A., and Barnett, P.A., Blood-brain barrier disruption in the treatment of brain tumors. Animal studies, *In:* "Implications of the Blood-Brain Barrier and Its Manipulation," Vol II, E.A. Neuwelt, ed., Plenum Medical Book Company, New York and London, pp. 107-194, 1989.

8. Oldendorf, W.H., Lipid solubility and penetration of the blood-brain barrier, *Proc. Soc. Exp. Biol. Med.* 147:813-816, 1974.

9. Timmerman, W., and Westerink, B.H.C., Importance of the calcium content infused during microdialysis for the effects induced by d2 agonists on the release of dopamine in the striatum of the rat, *Neurosci. Lett.* 131:93-96, 1991.

CEREBRAL ENDOTHELIAL MECHANISMS IN INCREASED PERMEABILITY IN CHRONIC HYPERTENSION

Sukriti Nag

Department of Pathology
Queen's University
Kingston, Ontario
Canada K7L 3N6

INTRODUCTION

Our previous studies (4,5) demonstrated that increased cerebrovascular permeability to protein in acute hypertension was associated with reduced activity of the endothelial plasma membrane adenosine triphosphatases (ATPases). This study was undertaken to determine when during the course of chronic hypertension a change in the activity of the plasma membrane calcium-activated ATPase (Ca^{2+}-ATPase) and the ouabain-sensitive, K^+-dependent p-nitrophenylphosphatase (Na^+,K^+ATPase) occurs.

METHODS

Chronic hypertension was induced in male Wistar-Furth rats (120-140 gm) by constricting the left renal artery followed by a right nephrectomy as described previously (3). Control rats were sham-operated. One week post-surgery, rats were sacrificed at weekly intervals over a 6-week period. Prior to sacrifice, rats were anesthetized and the blood pressure and blood gases were measured as described previously (3). Experiments were terminated by perfusion of the fixative suitable for the demonstration of the required enzymes as described previously (4,5).

Brain slices were reacted for the demonstration of either Ca^{2+}-ATPase using the method reported by Ando et al. (1), or Na^+,K^+ATPase, using the method of Mayahara et al. (2). Blocks of brain containing vessels from the occipito-temporal cortex were processed for ultrastructural examination using techniques described previously (4). Ultrathin sections were stained using lead citrate and examined using a Hitachi H500 electron microscope at 75 kV.

Frontiers in Cerebral Vascular Biology: Transport and Its Regulation
Edited by L.R. Drewes and A.L. Betz, Plenum Press, New York, 1993

263

RESULTS

An elevation of the blood pressure was observed as early as 1 week following the surgery for the induction of hypertension. At the time of sacrifice, most test rats had blood pressures that were above 200 mm Hg.

Electron-dense reaction product indicative of enzyme activity was distributed in a discontinuous manner on the outer plasma membranes of endothelial, smooth muscle and adventitial cells of cerebral cortical arterioles. The amount of reaction product was less in the case of Na^+,K^+-ATPase (Figure 1a) as compared with Ca^{2+}-ATPase (Figure 1b). A striking finding was the localization of both enzymes along the plasma membranes of pinocytotic vesicles (Figures 1a and b).

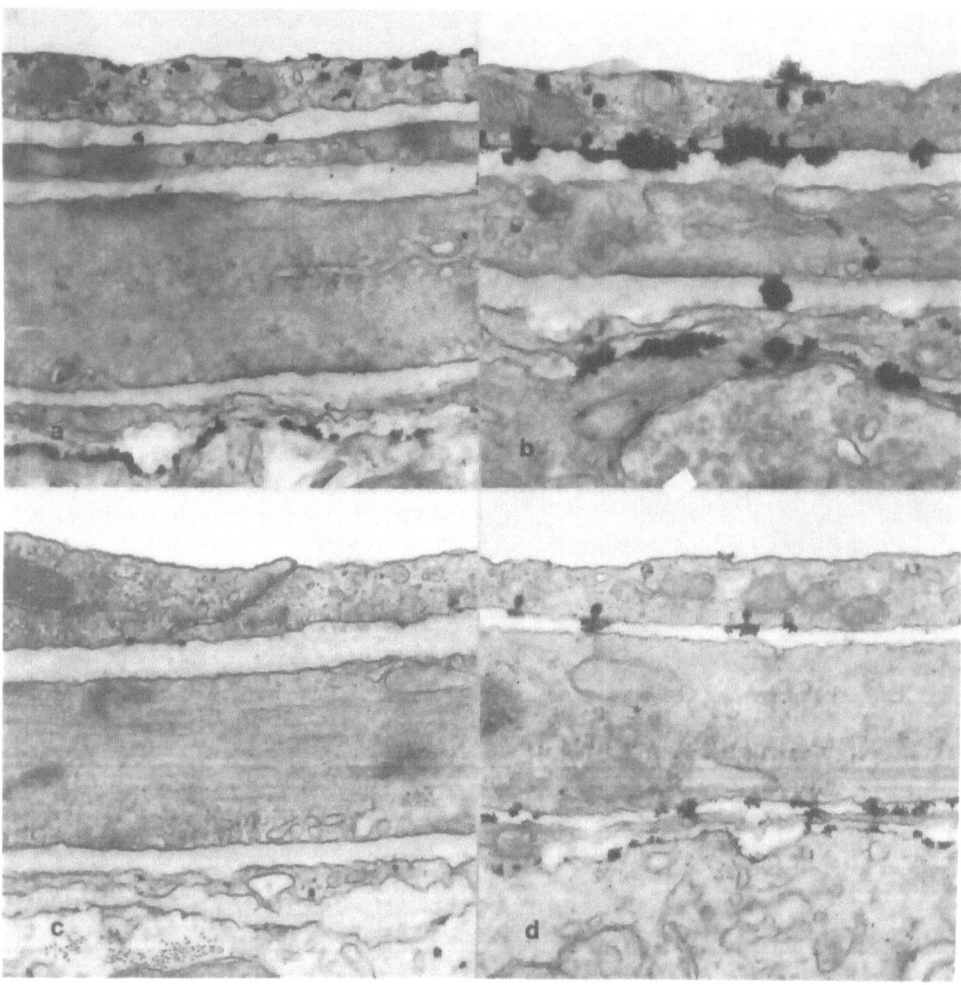

Figure 1. Segments of cerebral arterioles from normotensive rats showing localization of Na^+,K^+-ATPase (a) and Ca^{2+}-ATPase (b). Note the discontinuous distribution of electron-dense reaction product on the outer plasma membranes of endothelial, smooth muscle and adventitial cells. Enzyme is also localized along the plasma membranes of pinocytotic vesicles. There is reduced localization of Na^+,K^+-ATPase (c) and Ca^{2+}-ATPase (d) in arteriolar walls of hypertensive rats. a, X36,000; b, X45,000; c, X44,000; d, X32,000

Hypertensive rats demonstrated reduced localization of Ca^{2+}-ATPase in plasma membranes of cerebral endothelium as early as 1 week post-surgery while reduced Na^+,K^+-ATPase localization occurred about 4 weeks post-surgery (Figures 1c and d).

During the 4-6 week period post-surgery, some vessels with reduced enzyme localization were leaking endogenous serum proteins (Figure 2). The latter appeared as granular material that widened the basement membrane between different layers of the vessel wall. These permeable vessels showed increased endothelial pinocytotic vesicles that appeared prominent because of enzyme localization along the walls of these vessels. A rare finding was the presence of transendothelial channels that were formed by the fusion of pinocytotic vesicles (Figure 3).

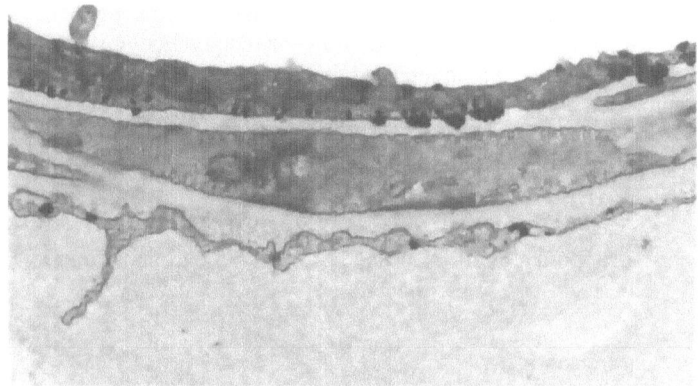

Figure 2. Segment of arteriole from a hypertensive rat showing leakage of endogenous serum proteins that appears as granular material distending the basement membrane. There is reduced localization of Ca^{2+}-ATPase in the plasma membranes of cells composing the vessel wall. Note the increased numbers of pinocytotic vesicles present. X19,300

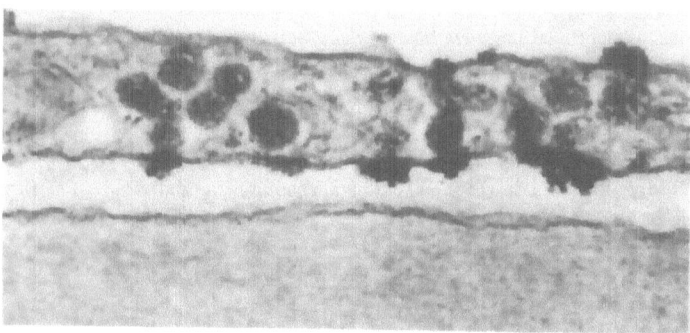

Figure 3. Segment of arteriolar endothelium from a hypertensive rat showing increased numbers of pinocytotic vesicles. Note the transendothelial channel formed by the fusion of vesicles. X88,000

DISCUSSION

Localization of the ATPases in cerebral cortical vessels of normotensive rats was similar to our previous observations (4,5). Arterioles of hypertensive rats showed a reduced amount of reaction product, indicating reduced enzyme activity. The latter was observed 1 week post-surgery in the case of Ca^{2+}-ATPase and occurred at 4 weeks in the case of Na^+,K^+-ATPase.

Increased cerebrovascular permeability to protein occurred during the 4-6 week period of observation. Permeable vessels showed reduced enzyme localization and increased numbers of pinocytotic vesicles. In previous studies using the acute hypertension model (4,5), it was uncertain whether enzyme alterations preceded or followed permeability alterations. The findings of this study suggest that alteration in activity of the endothelial plasma membrane ATPase precedes the permeability alterations. This study supports previous observations of the importance of enhanced pinocytosis and formation of transendothelial channels in increased cerebrovascular permeability to protein and edema formation (3,4).

Adenosine triphosphatases play an important part in calcium homeostasis; therefore, reduced activity of these enzymes could result in increases of intraendothelial ionic calcium. The latter could trigger increased permeability by enhanced pinocytosis. This hypothesis is supported by our recent study, which reported that cerebrovascular permeability to protein is reduced in the presence of calcium entry blocker (6).

ACKNOWLEDGMENTS

This work was supported by the Heart and Stroke Foundation of Ontario, Grant No. B2019.

REFERENCES

1. Ando, T., Fujimoto, K., Mayahara, H., Miyajima, H., and Ogawa, H., A new one-step method for the histochemistry and cytochemistry of Ca^{2+}-ATPase activity, *Acta Histochem. Cytochem.* 14:705, 1981.

2. Mayahara, H., Fujimoto, K., Ando, T., and Ogawa, K., A new one-step method for the cytochemical localization of ouabain-sensitive potassium-dependent p-nitrophenyl-phosphatase activity, *Histochem.* 67:125, 1980.

3. Nag, S., Cerebral cortical changes in chronic hypertension: Combined permeability and immunohistochemical studies, *Acta Neuropathol.* (Berl). 62:178, 1984.

4. Nag, S., Localisation of calcium-activated adenosine-triphosphatase (Ca^{2+}-ATPase) in intracerebral arterioles in acute hypertension, *Acta Neuropathol.* (Berl). 75:547, 1988.

5. Nag, S., Ultracytochemical localisation of Na^+,K^+-ATPase in cerebral endothelium in acute hypertension, *Acta Neuropathol.* (Berl.) 80:7, 1990.

6. Nag, S., Protective effect of flunarizine on blood-brain barrier permeability alterations in acutely hypertensive rats, *Stroke* 22:1265, 1991.

ALTERED FATTY ACID COMPOSITION OF ETHANOLAMINE PHOSPHOGLYCERIDE IN BRAIN MICROVESSELS FROM SENESCENT MOUSE

Wesley M. Williams and Stanley I. Rapoport

Laboratory of Neurosciences
National Institute on Aging
National Institutes of Health
Bethesda, MD 20892

INTRODUCTION

The cerebral microcirculation (small arterioles, capillaries and venules) is a primary intermediary for the maintenance of cellular homeostasis in the brain. This regulatory capacity, residing principally within the microvessel endothelial cell, encompasses functional aspects of the blood-brain barrier (1), production of vasoactive eicosanoids (2) and receptor binding (3). Stability of the endothelial cell membrane is, therefore, essential to cerebral homeostasis and proper neuronal function. Advanced cellular aging can be associated with destabilization of the plasma membrane as a result of changes in constituent lipids and proteins (4). Age-associated changes in phospholipid composition of neuronal membranes have been reported (5-7), and recently Tayarani et al. (8) have demonstrated age-related differences in the relative percent of mono and polyunsaturated fatty acids in the rat brain capillaries. Phospholipids and their constituent unsaturated fatty acids may be especially prone to catabolic processes within the endothelial cell as a result of persistent oxidative stress (9). The present study was, therefore, designed to assess whether microvessel membranes exhibit changes in phosphoglyceride composition and degree of fatty acid unsaturation with advancing age in the mouse.

EXPERIMENTAL METHODS

Cerebral microvessels were isolated from male, C57BL6/NNIA mice at 10, 20 and 27-30 months of age. The isolation procedure used has been described in detail by Williams et al. (10). Briefly, mice were killed by cervical dislocation without anesthesia, and the brains quickly removed (~ 30 sec) and immersed in cold (4°C) 0.4 M sucrose containing 20,000 I.U. sodium heparin/L. The meninges were removed and the cerebellum and cortical grey matter separated from the brain and placed in cold 0.4 M sucrose containing 0.1 mM $CaCl_2$, 0.1 mM ATP, 0.4 M MES buffer (pH 6.1), 1% bovine serum albumin (Fraction V), and

20,000 I.U. sodium heparin /L. The tissue was finely minced and homogenized with a 30-ml Teflon-on-glass Potter-Elvehjem tissue homogenizer. The homogenate was centrifuged at 2000 x g and the resulting pellet containing microvessels and parenchymal membranes resuspended in cold 15% dextran containing 0.1 mM $CaCl_2$, 0.1 mM ATP and 0.4 M MES buffer to remove myelin. After a second centrifugation at 5000 x g for 20 min, the resuspended tissue was passed through three glass bead columns to remove residual contamination of the microvessels by parenchymal membranes. Purified microvessels were subsequently concentrated by centrifugation and stored under nitrogen in HPLC grade methanol at -70°C.

Total lipid was extracted from cerebral microvessel membrane by a modified "Folch wash" (11). The diacyl and plasmalogen (alk-1-enyl acyl) fractions of ethanolamine phosphoglyceride were separated by a separation-reaction-separation thin-layer chromatographic procedure (12). Lipid spots were visualized with 1,6-diphenyl-2,3,5,-hexatriene (DPH). Silica gel containing the separated diacyl, alk-1-enyl acyl, and aldehyde moieties were methylated using boron trifluoride (BF3). The resulting methyl esters and aldehyde dimethyl acetals (DMA) were analyzed by gas chromatography (GC). Fatty acid and DMA peaks were identified by comparison of retention times with those of commercially prepared PUFA 1 and 2 fatty acid and DMA standards.

Phospholipid composition was quantitated by spectrophotometric determination of inorganic phosphate (13). Statistical significance at the p=0.05 level was determined by a Mann Whitney U test using a probability table for n=3 (14).

RESULTS

Phospholipid Composition (mol %) of Cerebral Microvessel Membrane

Overall, mol % for ethanolamine phosphoglyceride did not change significantly from 10 to 27-30 months of age. However, when alk-1-enyl acyl and diacyl EPG fractions were compared separately, a significant decrease (~38%) in mol % for alk-1-enyl acyl EPG was found in microvessels at 27-30 months of age relative to 10-month-old mice (Table 1).

Table 1. Comparison of phosphoglyceride composition (mol %) of cerebral microvessels from 10-, 20- and 30-month-old mice. [1]

Phosphoglyceride	10 month	20 month	30 month
alk-1-enyl acyl EPG	12.5±3.2	9.9±2.6	7.8±1.0*
diacyl EPG	12.9±2.5	9.7±2.0	10.0±2.1
Choline phosphoglyceride	45.7±2.9	51.7±5.8	57.5±5.7*
inositol phosphoglyceride	4.8±0.1	5.0±1.3	4.6±0.2
serine phosphoglyceride	8.2±0.6	8.1±1.2	6.8±1.6
sphingomyelin	15.8±3.2	15.5±2.3	13.3±1.3

[1] All analyses were done in duplicate on 3 separate isolations, mean ±S.D. *p=0.05

Comparative Acyl Composition in Ethanolamine Phosphoglyceride from Cerebral Microvessel Membrane

It is apparent from the data in Tables 2 and 3 that aging of cerebral microvessel membrane is accompanied by an elevation in saturated fatty acids concomitant with reduced levels of unsaturated species. These changes, apparent at both 20 and 27-30 months of age, are most evident in diacyl EPG. Alk-1-enyl acyl EPG also exhibits an increase in saturated fatty acids, but without a significant change in unsaturation content. Elevated saturation for both alk-1-enyl acyl and diacyl EPG results from an increase in both palmitic (16:1) and stearic acid (18:1) levels. The small, but statistically significant, loss in total unsaturated fatty acid level results from a reduction in both mono and polyunsaturated fatty acids. The decrease in monounsaturation for diacyl EPG is accounted for by a loss in 16:1 and 18:1 in both 20 and 27- to 30-month-old age groups, and in the 27- to 30-month-old age group for alk-1-enyl acyl EPG. Decreased levels of specific polyunsaturated fatty acids (18:2n-6, 20:3n-6, 22:5n-3) occur in both EPG species, but the reduction in total polyunsaturation is statistically significant only for diacyl EPG. Interestingly, the levels of arachidonic acid (20:4) and docosahexaenoic acid (22:6n-3) remain virtually unchanged, with the reduction in polyunsaturation occurring only in the relatively minor fatty acids of both the n-3 and n-6 series. Unlike other polyunsaturates, which remain unchanged or decrease, linolenic acid (18:3 n-3) increases with age in both alk-1-enyl acyl and diacyl EPG.

Table 2. Composition of constituent fatty acids (wt%) of alk-1-enyl acyl ethanolamine phosphoglyceride from mouse cerebral microvessels. [1]

Fatty acid	10 month	20 month	30 month
16:0	3.68±0.96	6.43±1.20*	6.27±0.99*
16:1	0.72±0.17	0.75±0.06	0.58±0.05
18:0	3.18±1.49	6.26±2.35*	5.02±0.92*
18:1	21.54±2.37	19.36±2.08	18.14±1.59
18:2n-6	1.91±0.21	1.16±0.44	1.45±0.24
18:3n-3	0.65±0.24	0.78±0.06	1.41±0.25*
20:3n-6	0.95±0.09	0.58±0.03*	0.59±0.08*
20:4n-6	42.40±3.96	43.85±5.33	43.43±4.66
20:5n-3	0.83±0.17	0.60±0.44	0.57±0.08
22:5n-3	0.77±0.06	0.59±0.16	0.55±0.02*
22:6n-3	10.66±1.69	9.43±1.25	11.35±1.00

[1] Analyses were made in duplicate on 3 separate isolations, mean±S.D. *p=0.05.

Alkenyl Group Composition of Ethanolamine Phosphoglyceride from Cerebral Microvessel Membrane

The major alkenyl groups present in cerebral microvessels are palmitate, stearate and oleate. As in alk-1-enyl acyl and diacyl EPG the proportion of saturation (18:0) increases with age, while the level of unsaturation (18:1) decreases. Both changes reach statistical significance in 27- to 30-month-old mice (Table 4).

Table 3. Composition of constituent fatty acids (wt%) of diacyl ethanolamine phosphoglyceride from mouse cerebral microvessels. [1]

Fatty acid	10 month	20 month	30 month
16:0	9.07±0.83	11.80±1.59*	11.26±0.47*
16:1	0.65±0.15	0.41±0.08*	0.35±0.12*
18:0	28.08±1.28	31.45±0.85*	33.38±1.01*
18:1	24.02±1.68	21.11±0.50*	18.38±1.98*
18:2n-6	4.61±0.28	3.95±0.52	3.23±0.43*
18:3n-3	0.60±0.08	0.72±0.07	0.83±0.10*
20:3n-6	0.94±0.04	0.85±0.31	0.50±0.05*
20:4n-6	20.79±1.87	18.69±1.65	19.45±1.57
20:5n-3	0.41±0.07	0.29±0.09	0.24±0.10*
22:5n-3	0.36±0.01	0.29±0.08	0.23±0.02*
22:6n-3	8.91±1.57	8.11±1.23	8.86±0.65

[1] All analyses were made in duplicate on 3 separate isolations, mean±S.D. *p=0.05.

Table 4. Major alkenyl groups (wt%) of ethanolamine phosphoglyceride from mouse cerebral microvessels. [1]

Alkenyl Group	10 month	20 month	30 month
16:0	34.43±1.50	36.52±2.00	35.83±3.37
18:0	40.98±4.04	43.84±3.83	48.24±3.00*
18:1	24.60±5.53	21.23±4.80	15.93±1.85*

[1] All analyses were made in duplicate on 3 separate isolations, mean±S.D. *p=0.05.

DISCUSSION

The findings of the present study suggest that with aging in the mouse, the fatty acid composition of the cerebral microvessel endothelial cell membrane becomes increasingly more saturated and less unsaturated. Next to choline phosphoglyceride, EPG is the most abundant phospholipid in the microvessel membrane, accounting for approximately 25% of total phospholipids in mouse brain at 10 months of age. EPG is also highly unsaturated containing large amounts of arachidonic and docosahexaenoic acids. Although the relative percentages of these two major fatty acids are not found to change in EPG with aging, the decrease in mol % for alk-1-enyl acyl EPG suggests that, in fact, a significant loss of these metabolically active fatty acids has occurred. Moreover, the apparent decreases in 18:2n-6 and 20:3n-6, both precursors of arachidonic acid, may suggest altered metabolism of these fatty acids and subsequently altered content of arachidonic acid in both alk-1-enyl acyl and diacyl EPG.

The cause or causes underlying these age-related changes in membrane fatty acid composition are not readily apparent. Several mechanisms may be proposed including: 1) altered absorbtion of essential fatty acids (arachidonate, linoleate and linolenate) from the gut or cerebral microcirculation; 2) altered phospholipase A_2 and/or phospholipase C activity; and 3) chronic, progressive oxidative degradation of membrane unsaturated fatty acids.

Results of the present study indicate that aging of the brain, at least in the mouse, is accompanied by changes in the fatty acid composition of ethanolamine phosphoglyceride from cerebral microvessel membranes. These changes could have a substantive effect upon physicochemical properties of the cell membrane and ultimately upon physiological processes within the endothelial cell.

ACKNOWLEDGMENT

This work was done while the author held a National Research Council-Laboratory of Neurosciences, N.I.H. Research Associateship.

REFERENCES

1. Rapoport, S.I., Sites and functions of the blood-brain barrier, *in:* "Blood-brain Barrier in Physiology and Medicine," Raven Press, New York, 1976.

2. Moore, S.A., Spector, A.A., and Hart, M.N., Eicosanoid metabolism in cerebromicrovascular endothelium, *Amer. J. Physiol.* 254:C37, 1988.

3. Homayoun, P., and Harik, S., Bradykinin receptors of cerebral microvessels stimulate phosphoinositide turnover, *J. Cereb. Blood Flow Metab.* 11:557, 1991.

4. Shinitzky, M., Heron, D.S., and Samuel, D., Restoration of membrane fluidity and serotonin receptors in the aged mouse brain, *in:* "Aging of the Brain," Vol 22, D. Samuel, S. Gershon, S. Algeri, and V.E. Grimm, eds., Raven Press, New York, 1983.

5. Calderini, G., and Toffano, G., Phospholipid methylation, ^{3}H-diazepam and ^{3}H-GABA binding in the cerebellum of aged rats, *in:* "The Aging Brain: Cellular and Molecular Mechanism of Aging in the Nervous System," Vol 20, E. Giacobini, G. Filogamo, G. Giacobini, and A. Vernadakis, eds., Raven Press, New York, 1982.

6. Cimino, M., Curatola, G., Pezzoli, C., Stramentinoli, G., Vantini, G., and Algeri, S., Age-related modification of dopaminergic and beta adrenergic receptor system: restoration of normal activity by modifying membrane fluidity with S-adenosyl methionine, *in:* "Aging Brain and Ergot Alkaloids," A. Agnoli, G. Crepalde, P.F. Spano, and M. Trabucchi, eds., Raven Press, New York, 1983.

7. Crews, F.T., Calderini, G., Battistella, A., and Toffano, G., Age dependent changes in the methylation of rat brain phospholipids, *Brain Res.* 229:256, 1981.

8. Tayarani, I., Cloez, I., Clement, M., and Bourre, J.-M., Antioxidant enzymes and related trace elements in aging brain capillaries and choroid plexus, *J. Neurochem.* 53:817, 1989.

9. Halliwell, B., and Gutteridge, J.M.C., "Free Radicals in Biology and Medicine," 2nd Edition, Clarendon Press, Oxford, 1989.

10. Williams, W.M., Richmann, M., and McNeill, T.H., Cerebral microvascular and parenchymal phospholipid composition in the mouse, *Neurochem. Res.* 13:743, 1988.

11. Ways, P., and Hanahan, D.J., Characterization and quantification of red cell lipids in normal man, *J. Lipid Res.* 5:318, 1964.

12. Sun, G.Y., Preparation and analysis of acyl and alkenyl groups of glycerophospholipids from brain subcellular membranes, *in:* "Neuromethods," Vol 7, A.A. Boulton, G.B. Baker, and L.A. Horrocks, eds., Human Press, Clifton, 1990.

13. Ames, B.N., Assay of inorganic phosphate, total phosphate and phosphotases, *Meth. Enzymol.* 8:115, 1966.

14. S. Siegal, "Nonparametric Statistics," McGraw-Hill, New York, 1956.

PREIRRADIATION OSMOTIC BLOOD-BRAIN BARRIER DISRUPTION PLUS COMBINATION CHEMOTHERAPY IN GLIOMAS: QUANTITATION OF TUMOR RESPONSE TO ASSESS CHEMOSENSITIVITY

Jeffery A. Williams[1], Simon Roman-Goldstein[2], John R. Crossen[3], Anthony D'Agostino[4], Suellen A. Dahlborg[5], Edward A. Neuwelt[6]

[1]Division of Neurosurgery, University of Oklahoma Health Sciences Center,
[2]Department of Diagnostic Radiology, Oregon Health Sciences University,
[3]Department of Psychology, Oregon Health Sciences University,
[4]Department of Pathology, Oregon Health Sciences University,
[5]Department of Neurology, Division of Neurosurgery, Oregon Health Sciences University,
[6]Departments of Neurology, Surgery (Neurosurgery), Biochemistry and Molecular Biology, Oregon Health Sciences University

INTRODUCTION

The blood-brain barrier (BBB) may prevent increased delivery of chemotherapeutic agents despite increased systemic blood concentrations. To increase drug delivery intra-arterial (i.a.) cisplatin has been administered at tumor recurrence by Mahaley et al. (20) and concurrent with radiotherapy by Calvo et al. (4). More recently, Dropcha gave i.a. cisplatin prior to radiation to better assess drug sensitivity in glioma patients (6). Osmotic disruption of the BBB (BBBD) further increases transfer of intravascularly administered chemotherapeutic agents to both tumor and brain around tumor (12,23-25).

At present, adjuvant chemotherapy is administered following radiotherapy in the treatment of human brain tumors (10,11,38,39,43,44). However, such a sequence may obscure the potential benefit of initial administration of chemotherapy. Following radiotherapy, both accelerated repopulation of the tumor (13) and enhanced DNA repair capacity (14,15) may render the tumor less responsive to subsequent chemotherapy. Initial cranial radiation may cause capillary endothelial hyperplasia (5) and may even compromise delivery of chemotherapeutic agents after successful osmotic BBBD (M.K. Gumerlock, manuscript in preparation, 1992). Additionally, sequencing radiation before chemotherapy may increase the risk of neurotoxicity. Leukoencephalopathy has been reported when methotrexate (MTX) is administered i.v. or i.a. after radiation therapy but is rarely seen if administered before radiation therapy or as the sole therapeutic modality (1,21). There is also

Frontiers in Cerebral Vascular Biology: Transport and Its Regulation
Edited by L.R. Drewes and A.L. Betz, Plenum Press, New York, 1993

273

evidence that cranial radiation is associated with lowered cognitive function, hormonal and psychosocial dysfunction, and cerebral atrophy (3,8,16-19,36,37).

Albeit not adequate to evaluate survival, this report presents 12 patients with supratentorial gliomas treated with osmotic BBBD and combination chemotherapy prior to radiotherapy to assess the chemotherapeutic responsiveness of astrocytomas to methotrexate and Cytoxan. Serial neuropsychological testing was conducted in patients who attained a near complete tumor response. This minimized the impact of residual tumor on test results and allowed evaluation of higher cortical function to assess neurotoxicity (45).

MATERIALS AND METHODS

Patients

Between October 1984 and April 1990, 12 adults (11 male, 1 female; Karnofsky Performance Status >70%) underwent craniotomy (8 patients) or biopsy (4 patients) with resultant diagnosis of Grade IV (7 patients), Grade III (3 patients) or Grade II (2 patients) glioma by Kernohan's classification (Table 1). Follow-up data was collected through April 1, 1991. Histologic grading of these tumors was confirmed by a single neuropathologist (A.D.). All patients underwent neuropsychologic testing on inclusion in the protocol (before osmotic BBBD chemotherapy). Follow-up testing was obtained on those patients completing a full course of treatment, one year or more after baseline assessment. Testing included Wechsler Adult Intelligence Scale - Revised (WAIS-R), Wechsler Memory Scale - Revised (WMS-R), Trail Making Test: Parts A and B (Trails A & B), Rey Auditory Verbal Learning Test (RAVLT) or California Verbal Learning Test (CVLT), Rey-Osterrieth Complex Figure Test (ROCFT), Finger Tapping Speed (TAP), and Grip Strength (GRIP) (9,30). The baseline and follow-up test results are reported only for those two patients who achieved 95% tumor response without recurrence for one year. These criteria are necessary to assess the risk of neurotoxicity from this treatment without confusing variables of tumor persistence, recurrence, or cranial irradiation that might render the results uninterpretable.

Table 1. Patient Characteristics and Blood-brain Barrier Disruption Treatment Summary

Patient Number	Age	Sex	Procedure	Path Grade	Chemotherapy Treatments Time (months) Dx-1st Rx	Time (months) 1st-last Rx.	Number Rx	Subsequent Radiation Therapy (Rad)
1	51	M	craniotomy	4	1.4	15.7	26	6100
2	44	M	biopsy	4	0.7	2.1	5	6480
3	41	M	craniotomy	3	0.4	12.0	25	0
4	45	M	biopsy	2	0.2	15.1	24	0
5	55	F	biopsy	2	1.0	7.5	13	7000
6	52	M	craniotomy	4	1.1	11.6	22	3400
7	23	M	craniotomy	3	1.5	10.8	22	0
8	50	M	craniotomy	4	0.7	15.3	25	0
9	67	M	biopsy	3	1.2	2.9	8	3600
10	42	M	craniotomy	4	1.4	2.4	6	0
11	42	M	craniotomy	4	1.7	1.0	3	0
12	57	M	craniotomy	4	1.4	2.1	6	6000

Osmotic BBBD and Chemotherapy Administration

Osmotic BBBD and chemotherapy infusions were administered as previously described (26). During each hospitalization, patients undergo two disruptions 24 hours apart, each followed by the administration of chemotherapeutic agents. Patients are placed under light general anesthesia. A catheter is inserted via the femoral artery into either an internal carotid artery at the C_{1-2} level or a vertebral artery at the C_6 level. Osmotic BBBD is performed by infusing warmed, 25% mannitol into either the internal carotid or vertebral artery, whichever provides the vascular supply to the tumor. Flow rates for mannitol varied from 6 to 12 cc/sec. The infusion rate is selected to achieve reflux into the external carotid artery for internal carotid artery infusions or contralateral vertebral artery for vertebral artery infusions, indicating complete filling of that vascular tree. Chemotherapy consisted of intravenous cyclophosphamide, 15 mg/kg, administered 10 to 20 minutes prior to osmotic BBBD; intra-arterial MTX, 2500 mg, injected immediately after osmotic BBBD; and procarbazine given orally, 100 mg/day for 14 days beginning 24 hours after the second BBBD. Patients were disrupted in two of the three vascular distributions 24 hours apart. Leucovorin is started 36 hours after the first barrier disruption (50 mg p.o. q.i.d. for 5 days after an 80 mg i.v. loading dose). These sequential osmotic BBBD procedures are repeated monthly for an anticipated 12-month course. Following chemotherapy, conventional radiotherapy was offered to those patients with documented tumor recurrence. Six of the nine such patients elected to receive radiation.

Radionuclide Assessment of BBBD

In order to assess the degree and extent of the BBBD, ^{99m}Tc-radiolabeled glucoheptonate, 20 mCi, was administered i.v. immediately following disruption. Planar images (frontal, occipital and vertex) were obtained approximately 3 hours after BBBD and allowed quantification of hemispheric and posterior fossa activity ratios. Mannitol infusion rates were increased in subsequent treatments if activity ratios (perfused vs. unperfused hemisphere) indicated poor disruption (31).

Computerized Tomographic Scan Volumetrics

Contrast-enhanced computerized tomography (CT) scans were obtained for each patient upon inclusion into the protocol and prior to each subsequent course of treatment. The product of the maximum transverse dimensions (length x width) and the number of CT scan cuts showing tumor (i.e., height) was calculated. The initial product L x W x H (cm^3), which is proportional to volume, was defined as V0, the initial volume index. Subsequent tumor volume indices, V, during treatment allowed calculation of the ratio V/V0. A value of V/V0 $\leq$ 0.5 was defined as a response to treatment. This corresponded to a reduction of 50% or more of enhancing volume. For each patient, the initial volume index (V0), lowest V/V0, and time (months) to achieve both the lowest V/V0 and any subsequent return to the initial baseline value (relapse, V/V0 $\geq$ 1) were recorded. A reduction of <50% in the enhancing volume was defined as no response. In these patients the time to recurrence was defined as 0 and lowest V/V0 was defined as 1 (Table 2).

Intervals

The following times (months) and treatments were calculated for each patient: (a) diagnosis to first BBBD and chemotherapy, (b) number of treatments, (c) interval from first to last treatment, (d) time to lowest V/V0 (see above), (e) diagnosis to progression (V/V0 $\geq$ 1) and (f) diagnosis to death.

RESULTS

Enhancing Volume Changes

The average initial tumor volume index (L x W x H) (cm^3) was 23 with the range from 0 (no measurable disease on contrast-enhanced CT scan) to 112 (Table 2). The average time (months) from diagnosis to beginning of treatment was 1 (range 0.2 to 1.7) (Table 1). The total number of treatments for all patients was 185; the average was 15 (range 3 to 26). The average interval (months) from first to last infusion was 8.2 (range 1.0 to 15.7).

Two patients had no enhancing tumor on entry into the protocol (Table 2). Patient #3 demonstrated only diffusely increased signal intensity on a T_2-weighted image (WI) magnetic resonance image (MRI) in the supratentorial white matter but no enhancement on CT or T_1-WI MRI. Patient #8 demonstrated no enhancing tumor after resection. These patients were, therefore, not included in the volumetric analysis. Four of the 10 patients with enhancing tumor demonstrated a response (V/V0 ≥ 0.5) to treatment (Table 2). Three of these four patients were on no steroids on inclusion in the protocol nor at the time of maximum tumor response. The fourth was tapered off steroids by the time of maximum response. The corresponding grades for patients showing response were 2 of 5 Grade IV, 1 of 2 Grade III, and 1 of 2 Grade II.

Table 2. Results of Chemotherapy Treatments

| | | | | Volumetric Analysis | | |
| | | Tumor Volume* | | | Time (Months) to | |
Pt. #	Initial Volume (V0) cm^3	Response (V/V0 ≥ 0.5)	Lowest V/V0	Nadir (Lowest V/V0)	Progression (V/V0 = 1)	Diagnosis to Death
1	65	yes	0.01	9.8	24.9	33
2	1	no	1	0	0	14.6
3	0	NA†	NA	NA	NA	13.2§‡
4	36	yes	0.03	8.6	25.4	28
5	13	no	1	0	0	14
6	2	yes	0.5	1.3	11.6	23.7
7	12	yes	0.08	5.9	NR‡	12.2§
8	0	NA	NA	NA	NA	14.2§
9	112	no	1	0	0	6.3
10	60	no	1	0	0	10.3
11	25	no	1	0	0	5.4
12	4	no	1	0	0	4.0

* V0 = initial tumor volume L x W x H (cm^3); V = subsequent tumor volume L x W x H (cm^3) see Materials and Methods

† NA, not applicable (i.e., no measurable tumor on inclusion in protocol - see Results)

‡ NR, no recurrence

§ Patient still alive.

Toxicity

Treatment resulted in no mortality nor permanent neurological deficit (Table 3). Focal motor or generalized seizures were observed either during or shortly after mannitol infusion in one-half of the patients, but occurred in only 16 of 185 total procedures. These seizures were readily controlled with i.v. medication and were not associated with the development of a chronic seizure disorder. By contrast to the prior reported study of BBB modification and chemotherapy in glioblastoma (31), sustained neurological deficit was not observed as a complication in the current series. Occasionally patients had significant stomatitis. The chemotherapy regimen caused moderate decreases in WBC and hematocrit values. No patient, however, developed sepsis as a result of decreased white blood cell count. In concordance with a prior study (31), thrombocytopenia (platelets less than 25,000) was not observed.

Table 3. Neurologic and Non-neurologic Complications

	Neurological Complications	
	Patients	Procedures
Mortality	0/12	0/185
Seizures	7/12	16/185
Permanent Deficit	0/12	0/185

	Hematologic Complications			
	Initial		Nadir	
	Average	Range	Average	Range
WBC	10	5.4 - 15.1	1.57	0.3 - 7.3
HCT	41.6	35.6 - 45	34.4	23.9 - 45.4

Neuropsychological Testing

All patients underwent initial neuropsychological evaluation. Two cases (patients # 1 and 4) illustrate the preservation of higher cortical functioning when treatment was successful in both reducing radiographic evidence of disease by over 95% and preventing tumor progression for at least one year (see Table 4 and Case Reports below). The case selection criteria permitted evaluation of potential neurotoxicity of BBBD and intra-arterial chemotherapy without the confounding factors of tumor progression or radiation (Figures 1 and 2). Both patients maintained normal levels of intellectual functioning and showed some areas of memory improvement beyond that expected on the basis of a test-retest practice effect. Although some deficits persisted at follow-up in attention/perception and motor areas, there was no significant decline greater than one standard deviation from pretreatment baseline.

Table 4. Neuropsychologic Test Results in Non-irradiated Patients With More Than 95% Volumetric Tumor Response[a].

	Normal Range[b]	Case # 1 (age 51)		Case # 4 (age 45)	
		Pre-tx	S/P 16 mo	Pre-tx	S/P 12 mo
Intelligence					
WAIS-R: VIQ	85-115	91	93	84	90
WAIS-R: PIQ	85-115	96	94	92	98
WAIS-R: FSIQ	85-115	92	93	86	91
Memory					
WMS-R: Visual	85-115	-	-	62	71
WMS-R: Verbal	85-115	-	-	93	108
WMS-R: Delayed	85-115	-	-	67	89
RAVLT: Trials 1-5	39-55	41	42	-	-
RAVLT: Recall	7-12	10	11	-	-
RVALT: Recognition	13-15	12	14	-	-
ROCFT: Recall	14-25	18	24	-	15.5
Attention/Perception					
WMS-R: Attention	85-115	-	-	80	95
Trials A (Sec.)[c]	24-46	52	36	-	58
Trials B (Sec.)[c]	39-110	97	115	-	126
ROCFT: Copy	32-36	25	35	-	36
Motor					
TAP: Dominant (taps/10 sec)	35-51	47	39	-	46
TAP: Nondominant (taps/10 sec)	34-46	-	-	-	45
GRIP: Dominant (kg)	41-55	37	39	-	38
GRIP: Nondominant (kg)	37-55	0	4.5	-	33

[a] Selected representative test results are reported

[b] Normal scores range from 1 S.D. above to 1 S.D. below the mean (see Methods)

[c] Longer completion times for Trails A and B reflect greater degrees of impaired functioning, in contrast to other variables where higher values reflect better performance.

CASE REPORTS

Patient #1

This 51-year-old male experienced a left upper extremity focal motor seizure in January 1986. By March 1986, physical exam documented left arm weakness and a concurrent CT scan revealed an enhancing right parietal mass with an initial volume (V0) of 65 cm^3. A right parietal craniotomy in April 1986 confirmed a Grade IV glioblastoma multiform. Osmotic BBB modification with combination chemotherapy was initiated and continued for a total of 26 treatments. Over the course of treatment, the patient's tumor first gradually decreased in size (Figure 1) and was, one year after initiation of treatment, <1% of its postoperative volume. He sustained brief, focal seizures during the sixth and fifteenth osmotic BBB infusions that were controlled with additional i.v. medication. No hematologic or permanent neurologic toxicity was observed. His serial neuropsychological tests results showed stable functioning 16 months after pretreatment baseline (Table 4). The extreme weakness in his nondominant grip and the inability to do the Finger Tapping test appear to reflect the pretreatment impact of his right parietal tumor location.

The tumor recurred 2 years after initiation of treatment, and the patient underwent a full course of external beam irradiation (6100 rad). He expired in January 1989, 33 months from diagnosis.

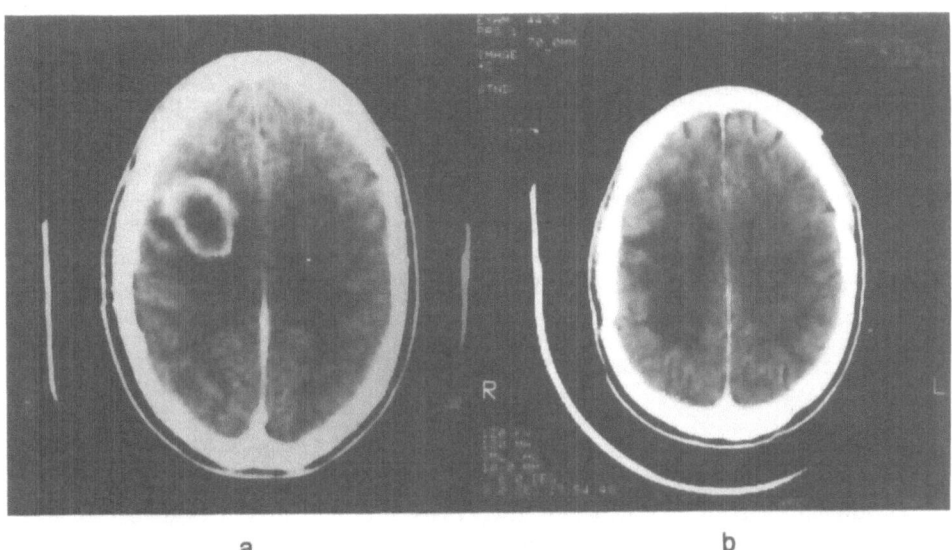

a   b

Figure 1. Patient #1: 51-year-old male with glioblastoma diagnosed at craniotomy (a) postoperative enhanced computerized tomographic (CT) scan obtained prior to first course of blood-brain barrier disruption (BBBD) chemotherapy showing large ring enhancing lesion in right parietal region. (b) Enhanced CT scan obtained nine months after "a" and prior to his 8th course (infusions 15 and 16) of BBBD chemotherapy. Note the dramatic decrease in the enhancing lesion.

Patient #4

This 45-year-old male experienced a grand mal seizure in January 1988. Physical exam revealed aphasia, right central seventh nerve palsy, and diminished muscle strength reflexes on the right. CT scan demonstrated a low density area in the left posterior parietal region that enhanced after i.v. contrast. CT-guided needle biopsy revealed a Grade II astrocytoma. At initiation of osmotic BBB modification with combination chemotherapy in March 1988, the contrast-enhancing volume of the tumor measured 36 cm^3 (Figure 2A). Over the ensuing eight months, the tumor gradually decreased in size by more than 90% of its original volume as seen on enhanced CT scan. The patient underwent 24 infusions through May 1989 (Figure 2B). He sustained no seizures nor hematological toxicity during treatment, and the seventh nerve palsy resolved but mild dysphasia persisted. Follow-up neuropsychological assessment showed improvements in cognitive test scores that, in some areas, were greater than expected in equivalent test-retest situations (Table 4). No evidence of neurotoxicity was detected.

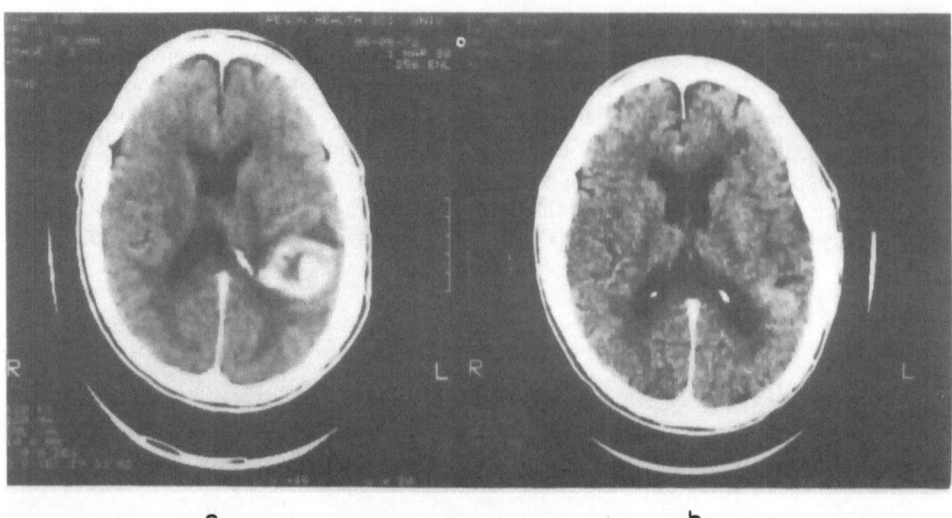

a　　　　　　　　　　　　　b

Figure 2. Patient #4: 45-year-old male with needle biopsy proven Grade II astrocytoma (a) enhanced computerized tomographic (CT) scan obtained prior to first course of blood-brain barrier disruption (BBBD) chemotherapy showing large enhancing lesion in left posterior parietal region. (b) Enhanced CT scan obtained 14 months after "a" and prior to his last course (24th infusion) of BBBD chemotherapy. The former enhancing lesion is barely detectable.

Follow-up CT scan in March 1990 (two years after initiation of treatment) revealed tumor recurrence. The patient declined further treatment and expired four months later, 28 months from diagnosis.

DISCUSSION

Sequencing of Radiation Therapy and Chemotherapy

The traditional sequencing of radiation and chemotherapy after surgery (22) may obscure the potential benefit of chemotherapy given initially. Drug resistance is often encountered in patients bearing a variety of disparate neoplasms and who previously received radiotherapy (7,15,35,41). In tumor models, external beam irradiation induces striking increases in cellular capacity for repair of DNA damage caused by chemotherapy. Following sequential fractionated external beam radiotherapy and exposure to cisplatin in vitro (15), accelerated excision and repair of platinum-DNA adducts compared with nonirradiated controls in a human teratoma model was observed. Similarly, Sharma et al. (40) demonstrated a 2000-fold increase in frequency of MTX-resistant Chinese hamster ovary and mouse 3T6 cells after a single 1000 rad fraction. Further treatment with any cytotoxic agent, including radiation, can trigger surviving cells (clonogens) in a tumor to divide faster than before, a process known as accelerated repopulation (13). In such a setting, the concomitant decrease in cell cycle duration reduces the available time for an agent to cause DNA damage, perhaps reflected in decreased treatment efficacy.

Osmotic BBBD

Pharmacological studies in normal and tumor-bearing rodents, in canines, and extrapolation in clinical studies, showed that BBB modification resulted in increased drug concentrations both in tumor and in those portions of brain supplied by the infused artery (25,28,29,31-34,43). The initial treatment of primary central nervous system (CNS) lymphoma with such cyclic, intra-arterial osmotic BBB modification and chemotherapy resulted in improved survival to 45 months when compared to patients in an historical control series who were treated initially with radiotherapy (median survival 15 months) (27,30). By contrast, when radiotherapy was administered prior to osmotic BBBD and chemotherapy, survival was comparable to established regimens employing initial radiotherapy with or without systemic chemotherapy.

Response Factors

In enhanced CT imaging of human gliomas, contrast enhancement in the region of the tumor reflects regional breakdown of the BBB. Such loss of BBB integrity may indicate the presence of tumor, although viable tumor cells may exist some distance from any radiographic abnormality (2). Contrast-enhanced CT scans are useful for evaluation of macroscopic growth or regression of large tumor masses.

Although the number of treated glioma patients is too small to assess survival, several observations deserve comment. The type of surgery in this study (craniotomy with debulking vs. biopsy) did not appear to affect survival. Patients demonstrating the greatest decreases in the volume index, V/V0, demonstrated a longer survival. Patients who received radiotherapy following tumor recurrence did sustain a moderate increase in survival.

Neuropsychologic Evaluation

In the two case examples demonstrating minimal measurable disease one year after treatment, no neuropsychologic evidence of either global dementia or focal cognitive deficits was observed. As a result, patients continued their normal life style without significant compromise of their intellect, memory, new learning or ability to concentrate. Systematic baseline and follow-up neuropsychological assessment can contribute substantially to the evaluation of quality of life during and after brain tumor treatments.

CONCLUSIONS

This report suggests that sequencing chemotherapy before radiation therapy may afford an improved method to assess glioma chemosensitivity, maximize tumor cell cytotoxicity and minimize neurotoxicity (42). As tumor cells appear more sensitive to initial chemotherapy, administering radiation therapy subsequent to chemotherapy may avoid such problems as capillary endothelial hyperplasia, radiation induced tumor cell resistance and radiation neurotoxicity. Neuropsychological testing demonstrates preservation of cognitive, psychosocial function in patients responding to chemotherapy associated with osmotic BBBD. This technique increases the concentration of chemotherapeutic agents to both the tumor infiltrated brain around tumor as well as to tumor when compared with other routes of administration. A similar trial in gliomas has been initiated with preirradiation carboplatin and etoposide given in association with osmotic BBBD to assess initial drug response (46). The increased delivery of chemotherapeutic agents to brain tumor, coupled with increased susceptibility of tumor cells to initial chemotherapy may improve tumor responses while maintaining cognitive function. This study offers a new and unique means to assess both

CNS toxicity and tumor responsiveness, which may lead to protocols that also prolong survival in glioma patients.

SUMMARY

Twelve adults (11 male, 1 female), diagnosed as having supratentorial gliomas, were treated with osmotic blood-brain barrier disruption and chemotherapy (intra-arterial methotrexate, 2500 mg/infusion; intravenous Cytoxan, 15 mg/kg/infusion; and oral procarbazine, 100 mg daily x 14 days) prior to radiotherapy. To assess higher cortical function, all patients underwent neuropsychological testing. Volumetric analysis of pretreatment and serial enhanced computerized tomographic scans were recorded. Four of ten patients with enhancing tumor showed radiographic tumor response, defined as 50% reduction of enhancing tumor volume. These four patients received no steroids at the time of maximum tumor response. Osmotic blood-brain barrier disruption and chemotherapy administered prior to radiotherapy can result in significant objective tumor responses with maintenance of cognitive function. It also offers a new and unique means to assess chemosensitivity, which may lead to improved treatment protocols.

ACKNOWLEDGMENTS

This work was supported by the Veterans Administration Merit Review Grant, and the National Institutes of Health, Grant # 31770.

REFERENCES

1. Bleyer, W.A., and Griffin, T.W., White cell necrosis, mineralizing microangiopathy and intellectual abilities in survivors of childhood leukemia, in: "Radiation Damage to the Nervous System," H.A. Gilbert, and A.R. Kagna, eds., Raven Press, New York, pp 155-173, 1980.
2. Burger, P.C., and Kleihues, P., Cytologic composition of the untreated glioblastoma with implications for evaluation of needle biopsies, Cancer 63:2014-2023, 1989.
3. Burger, P.C., Mahaley, M.S., and Dudka, L., The morphologic effects of radiation administered therapeutically for intracranial gliomas: A postmortem study of 25 cases, Cancer 44:1256-1272, 1979.
4. Calvo, F., Dy, C., and Henriquez, I., Postoperative radical radiotherapy with concurrent weekly intra-arterial cis-platinum for treatment of malignant glioma: A pilot study, Radiother Oncol 14:83-88, 1989.
5. Caveness, W.F., Experimental observations: Delayed necrosis in normal monkey brain, in: "Radiation Damage to the Nervous System," H.A. Gilbert, and A.R. Kagna, eds., Raven Press, New York, 1980, pp 1-38.
6. Dropcha, E.J., Rosenfeld, S.S., and Morawetz, R.B., Preradiation intracarotid cisplatin treatment of newly diagnosed anaplastic gliomas, J Clin Oncol 10:452-458, 1992.
7. Einhorn, L.H., and Williams, S.D., The management of disseminated testicular cancer, in: "Principles of Cancer Treatment," S.K. Carter, E. Glatstein, and R.B. Livingston, eds., McGraw-Hill, New York, pp 605-612, 1982.
8. Frytak, S., Shaw, J.N., and O'Neill, B.P., Leukoencephalopathy in small cell lung cancer patients receiving prophylactic cranial irradiation, Am J Clin Oncol 12:27-33, 1989.
9. Geffen, G., Moar, K.J., and O'Hanlon, A.P., et al., Performance measures of 16- to 86-year-old males and females on the auditory verbal learning test, Clin Neuropsychol 4:45-63, 1990.

10. Green, S.B., Byar, D.P., and Strike, T.A., et al., Randomized comparisons of BCNU, streptozotocin, radiosensitizer, and fractionation of radiotherapy in the postoperative treatment of malignant glioma (Study 7702), *Proc Am Soc Clin Oncol* 3:260, 1984(abstr).

11. Green, S.B., Byar, D.P., Walker, M.D., Pistenmaa, D.A., Alexander, E., Batzdorf, U., Brooks, W.H., Hunt, W.E., Mealey, J., Odom, G.L., Paoletti, P., Ransohoff, J., Robertson, J.T., Seller, R.G., Sahpiro, W.R., Smith, K.R., Wilson, C.B., and Strike, T.A., Comparisons of carmustine, procarbazine, and high-dose methylprednisolone as additions to surgery and radiotherapy for the treatment of malignant glioma, *Cancer Treat Rep* 67:121-132, 1983.

12. Gumerlock, M.K., Belshe, B.D., and Madsen, R., et al., Osmotic blood-brain barrier disruption and chemotherapy in the treatment of high grade malignant glioma: Patient series and literature review, *J Neurooncol*, 1991 (In press).

13. Hall, E.J., "Radiobiology for the Radiologist," J.B. Lippincott, Philadelphia, 1988.

14. Hill, B.T., *In vitro* drug-radiation interactions using fractionated X-irradiation regimens, *in:* "Antitumor Drug-radiation Interactions," B.T., Hill, and A.S. Bellamy, eds., CRC Press, Boca Raton, pp 207-222, 1989.

15. Hill, B.T., Shellard, S.A., Hosking, L.K., Fichtinger-Schepman, A.M.J., and Bedford, P., Enhanced DNA repair and tolerance of DNA damage associated with resistance to cis-diamine-dichloroplatinum (II) after *in vitro* exposure of a human teratoma cell line to fractionated x-irradiation, *Int J Radiat Oncol Biol Phys* 19:75-83, 1990.

16. Johnson, B.E., Becker, B., Goff, W.B., Petronas, N., Krehbiel, M.A., Makuch, R.W., McKenna, G., Glatstein, E., and Ihde, D.C., Neurologic, neuropsychologic, and computed cranial tomography scan abnormalities in 2- to 10-year survivors of small-cell lung cancer, *J Clin Oncol* 3:1659-1667, 1985.

17. Johnson, B.E., Patronas, and N., Hayes, W., et al., Neurologic, computed cranial tomographic, and magnetic resonance imaging abnormalities in patients with small-cell lung cancer: Further follow-up of 6- to 13-year survivors, *J Clin Oncol* 8:48-56, 1990.

18. Lee, J.S., Umswadi, T., and Lee, Y.Y., et al., Neurotoxicity in long-term survivors of small-cell lung cancer, *Int J Oncol Biol Phys* 12:313-321, 1986.

19. Levin, H.S., Neurobehavioral sequelae of head injury, *in:* P.R. Cooper, ed, "Head Injury," Williams & Wilkins, Baltimore, pp 442-463, 1987.

20. Mahaley, M.S., Hipp, S., and Dropcho, E., et al., Intracarotid cisplatin chemotherapy for recurrent gliomas, *J Neurosurg* 70:371-378, 1989.

21. Mahler, P., Griffin, B., and Geyer, J.R., et al., Chemoradiotherapy interactions and the blood-brain barrier, *in:* E.A. Neuwelt, ed., "Implications of the Blood-Brain Barrier and Its Manipulation: Vol. 1. Basic Science Aspects," Plenum Press, New York, pp 373-387, 1989.

22. Nazzaro, J.M., and Neuwelt, E.A., The role of surgery in the management of supratentorial intermediate and high grade astrocytomas in adults, *J Neurosurg* 73:331-344, 1990.

23. E.A. Neuwelt, ed., "Implications of the Blood-Brain Barrier and Its Manipulation: Vol. 1. Basic Science Aspects," Plenum Press, New York, 1989.

24. E.A. Neuwelt, ed., "Implications of the Blood-Brain Barrier and Its Manipulation: Vol. 2. Clinical Implications," Plenum Press, New York, 1989.

25. Neuwelt, E.A., Barnett, P., McCormick, C., Frenkel, E.P., and Minna, J.D., Osmotic blood-brain barrier modification: Monoclonal antibody, albumin and methotrexate delivery to CSF and brain, *Neurosurg* 17:419-423, 1985.

26. Neuwelt, E.A., Frenkel, E.P., Diehl, J., Vu, L.H., Rapoport, S., and Hill, S., Reversible osmotic blood-brain barrier disruption in humans: Implications for the chemotherapy of malignant brain tumors, *Neurosurg* 7:44-52, 1980.

27. Neuwelt, E.A., Frenkel, E.P., Gumerlock, M.K., Braziel, R., Dana, B., and Hill, S.A., Developments in the diagnosis and treatment of primary CNS lymphoma, *Cancer* 58:1609-1620, 1986.

28. Neuwelt, E.A., Frenkel, E.P., Rapoport, S., and Barnett, P., Effect of osmotic blood-brain barrier disruption on methotrexate pharmacokinetics in the dog, *Neurosurg* 7:36-43, 1980.

29. Neuwelt, E.A., Glasberg, M., Diehl, J., Frenkel, E.P., and Barnett, P., Osmotic blood-brain barrier disruption in posterior fossa of the dog, *J Neurosurg* 55:742-748, 1981.

30. Neuwelt, E.A., Goldman, D., Dahlborg, S.A., Crossen, J., Ramsey, F., Goldstein, S.M., Braziel, R., and Dana, B., Primary central nervous system lymphoma treated with osmotic blood-brain barrier disruption and combination chemotherapy: Prolonged survival and preservation of cognitive function, *J Clin Oncol* 9:1580-1590, 1991.

31. Neuwelt, E.A., Howieson, J., Frenkel, E.P., Specht, D., Weigel, R., Buchan, C.G., and Hill, S.A., Therapeutic efficacy of multiagent chemotherapy with drug delivery enhancement by blood-brain barrier modification in glioblastoma, *Neurosurgery* 19:573-582, 1986.

32. Neuwelt, E.A., Maravilla, K.R., Frenkel, E.P., Barnett, P., Hill, S., and Moore, R.J., Use of enhanced computerized tomography to evaluate osmotic blood-brain barrier disruption, *Neurosurgery* 6:49-56, 1980.

33. Neuwelt, E.A., Maravilla, K.R., Frenkel, E.P., Rapoport, S.I., Hill, S.A., and Barnett, P.A., Osmotic blood-brain barrier disruption: Computerized tomographic monitoring of chemotherapeutic agent delivery, *J Clin Invest* 64:684-688, 1979.

34. Neuwelt, E.A., Pagel, M., Barnett, P., Glasberg, M., and Frenkel, E.P., Pharmacology and toxicity of intracarotid Adriamycin administration following osmotic blood-brain barrier modification, *Cancer Res* 41:4466-4470, 1981.

35. Price, L.A., and Hill, B.T., A kinetically based logical approach to chemotherapy of head and neck cancer, *Clin Otolaryngol* 2:339-345, 1977.

36. Salazar, O.M., Rubin, P., and Fieldstein, M.L., et al., High dose radiation therapy in the treatment of malignant gliomas: final report, *Int J Radiat Oncol Biol Phys* 5:1733-1748, 1979.

37. Schiffer, D., Giordana, M.T., and Soffietti, R., et al., Radio- and chemotherapy of malignant gliomas: Pathological changes in the normal nervous tissue. *Acta Neurochir (Wein)* 58:37-58, 1981.

38. Shapiro, W.R., and Ausman, J.I., The chemotherapy of brain tumors: A clinical and experimental review, *in:* "Recent Advances in Neurology," F. Plum, ed., FA Davis Co., Philadelphia, pp 150-235, 1969.

39. Shapiro, W.R., Green, S.B., and Burger, P.C., et al., Randomized trial of three chemotherapy regimens and two radiotherapy regimens in postoperative treatment of malignant glioma, *J Neurosurg* 74:1-9, 1989.

40. Sharma, R.C., and Schimke, R.T., Enhancement of the frequency of methotrexate resistance by gamma radiation in Chinese hamster ovary and mouse 3T6 cells, *Can Res* 49:3861-3866, 1989.

41. Thigpen, T., Vance, R.B., and Balducci, L., et al., Chemotherapy in the management of advanced or recurrent cervical or endometrial carcinoma, *Cancer* 48:658-665, 1981.

42. Turrisi, A.T., Brain irradiation and systemic chemotherapy for small-cell lung cancer: Dangerous liaisons? *J Clin Oncol* 8:196-199, 1990.

43. Walker, M.D., Alexander, E., Jr, and Hunt, W.E., et al., Evaluation of BCNU and/or radiotherapy in the treatment of anaplastic gliomas. A cooperative clinical trial, *J Neurosurg* 49:333-343, 1978.

44. Walker, M.D., Green, S.B., and Byar, D.P., et al., Randomized comparisons of radiotherapy and nitrosoureas for the treatment of malignant glioma after surgery, *N Engl J Med* 303:1323-1329, 1980.

45. Weintraub, S., and Mesulam, M., Mental state assessment of young and elderly adults in behavioral neurology, *in:* "Principles of Behavioral Neurology," M. Mesulam, ed., F.A. Davis Co., Philadelphia, 1985.

46. Williams, P.C., Neuwelt, E.A., Hogan, R.L., Dana, B.W., and Roman-Goldstein, S., Toxicity and efficacy of carboplatin and etoposide in conjunction with blood-brain barrier modification in the treatment of intracranial neoplasms, *Am Assoc Cancer Res* meeting, San Diego, May, 1992(abst).

THE UPTAKE OF THYROXINE BY THE ISOLATED
PERFUSED CHOROID PLEXUS OF THE SHEEP

Malcolm B. Segal and Jane E. Preston

Sherrington School of Physiology, UMDS, St. Thomas's Hospital
London, England, SE1 7EH

INTRODUCTION

Thyroxine (T_4) can enter the brain either directly across the blood-brain barrier (BBB) or secondarily via the choroid plexus (CP) into the cerebrospinal fluid (CSF) and then by diffusion from the CSF into the extracellular fluid surrounding the neurones (3-5). The concentration of both T_4 and tri-iodothyronine (T_3) as the free hormones in CSF is in fact greater than in plasma, and it is difficult to envisage how this adverse gradient in CSF can be established without some sort of transport mechanism. Schreiber and his colleagues have shown that the T_4 binding hormone transthyretin (TTR) is synthesized by the choroid plexus, and together we have shown that TTR is secreted into the CSF and not into the plasma (2,7). Studies by this group on the uptake of T_4 from the blood using both in vivo and in vitro techniques failed to show a carrier-mediated uptake by the choroid plexus. Earlier in vivo studies by Hagen and Solberg (4) did show the entry into the CSF could be saturated, but in vivo it is not possible to separate the entry across the BBB from that across the CP. In vitro, using the non-perfused choroid plexus, it is also not possible to study the uptake by the blood side of this tissue. To overcome these problems, we have used the isolated perfused choroid plexus of the sheep. With this preparation, we have recently shown that the uptake of T_3 by the blood side of the CP is carrier-mediated, although the rate of entry of this more lipid soluble hormone into the CSF is less than that of T_4 (2,6). Since T_4 is much less lipid soluble than T_3, a carrier-mediated uptake would appear likely.

METHODS

Sheep (Clun Forest 20-30 kg) were anesthetized with thiopentone sodium (20 mg.kg^{-1} I.V.), heparinized and exsanguinated. The vault of the skull was reflected and the brain carefully removed. The internal carotid arteries on both sides were cannulated and perfused with a modified Ringer solution at 0.5-1.5 ml.min^{-1}. All other vessels on the circle of Willis were tied off to direct the perfusate into the anterior choroidal arteries. The roof of both lateral ventricles was opened and the exposed choroid plexuses kept moist by

Frontiers in Cerebral Vascular Biology: Transport and Its Regulation
Edited by L.R. Drewes and A.L. Betz, Plenum Press, New York, 1993

superperfusion with an artificial CSF. The great vein of Galen was cannulated to collect the outflow from both choroid plexuses.

The perfusion fluid had the following composition in mM: Na^+ 145.8, K^+ 5.4, Cl^- 119.7, HCO_3^- 25, $HPO_4^=$ 1.2, Ca^{++} 2.35, Mg^{++} 1.13 and glucose 5 mM; 50% Dextran 70 in saline as colloid agent and 0.05 g.dl^{-1} bovine serum albumin (BSA, Sigma Fractions). The perfusate was gassed with 95% O_2 and 5% CO_2 and warmed to 37°C. The mock CSF contained in mM: Na^+ 148, K^+ 2.9, Ca^{++} 2.5, Mg^{++} 1.8, Cl^- 135, HCO_3^- 26 and $HPO_4^=$ 0.25; gassed and prewarmed as above. For low sodium experiments the NaCl and NaHCO$_3$ were replaced by the appropriate choline salts.

The uptake of ^{125}I-L-thyroxine (T_4) by the basolateral face of the choroid plexus was studied with reference to the passage and recovery of ^3H labelled mannitol. A 100-µl bolus was introduced into a side loop that contained 0.5 µCi ^{125}I-L-T_4 and 2 µCi ^3H-mannitol; then by a system of taps, the bolus could be introduced into the inflow of either choroid plexus without any change in pressure. After allowing for clearance of the dead space of the system, 20 sequential one-drop samples of the various effluent were collected and added to scintillation fluid. The activities of both isotopes as dpm in each sample were determined using an LKB Rack Beta Spectral 1219.

The uptake of the ^{125}I-L-T_4 in each drop was related to the recovery of ^3H-mannitol, a passive marker, which passed into and returned from the extracellular space of the choroid plexus. The recovery of T_4 was always less than that of mannitol, indicating uptake by the cells of the choroid plexus; however, the peak recovery of both isotopes was simultaneous (see Figure 1).

The uptake (U%) for each drop is calculated as:

$$U\% = \frac{\% \ ^3H\text{-mannitol recovered} - \% \ ^{125}I\text{-}T_4 \text{ recovered}}{\% \ ^3H\text{-mannitol recovered}}$$

The U% value for the samples containing the highest levels of radioactivity were averaged to give an index of the maximal cellular uptake, $U_{max}\%$.

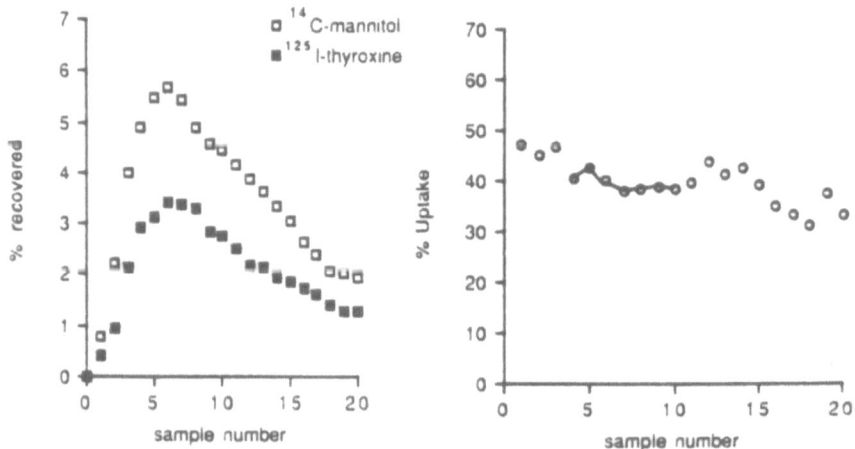

FIGURE 1. On the left is shown the % uptake of mannitol and thyroxine from a 100µl bolus injected into the arterial inflow of the perfused sheep choroid plexus and collected as single drops from the great vein of Galen in a typical experiment. The recovery of T_4 is less than that of the mannitol (a non-transported molecule) and represents the uptake of T_4 by the cells of the choroid plexus. On the right is shown the % uptake of T_4 in each drop. The maximal uptake of T_4 ($U_{max}\%$) is taken from the record where the maximum uptake of radioactivity has occurred (joined circles).

RESULTS

Table 1 shows that the uptake of ^{125}I-L-thyroxine in the absence of any unlabelled hormone was 34%. This uptake was reduced by 54% when 600-800 μM of unlabelled T_4 was added to the bolus. A similar inhibition was found when 800 μM of unlabelled D-T_4 or 600 μM of L-tri-iodothyronine (L-T_3) was added to the bolus.

TABLE 1. The maximal uptake (U_{max}%) of ^{125}I-L-thyroxine by the blood side of the isolated perfused sheep choroid plexus.

Content of Bolus	U_{max}%	Inhibition	p
^{125}I.L-T_4 only	34.4±2.8 (21)		
+ unlabelled L-T_4 600μM	16.0±1.8 (4)	53.5%	<0.01
+ unlabelled D-T_4 800μM	17.4±1.8 (4)	49.4%	<0.05
+ unlabelled L-T_3 600μM	14.9±2.2 (5)	56.7%	<0.01

Values are mean ± SE, n in parentheses, p = difference from control U_{max}%.

Table 2 shows the effect of replacing the sodium in the perfusate and bolus on the uptake of ^{125}I-L-T_4 and, although there was a slight reduction in uptake, this failed to reach significance.

In addition in Table 2 is shown the effect of the presence of 5 mM sodium perchlorate on the uptake of ^{125}I-L-T_4. Again no significant change in the uptake was found.

TABLE 2. Effect of sodium replacement with choline and of the presence of 5 mM sodium perchlorate in the bolus on the maximal uptake (U_{max}%) of L-T_4 by the blood side of the isolated perfused choroid plexus.

	U_{max}%	p
Normal sodium	35.8±4.2 (11)	
Sodium free	30.6±0.7 (5)	>0.05
Normal sodium	34.4±2.8 (21)	
+5mM sodium perchlorate	37.0±1.4 (6)	>0.05

Values are mean ± SE, n in parentheses, p = difference from control.

CONCLUSIONS

The uptake (U_{max}) of thyroxine by the basolateral face (blood side) of the choroid plexus was 34% and this uptake was inhibited about 50% by the presence of unlabelled T_4, which would indicate that the process is partially carrier mediated. The non-saturable uptake most probably reflects passive diffusional uptake related to lipid solubility of this molecule. A similar degree of inhibition was observed in the presence of unlabelled D-T_4 and L-T_3, which would suggest that the carrier-mediated component of the uptake of T_4 has a low stereospecificity. The lack of effect of low sodium on the uptake of T_4 suggests that this uptake is not sodium dependent and the small non-significant reduction in T_4 uptake in the low sodium solutions may be the consequence of the reduction in CSF secretion caused by the absence of sodium in the perfusate (1). The lack of effect of sodium perchlorate on the uptake of T_4 would suggest that there is little free iodide released from the T_4 as it transits the CP. Therefore, the observed uptake is of the intact hormone.

These studies have shown that the uptake of T_4 into the cells of the choroid plexus from the blood is partly carrier mediated and, if this entry is in series with the synthesis and secretion of TTR, the adverse gradient of free T_4 between plasma and CSF can be explained.

ACKNOWLEDGMENTS

The authors thank the Wellcome Trust whose support made this work possible.

REFERENCES

1. Davson, H., and Segal, M.B., The effect of some inhibitors and accelerators of sodium transport on the turnover of [22]Na in the cerebrospinal fluid and the brain, *J. Physiol.* 209:131-153, 1969.
2. Dickson, P.W., Aldred, A.R., Menting, J.G.T., Morley, P.D., Sawyer, W.H., and Schreiber, G., Thyroxine transport in choroid plexus, *J. Biol. Chem.* 262:13907-13915, 1987.
3. Dratman, M.B., Crutchfield, F.L., and Schoenhoff, M.B., Transport of iodothyronines from bloodstream to brain: contributions by blood:brain and choroid plexus: cerebrospinal fluid barriers, *Brain Res.* 554:229-236, 1991.
4. Hagen, G.A., and Solberg, L.A., Brain and cerebrospinal fluid permeability to intravenous thyroid hormones, *Endocrin.* 95:1398-1410, 1974.
5. Pardridge, W.M., Carrier-mediated transport of thyroid hormones through the rat blood-brain barrier: Primary role of albumin-bound hormone, *Endocrin.* 105:605-612, 1979.
6. Preston, J.E., and Segal, M.B., Saturable uptake of [125]I.L-tri-iodothyronine at the basolateral (blood) and apical (cerebrospinal fluid) sides of the isolated perfused sheep choroid plexus, *Brain Res.* in press.
7. Schreiber, G., Aldred, A.R., Jaworowski, A., Nelsson, C., Achen, M.G., and Segal, M.B., Thyroxine transport from blood to brain via transthyretin synthesis in choroid plexus, *Am. J. Physiol.* 258:R338-R345, 1990.

PARTICIPANTS

Marc G. Achen, Abteilung Neurochemie, Max-Planck-Institute für Psychiatrie, Planegg-Martinsried, Germany

Takao Asano, Department of Neurosurgery, Saitama Medical Center, Saitama, Japan

Richard Béliveau, Département de chimie, Laboratoire de membranalogie moléculaire, Université du Québec à Montréal, Montréal, (Québec), Canada

A. Lorris Betz, Departments of Pediatrics, Surgery (Neurosurgery) and Neurology, University of Michigan, Ann Arbor, Michigan

Nancy D. Borson, Department of Biochemistry and Molecular Biology, University of Minnesota, Duluth, Minnesota

Peter Brust, Forschungszentrum Rossendorf, Dresden, Germany

Roméo Cecchelli, Institut Pasteur de Lille, France

Pierre-Olivier Couraud, Immunopharmacologie, Institut Cochin de Genetique Moleculaire, Paris, France

A. G. de Boer, Center for Biopharmaceutical Sciences, Leiden, The Netherlands

Elizabeth C.M. de Lange, Center for Biopharmaceutical Sciences, Leiden, The Netherlands

H.E. de Vries, Center for Biopharmaceutical Science, Leiden, The Netherlands

Paula Dore-Duffy, Department of Neurology, Wayne State University School of Medicine, Detroit, Michigan

Lester R. Drewes, Department of Biochemistry and Molecular Biology, University of Minnesota, Duluth, Minnesota

Catherine Dubertret, Biotechnology Department, Rhone Poulenc Rorer, France

Steven R. Ennis, Department of Surgery, University of Michigan, Ann Arbor, Michigan

James R. Ewing, Department of Neurology, Henry Ford Hospital, Detroit, Michigan

James H. Fitzpatrick, Department of Anesthesiology, University of Wisconsin, Madison, Wisconsin

Ronald D. Franks, School of Medicine, University of Minnesota, Duluth, Minnesota

Phillip Friden, Alkermes, Inc., Cambridge, Massachusetts

David Z. Gerhart, Department of Biochemistry and Molecular Biology, University of Minnesota, Duluth, Minnesota

David D. Gilboe, Department of Neurosurgery, University of Wisconsin, Madison, Wisconsin

Gary W. Goldstein, Kennedy Krieger Institute, Baltimore, Maryland

Rolf Gruetter, Department of Molecular Biophysics, Yale University, New Haven, Connecticut

Mary Kay Gumerlock, Department of Surgery, University of Oklahoma, Oklahoma City, Oklahoma

Jay M. Hemmila, Department of Biochemistry and Molecular Biology, University of Minnesota, Duluth, Minnesota

Sami I. Harik, Department of Neurology, Case Western Reserve University, Cleveland, Ohio

R. C. Janzer, Division of Neuropathology, Institute of Pathology, Lausanne, Switzerland

Hiroo Johshita, Saitama Cardiovascular Center, Saitama, Japan

Hazel C. Jones, Department of Pharmacology and Therapeutics, University of Florida, Gainesville, Florida

Ferenc Joó, Laboratory of Molecular Neurobiology, Institute of Biophysics Biological Research Center, Szeged, Hungary

Richard F. Keep, Department of Surgery, University of Michigan, Ann Arbor, Michigan

Robert Klein, Department of Anatomy, University of Kansas, Kansas City, Kansas

Peter Kollros, Department of Pediatrics, Thomas Jefferson University, Philadelphia, Pennsylvania

Jeanne-Marie Lafauconnier, Hôpital Fernand Widal, Institut Cochin de Genetique Moleculaire, Paris, France

John Laterra, Kennedy Krieger Institute, Baltimore, Maryland

Dorothee Krause, Institute of Anatomy, University of Regensburg, Germany

Paul G. M. Luiten, University of Groningen, Department of Animal Physiology, Haren, The Netherlands

Frances Maher, EDMNS/DB/NIDDK, National Institutes of Health, Bethesda, Maryland

Michael Man, Department of Biochemistry and Molecular Biology, University of Minnesota, Duluth, Minnesota

Graeme Mason, Department of Molecular Biophysics and Biochemistry, Yale University School of Medicine, New Haven, Connecticut

Toru Matsui, Department of Neurosurgery, Saitama Medical Center, Saitama, Japan

Richard McCarron, NINDS/LNNS, National Institutes of Health, Bethesda, Maryland

J. Gordon McComb, Department of Neurosurgery, Childrens Hospital Los Angeles, USC School of Medicine, Los Angeles, California

James C. McKenzie, Department of Anatomy, Howard University, Washington, DC

Steven A. Moore, Department of Pathology, University of Iowa, Iowa City, Iowa

Sukriti Nag, Department of Pathology (Neuropathology), Queen's University, Kingston, Ontario, Canada

Edward A. Neuwelt, Departments of Neurology, Neurosurgery, Biochemistry and Molecular Biology, Oregon Health Sciences University, Portland, Oregon

Olaf B. Paulson, Department of Neurology, Rigshospitalet, Copenhagen, Denmark

John P. Pirro, Radiopharmaceuticals Department, Bristol-Myers Squibb Pharmaceutical Research Institute, New Brunswick, New Jersey

Olivier Rabin, Hôpital Fernand Widal, Institut Cochin de Genetique Moleculaire, Paris, France

Francoise Roux, Unité de Neurotoxocologie, Tissue Culture Laboratory, Hôpital Fernand Widal, Institut Cochin de Genetique Moleculaire, Paris, France

Thomas N. Sato, Department of Neurosciences, Roche Institute of Molecular Biology, Nutley, New Jersey

Gerald P. Schielke, Warner-Lambert/Parke Davis Pharmaceuticals, Ann Arbor, Michigan

Ursula Schindler, Department of CNS-Pharmacology, Cassella AG, Frankfurt, Germany

Malcolm B. Segal, Department of Physiology UMDS, St. Thomas' Hospital, London , England

Quentin R. Smith, Laboratory of Neurosciences, NIA, National Institutes of Health, Bethesda, Maryland

Maria Spatz, NINDS/LNNS, National Institutes of Health, Bethesda, Maryland

Danica B. Stanimirovic, NINDS/LNNS, National Institutes of Health, Bethesda, Maryland

William Taylor, Department of Biochemistry and Molecular Biology, University of Minnesota, Duluth, Minnesota

Hiroshi Tenjin, Neurosurgery, Mayo Clinic, Rochester, Minnesota

Susan J. Vannucci, Division of Pediatric Neurology, Pennsylvania State University, The Milton S. Hershey Medical Center, Hershey, Pennsylvania

Kishena C. Wadhwani, National Institutes on Aging, Bethesda, Maryland

Devin Welty, Warner-Lambert/Parke-Davis, Ann Arbor, Michigan

Mathew Whittico, Pharmacy, University of Michigan, Ann Arbor, Michigan

Wesley Williams, NIA, National Institutes of Health, Bethesda, Maryland

Joanna Woyciechowska, The Duluth Clinic, Ltd., Duluth, Minnesota

P. Wülforth, Department of Pharmacology, Merz + Co., Frankfurt , Germany

Berislav V. Zlokovic, Department of Neurosurgery, Childrens Hospital Los Angeles, USC School of Medicine, Los Angeles, California

INDEX

Histamine, 156, 160
HT7 antigen, 211, 217, 225
Hyperglycemia, 38
Hyperinsulinemia, 26
Hypernatremia, 66
Hyperosmotic BBB disruption, 257, 273
Hypertension, 263
Hypoglycemia, 26
Hypoxia, 161

Immunocytochemistry, 13, 132, 150, 202, 211,
    218, 224, 238, 244, 250
Immunoglobulin superfamily, 213, 226
Inflammation, 223, 237, 243
Inositol triphosphate, 168, 173
Integrins, 237
Intercellular adhesion molecule-1 (ICAM-1), 237,
    244
Interferon, 208, 238, 250
Interleukin-1α, 238
Intra-uterine growth, 107
Intravenous double-indicator method, 26

Leukocytes, 237, 243
Linolenic acid, 229
Lipids, 267
Lipoxygenase, 165
Lock-docking primer, 21
Luminal surface, 214
Lumped constant, 27
Lymphocytes, 237, 249

Melphalan, 88
Membrane lipids, 267
Methotrexate, 257, 273
Methyl-prednisolone, 117
MHC antigens, 205, 243
Microdialysis, 257
Microvessels, 10, 13, 45, 56, 61, 66, 76, 113,
    121, 149, 159, 189, 211, 244, 267
Migration, 189
Monoclonal antibodies, 149, 244, 249
Morphogenesis, 189
Multidrug resistance genes, 121
Multiple sclerosis, 237, 243, 249
Myoinositol, 180

Natriuretic peptide, 206
Nerve growth factor, 130
Neural tube grafts, 226
Neurons, 229
Neurothelin, 217
Nimodipine, 117
Nitric oxide, 205

NMR spectroscopy, 29, 35

Octanol-water partition coefficient, 89
Ouabain, 47, 49, 58, 68, 73

P-glycoprotein, 121
Patch clamp, 50
Perfused brain, 117, 143
Pericytes, 137, 149, 189
Permeability, regulation of, 155
Permeability-surface area product, 44, 87, 102
Phorbol esters, 61, 159, 177, 195
Phosphatidylcholine, 180
Phosphatidylethanolamine, 180
Phospholipases
    Phospholipase A$_2$, 165, 166, 271,
    Phospholipase C, 169, 174, 181, 271
Phospholipids, 180, 267
Phosphorylation, 181
Pial vessels, 71
Pinocytosis, 156, 264
Plasmalemma, 141
Plasminogen activator, 192, 195
Polymerase chain reaction, 19
Positron emission tomography, 97
Postnatal, 3
Prodrug, 84, 86
Proliferation, 189
Prostaglandins, 165, 208
Protein kinase C, 61, 159, 177, 181
Protein phosphorylation, 160

Quinidine, 67

Radiotherapy, 273
Receptors, 183
    transferrin, 129, 244

Sciatic nerve, 101
Second messengers, 155
Sodium, potassium-ATPase, 46, 55, 61, 73, 263
Staurosporine, 62
Streptozotocin, 9
Subarachnoid hemorrhage, 177

Thromboxane, 168
Thyroxine, 285
Transcytosis, 137
Transfer coefficient, 108
Transferrin, 129
    transferrin receptor, 129, 244
Transforming growth factor-β, 238
Transport
    amino acid, 88, 95, 101, 108